Additive Manufacturing for Chemical Sciences and Engineering

Additive Manufacturing for Chemical Sciences and Engineering

Editor-in-chief

Suresh K. Bhargava
Member of Order of Australia

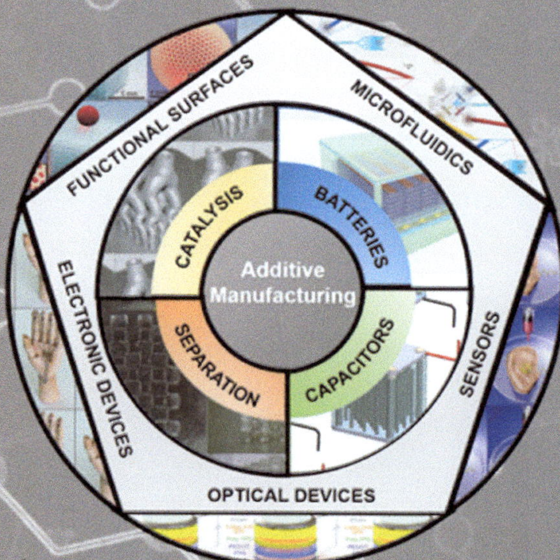

Editors

Seeram Ramakrishna
Milan Brandt
PR. Selvakannan

Suresh K. Bhargava · Seeram Ramakrishna ·
Milan Brandt · PR. Selvakannan
Editors

Additive Manufacturing for Chemical Sciences and Engineering

Springer

Editors
Suresh K. Bhargava
Centre for Advanced Materials
and Industrial Chemistry
RMIT University
Melbourne, VIC, Australia

Milan Brandt
Centre for Additive Manufacturing
RMIT University
Melbourne, VIC, Australia

Seeram Ramakrishna
Centre for Nanofibers and Nanotechnology
National University of Singapore
Singapore, Singapore

PR. Selvakannan
Centre for Advanced Materials
and Industrial Chemistry
RMIT University
Melbourne, VIC, Australia

ISBN 978-981-19-2295-4 ISBN 978-981-19-2293-0 (eBook)
https://doi.org/10.1007/978-981-19-2293-0

This Springer imprint is published by the registered company Springer Nature Singapore Pte Ltd.
The registered company address is: 152 Beach Road, #21-01/04 Gateway East, Singapore 189721, Singapore

Students who were badly affected by COVID but accepted the challenges, became resilient and kept growing through their educational pathway.

The editor-in-chief and other editors have decided to donate the royalties of this book for the welfare of these students

Preface

Today we are witnessing the dawn of the fourth industrial revolution (4IR) which is fundamentally altering the way we live, eat, commute, work, and even celebrate. In its scale, outreach, and potential this revolution is unlike any other the humankind has seen before. Ushering a gradual metamorphosis of our modern society, the 4IR is rooted in powerful technological innovations ranging from Internet for global connectivity and information technology for data analytics and management to audacious technologies of the future such as Internet of Things, augmented reality, artificial intelligence, and additive manufacturing. This book analyzes one such technological innovation, *viz.,* additive manufacturing (AM) and elaborates upon the emerging paradigm shift brought about by AM in chemical sciences and engineering.

Chemical sciences and engineering play a paramount role in improving the quality of human life by manufacturing of essential commodities such as sanitizers, pharmaceuticals, textiles, plastics, rubbers, asphalt, and so on. Such manufacturing in turn relies on chemical technologies and processing equipment. While the automotive, aerospace, engineering, and biomedical industries have hastily adopted AM technologies for associated benefits, the chemical industries have been lagging in the uptake due to the lack of agility and adaptability in industrial chemical processes. AM is a manufacturing philosophy which uses incremental addition of material in a layer-by-layer fashion for fabrication of the object as per a digital design. As a result, use of AM engenders benefits such as complete design freedom, capability to manufacture sophisticated geometries, improved material utilization, tool-less fabrication, and shortened lead times. Due to the unmatched advantages of AM, it has tremendous potential in facilitating wide-scale upgradation, improvisation, and miniaturization of chemical processing equipment and plants in line with the ethics of Industry 4.0. Integration of AM with the chemical industry is anticipated to bring widespread process automation, improved feedstock and product inventories, and digital process control and process intensification. Future chemical reactors will be complex tailor-made devices with design optimized for fluid and particle flow, heat and mass transport, and reaction thermodynamics. This will lead to improved efficiencies of existing chemical processing technologies, improved process economics, environmental compatibility, reduced carbon footprint, and reduction in wastes and emissions.

The promise of AM in chemical sciences and engineering goes beyond the gradual reinvigoration and modernization of the chemical industry. AM is breaking the age-old barriers of localized engineering to enable decentralized engineering and production. It is now possible for design engineers to develop and analyze a model anywhere in the world and subsequently manufacture the matured model in an AM facility located in a different corners of the world altogether. Similarly, it is possible today to digitally store items and goods and mass produce at the point of high demand either in a large facility located in a developed nation or a temporary facility established in a developing nation facing an environmental, health, or political catastrophe. AM is also empowering mass customization and modular design movements which aim at tailor-designing of goods and services as per the requirements and demands of the customer. Thus, AM is a marvelous feat of modern technology and being hailed globally for enabling high standards of customer satisfaction, sustainability, resource utilization, and environmental protection.

This book elucidates all such unique aspects of AM and its enormous potential in shaping of the future of chemical sciences and engineering. Chapter 1 introduces AM to the chemical community followed by a description of the history and evolution of AM in Chap. 2. Chapters 3, 4, and 5 analyze the principles, challenges, and opportunities of different AM technologies in detail such as material extrusion, vat photopolymerization, laser powder bed fusion, and robocasting. Chapter 6 describes methods and technologies to impart chemical functionality to additive manufactured parts, components, reactors, and devices. Finally, the emerging technological applications of AM are described in Chap. 7 and the applications of AM in chemical engineering and micro-reaction technology are described in Chap. 8.

I sincerely hope that this book will channel the enthusiasm and focus of the chemical science and engineering community and the presented knowledge will empower them to build a better, greener, and brighter future for all humanity. As Abraham Lincoln once said, "*the best way to predict the future is to create it*". With this book, I want to give you some knowledge and tools of AM to create innovation in Chemical Engineering.

Melbourne, Australia

Suresh K. Bhargava
Editor-in-chief

Contents

About the Editors

Distinguished Professor Suresh K. Bhargava, AM, RMIT University, Melbourne.
Editor-in-chief
Distinguished Professor Bhargava is a world-renowned interdisciplinary scientist. He completed his Ph.D. from the University of Exeter, UK under the supervision of the late Professor E W Abel. He then joined the Australian National University in 1983 as a Research Fellow and today is the Dean-Research & Innovation and Founding Director of the Centre for Advanced Materials & Industrial Chemistry (CAMIC), STEM College, RMIT University. He has successfully translated his research excellence into innovative technologies for industrial application, championed gender equality in STEM, and strengthened academia-industry relationships at an international level. He has an i-10 index of 412, over 600 highly cited journal articles, 2 books, 16 book chapters, and 12 patents with a h-index of 76, and over 21100 citations. Holding distinguished professorships at top universities across 6 countries, he has created platforms for nurturing high-quality, globally agile Ph.D. students through collaborative research and international engagement. His outstanding contribution to academia and industry has been recognized through many prestigious national and international awards including the "CHEMECA Medal", the most prestigious award in engineering in Australia and New Zealand, and Honoris Causa D.Sc. Degree from Rajasthan University, India which was presented by the President of India. Quoted as being among the top 1% of world scientists in industrial chemistry, he provides

consultancy and advisory services to many government and industrial bodies across the world including BHP Billiton, Alcoa World Alumina, Rio Tinto, and Exxon Mobil. He is a fellow of six world academies, including the American Association for the Advancement of Science and The World Academy of Sciences, UNESCO. His main research interests focus on CO_2 utilization, hydrometallurgy, mineral characterization and processioning, heterogeneous and homogenous catalysis, metallodrugs, nanotechnology, and chemical sensors. He introduced the concept of 3G catalysis using 3D printing through his recent research endeavors at RMIT University.

Distinguished Professor Seeram Ramakrishna National University of Singapore, Singapore.
Professor Seeram Ramakrishna, FREng is the Chair of the Circular Economy Taskforce at the National University of Singapore. He is an editorial board member of Nature Scientific Reports and an elected Fellow of major professional societies and academies in Singapore, UK, India, and USA. He is named among the World's Most Influential Minds and the Top 1% Highly Cited Researchers in Materials Science by Thomson Reuters and Clarivate Analytics. He has co-authored 1000 journal papers, authored 8 books, received 110,000 citations, and has a h-index of 158. His research interests include innovations in sustainable materials and evaluation of circularity by life cycle assessment.

Distinguished Professor Milan Brandt RMIT University, Melbourne.
Distinguished Professor Milan Brandt is the Technical Director of the Digital Manufacturing Facility and Director of the Centre for Additive Manufacturing, RMIT University. He is the lead Australian Researcher in macro-machining and additive manufacturing with lasers and has conducted work over the last 37 years in laser cladding, cutting, drilling, welding, and more recently additive manufacturing. This has resulted in technological advancements, patents, research papers, and commercial products, which have been recognized nationally and internationally in both scientific and industrial circles. He is a Fellow of the Laser Institute of America, Honorary Fellow of Weld Australia,

Professorial Fellow of the Department of Medicine, Melbourne University, and Adjunct Prof. at the University of Waterloo, Canada. In 2018, he was the President of the Laser Institute of America, the largest international association of researchers and industry involved with lasers and laser additive technology and in 2019 he was named the Engineers Australia Centenary Hero for his research which led to Australia's first locally manufactured 3D printed spinal implant that was implanted into a patient.

Dr. PR. Selvakannan RMIT University, Melbourne.
Dr. PR. Selvakannan is a middle career researcher at RMIT University, under the mentorship of Professor Bhargava. His major fields of interest are CO_2 utilization, methane activation, endothermic fuels for high-speed flight vehicles, surface-enhanced Raman scattering, and additive manufacturing in catalysis. He has 75 publications in international peer-reviewed journals and a h-index of 34. Before joining RMIT, he worked as a Research Fellow at University of Paris-Sud, France (2007–2009) and as a Research Scientist at the Innovation Centre, Tata Chemicals Ltd, India (2005–2007). He received his Ph.D. (2005) from the National Chemical Laboratory (University of Pune), India. He graduated from the American College, India with Master's and Bachelor's degrees in Chemistry (1996–2001).

Chapter 1
An Introduction to the World of Additive Manufacturing

Milan Brandt and Suresh K. Bhargava

Contents

Abstract Science and engineering have been instrumental in advancing the progress of humanity and are the key enablers that are shaping our very present and the future, improving our health and living standards, allowing humanity to leave the planet, walk on the moon, and explore distant stars and galaxies. Additive manufacturing (AM) or 3D printing is the latest technological marvel in manufacturing that is compelling us to think differently about how we design and manufacture parts and components, the location where the parts are being manufactured, the supply chains, environmental impact of manufacturing and sovereign capability. AM is now being perceived as a disruptive technology because of the benefits and opportunities it offers over the long-established traditional manufacturing technologies. It is a relatively young technology, some 30 years since the invention of the first 3D printer, compared to conventional subtractive technologies that have been around for centuries. Mainstream AM has in fact only been around the last 5 to 10 years as both the 3D printing technology and feedstock materials have evolved and become more robust, cheaper, and reliable giving manufacturers the necessary level of confidence to adopt them in serial production. The level of activity in AM is at an all-time high as various stake holders, company executives, investors, researchers, and government agencies

M. Brandt
Centre for Additive Manufacturing, RMIT University, Melbourne, VIC, Australia
e-mail: milan.brandt@rmit.edu.au

S. K. Bhargava (✉)
Centre for Advanced Materials and Industrial Chemistry, RMIT University, Melbourne, VIC, Australia
e-mail: suresh.bhargava@rmit.edu.au

S. K. Bhargava et al. (eds.), *Additive Manufacturing for Chemical Sciences and Engineering*, https://doi.org/10.1007/978-981-19-2293-0_1

try to predict its trajectory. Global organizations of all types are attempting to understand the role they can play in the AM space. This chapter presents an overview of AM and its enormous potential in the advancement of chemical sciences and engineering. In-depth discussion of fundamental and applied aspects of each individual AM technology and how they relate to chemical sciences and engineering are presented in subsequent chapters. To date AM has seen greater adoption by the aerospace, defense, and biomedical industries and to somewhat lesser degree in the chemical manufacturing and processing industries. Introduction of AM practices in this sector can improve the efficiencies of existing chemical processing technologies, allow sophistication of existing processes for manufacturing of new products, improve process economics, and address complex challenges and demands of our modern society such as environmental compatibility, reduction in wastes and emissions and localized production. AM in chemical sciences has enormous potential in process intensification which enables efficient heat and mass transfer, large reduction in the footprint of chemical manufacturing and processing and reduce environmental pollution.

Keywords Additive manufacturing · 3D printing · Rapid prototyping · Customization · Chemical manufacturing · Process intensification · Energy storage · Reactors

Abbreviations

2D	Two-dimensional
3D	Three-dimensional
AM	Additive Manufacturing
ASB	Academic Sharp Brain
BJT	Powder Bed Inkjet printing
CAD	Computer Aided Design
EPBM	Electron Beam Melting
FDM	Fused Deposition Modelling
Ge	Germanium
LPBF	Laser Powder Bed Fusion
MJT	Material Jetting
Ni	Nickel
SLA	Stereolithography
SLM	Selective Laser Melting
SLS	Selective Laser Sintering
STL	Standard Tessellation Language
S	Sulfur
U	Uranium

1.1 Background to Additive Manufacturing

Additive Manufacturing (AM), commonly known as 3D printing, is a manufacturing methodology wherein the material is added to form a desired solid geometry instead of subtraction or casting as in conventional manufacturing processes. Other common names for AM include solid free form fabrication, direct digital manufacturing, desktop manufacturing, additive technology and layered additive manufacturing. The AM technologies were initially developed to meet the demand of rapid prototyping and testing in the industrial manufacturing of sophisticated objects and components. The fundamental working principle behind most AM technologies is essentially a layer-by-layer fabrication method. AM process typically starts with a digital design of the component using a computer aided design (CAD) software. The design is then converted into a stereolithography file format (.STL), wherein the component surfaces are tessellated into triangles, and a database of the triangle nodes stored. In the next step, layers are generated from the created database wherein each layer governs the deposition of material in two dimensions (2D) and the layer thickness and interweaving govern the deposition of material in the third dimension. All geometric information about the desired solid object to be manufactured is now available in the form of a g-code which can be subsequently fed into a material deposition device. When the object is manufactured, material deposition takes place in a layer-by-layer fashion wherein each layer is manufactured according to the deposition scan implementation in the g-code and is repeated until the entire component is built. Thus, the additive process flow in AM alleviates the constraints associated with traditional manufacturing methods and provides tremendous opportunities for the design of novel geometries and complex internal structures to enhance component properties such as its strength-to-weight ratio or cooling in comparison with traditional manufacturing methods.

AM also allows fabrication of 3D objects in a variety of material compositions such as metals, alloys, ceramics, polymers, tissues, and cells. However, each AM technology is accompanied by its own benefits and limitations which may constrain the design and manufacturing process flow. For instance, Fig. 1.1 illustrates the differences in design elements and manufacturing process flow involved in the realization of a design concept the ‘Academic Sharp Brain (ASB)’ into a 3D object composed of metal or plastic and fabricated using AM. At the design stage, the designer develops a complete layout and geometrical design of an object based on the project objective, availability of feedstock material, available funds and resources and desired characteristics of finished product such as weight, strength, visual appeal, and shelf life. This stage is depicted for metallic ASB in Fig. 1.1a and plastic ASB in Fig. 1.1e. Varying degree of infills are commonly employed in AM for reduction of material wastage since the desired strength and shelf-life can be easily achieved by fabrication of hollow objects with solid shells. For the metallic ASB, the designer gives a masculine form to ASB and imbibes the infill requirement directly in the design stage by designing a meshed object with interwoven triangular lattice. Such a design allows optimization of object strength by controlling the geometry of the mesh element

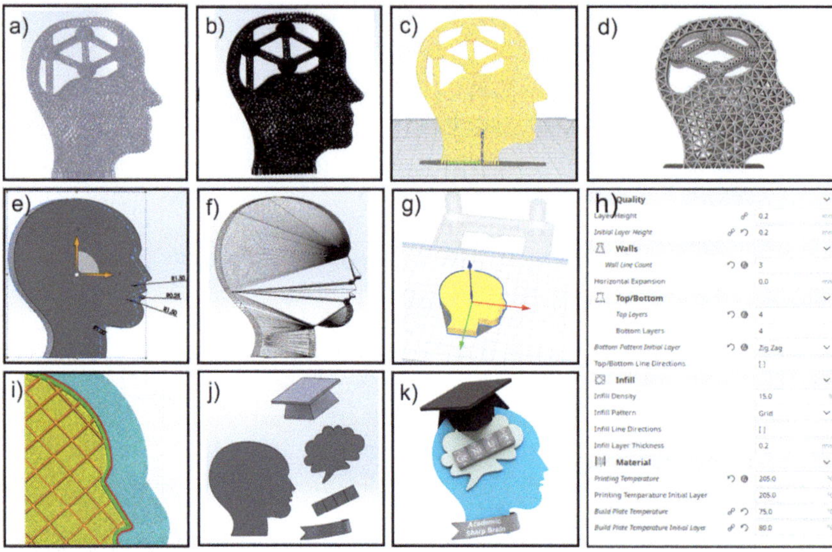

https://www.youtube.com/watch?v=ZS_vQ0ObqaY

Fig. 1.1 Illustration of different design elements and process flow involved in the realization of a bespoke design concept into a metallic (a–d) and plastic (e–k) 3D object 'Academic Sharp Brain Logo (ASB)' using additive manufacturing; **a** CAD design of the metallic ASB composed of triangular meshes to control the weight, reduce material wastage, reduce print times and improve visual appeal; **b** Tessellation of the CAD design file into STL file format for supplying the geometric information to laser position stage; **c** Adjusting the build orientation to vertical based on the requirements of SLM; **d** Image of the finished metallic ASB after post processing and finishing; **e** Design of a single component of the plastic ASB using Solidworks CAD software; **f** Conversion of the 3D model to a STL file wherein all component surfaces are tessellated into triangles and data stored in a g-file; **g** Import of g-file into an open-source and free to use process modelling software 'Cura' available under public license LGPLv3; **h** An example of different printing parameters that govern the quality of finished printed object such as layer height, wall thickness, number of top and bottom layers, infill type and density, printing speed, bed temperature etc.; **i** Use of 15% infill volumetric density and a grid infill pattern to reduce print times and consumption of feedstock material; **j** Assembly of the final object using different printed components; and, **k** The final object after painting and post processing. The Academic Sharp Brain event was hosted as a part of the Royal Australian Chemical Institute Centenary Congress in July 2017 (https://www.youtube.com/watch?v=ZS_vQ0ObqaY)

such as size, thickness, and shape. Moreover, the designer incorporates symbols of the four elements, Germanium (Ge), Nickel (Ni), Uranium (U) and Sulfur (S) to communicate 'GENIUS'. Figure 1.1b depicts the tessellation of the CAD design file into the STL file format. The STL file can then in turn be converted into a g-code which provides positioning coordinates to the fabrication device such as the laser positioning stage in selective laser melting (SLM). At this stage the fabricator must decide the orientation of the build which impacts the build quality as demonstrated in Fig. 1.1c for fabrication of the metallic ASB in a vertical build orientation. Finally, an image of the finished object after the SLM process and post manufacturing cleanup is depicted as Fig. 1.1d.

For the plastic ASB, the designer has given a feminine form to the design concept and decided to add décor such as a graduation hat and an ASB collar. Instead of printing the object as a whole, the designer decided to print each component separately and assemble in the last stage. Figure 1.1f depicts the tessellation of the CAD design into STL file format and data storage in a g-file. Considering the efficacy of the Fused Deposition Modelling (FDM) process, the fabricator decided to print in a horizontal build orientation as shown in Fig. 1.1g which was controlled through an open-source and free to use process modelling software 'Cura'. The process modelling software also allows control over printing parameters such as layer height, wall thickness, number of top and bottom layers, infill type and density, printing speed and bed temperature as depicted in Fig. 1.1h. In contrast to the metallic ASB where infill was incorporated directly into the design, for the plastic ASB the fabricator incorporates an infill of 15% volumetric density and a grid infill pattern in the process modelling stage as depicted in Fig. 1.1i. The design of different components of the desired object are shown in Fig. 1.1j and an image of the finished object after assembly, finishing, and painting is depicted in Fig. 1.1k. This example illustrates the efficacy of additive manufacturing in enabling total design freedom and free form fabrication.

Today AM is the fastest growing sector of manufacturing globally. Wohlers Associates' 2021 annual Report [1], the most recognized and comprehensive account of the current state of the global additive manufacturing industry, estimated that the global 3D printing industry has grown to a total of US $12.8 billion in 2020 as shown in Fig. 1.2. The 374-page report also discusses the impact of COVID-19 on the additive manufacturing industry and estimates that the annual growth was down

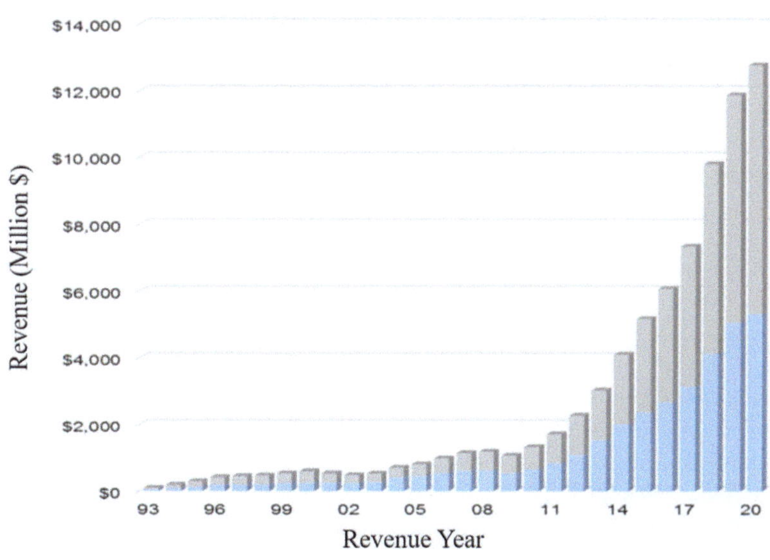

Fig. 1.2 Global revenue generated from AM products and services since 1993 over the years. The lower (blue) segment of the bars represents products, while the upper (grey) segment represents services. *Source* Wohlers Report 2021 [1]

considerably from an average growth of some 27% over the previous 10 years but nevertheless, the marked grew by some 7.5%. The report also includes commentary on 74 early-stage investments and 35 acquisitions and public offerings. Additive manufacturing startups and established companies have received substantial funding in 2020. For example, Desktop Metal, received $575 million as part of a merger with a special acquisitions company. After going public in December 2020, Desktop Metal's market capitalization exceeded $7.5 billion in February 2021.

Despite COVID-19, many engineering industrial giants such as Stryker, Audi, BMW, Boeing, Ford, General Electric, Hewlett-Packard, Nike, Airbus and Alcoa to name a few as well as many small and medium-sized enterprises globally, are on the path of partial to complete adoption of additive manufacturing technologies. In 2020, many companies quickly pivoted to additive manufacturing technologies in response to the shortage of personal protection equipment, testing equipment and ventilators caused by COVID-19 pandemic. Popular press, technical and trade journals are reporting new innovations and advancements in 3D printing almost every day such as launch of new and innovative printers, sophisticated parts additively manufactured in both metallic alloys as well as polymers and design and manufacturing of modern components that challenge the imagination. Many countries have initiated large national programs in the last few years to further accelerate the development of new AM technologies, feedstock materials and their applications.

In the modern age, the global shift from mechanical and analogue electronics to digital electronics is being called the digital revolution or the fourth industrial revolution. 3D printing is the manufacturing component of this digital revolution wherein manufacturing of a solid object from a digital design is being branded as direct digital manufacturing. The global inclination towards direct digital manufacturing can be traced back to some five years ago. COVID-19 has provided further impetus to this movement by posing a serious threat to supply chain logistics, sovereign capability, and national security. All these factors are changing the nature and economics of global manufacturing, particularly in "first-world" countries and will continue to do so in the future. Thus, it is critical for companies in "first-world" countries to leverage advanced technology and latest research for addressing the challenges of modern global economics and competitive manufacturing to remain in business and make profits. AM is being increasingly seen as 'the solution' to such challenges owing to the benefits it offers compared to conventional subtractive manufacturing. Unlike the early days of rapid prototyping, latest additive manufacturing technologies are capable of directly manufacturing functional polymers and metal components from computer generated models or based on measurements taken to re-engineer existing components. As such AM has truly bypassed traditional manufacturing processes such as cutting, milling, and grinding in most aspects. Various benefits and advantages of AM shown in Fig. 1.3, include: (1) capability to manufacture new and sophisticated designs not possible using conventional subtractive technology, (2) dramatic savings in time, materials, wastage, energy, and other costs in producing new components, (3) significant reductions in environmental impact, and (4) faster time to market for products.

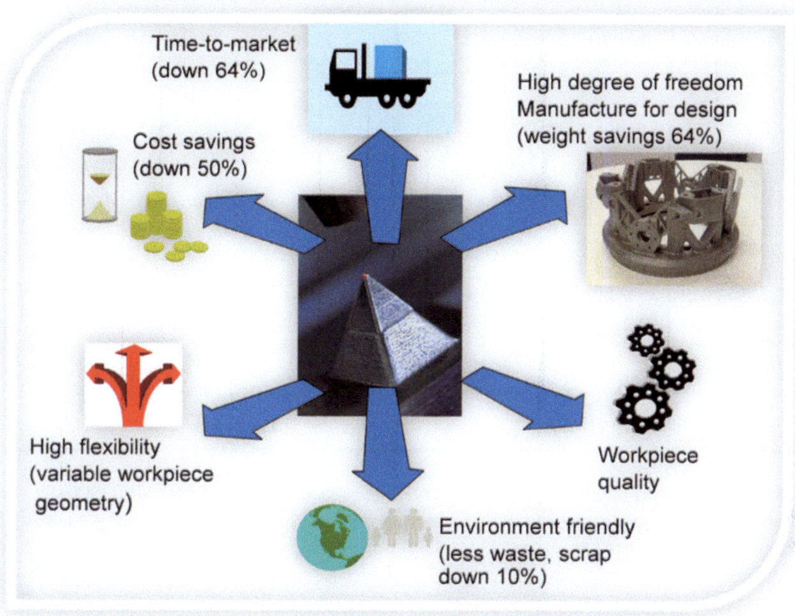

Fig. 1.3 Various benefits and advantages of additive manufacturing. Figure in brackets refers to savings in aerospace and defense sectors. Permission pending from [2]

1.2 Standardization of Additive Manufacturing Technologies

In common with the conventional subtractive technologies, there has been a push in recent years to standardize the AM technologies to ensure part quality, repeatability, and consistency across builds and machines. According to the American Society for Testing and Materials (ASTM F52900-15) [3] there are over 50 different AM technologies, which were classified into 7 standard processes as binder jetting, material jetting, material extrusion, vat photopolymerization, powder bed fusion, energy deposition and sheet lamination as shown in Fig. 1.4. Besides the standard classification, AM technologies are often classified based on feedstock materials such as metals, polymers and other materials and the method used to fuse the materials as shown in Fig. 1.5. A number of these technologies are now quite mature, such as fused deposition modeling (FDM), selective laser sintering (SLS), stereolithography (SLA), powder bed inkjet 3D printing (BJT), and their variants (Material jetting (MJT)) and laser (LPBF) and electron beam melting (EBM) [4–6].

The wide interest and enormous potential in the dynamic field of AM has led to development and commercialization of new technologies and feedstock materials rapidly and on a regular basis. In the recent years it has become possible to manufacture objects with an ever-expanding library of printable materials in shorter

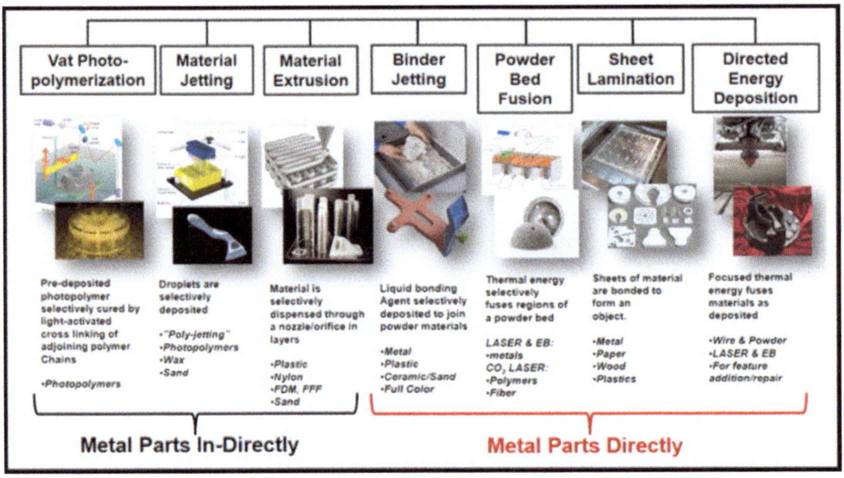

Fig. 1.4 Standard classification of AM technologies according to ASTMF52900-15. *Source* https://insights.globalspec.com/article/7447/factors-to-consider-when-3d-printing-or-additive-manufacturing-metal-parts [7]

production times and at lower cost which was not feasible before. A more detailed description of each technology, its benefits, and limitations in terms of print resolution, component manufacturability and part properties, examples of innovative component designs and range of current applications in relation to chemical science and engineering are presented in the following chapters.

1.3 Additive Manufacturing—Features and Characteristics

Like any manufacturing technology, all AM technologies have some common attributes which are characteristic of the direct digital and layer-by-layer manufacturing methodology such as the need of a CAD and process modelling software, processable feedstock materials and quality of manufactured objects in comparison to conventional manufacturing technologies. For additive manufacturing to become a global household phenomenon it is critical that CAD design and process modelling software are made available to the user with an interactive and intuitive user interface at an affordable price or free of cost. The AM community is already responding to this need and the latest generation open-source software are revolutionizing how 3D objects are designed and modelled for 3D printing. While the developments in commercially available AM software have been significant over the last five years in terms of their design and process modelling capability, they are still lacking in the ability to easily model or analyze multiple material geometries and their accompanying anisotropy.

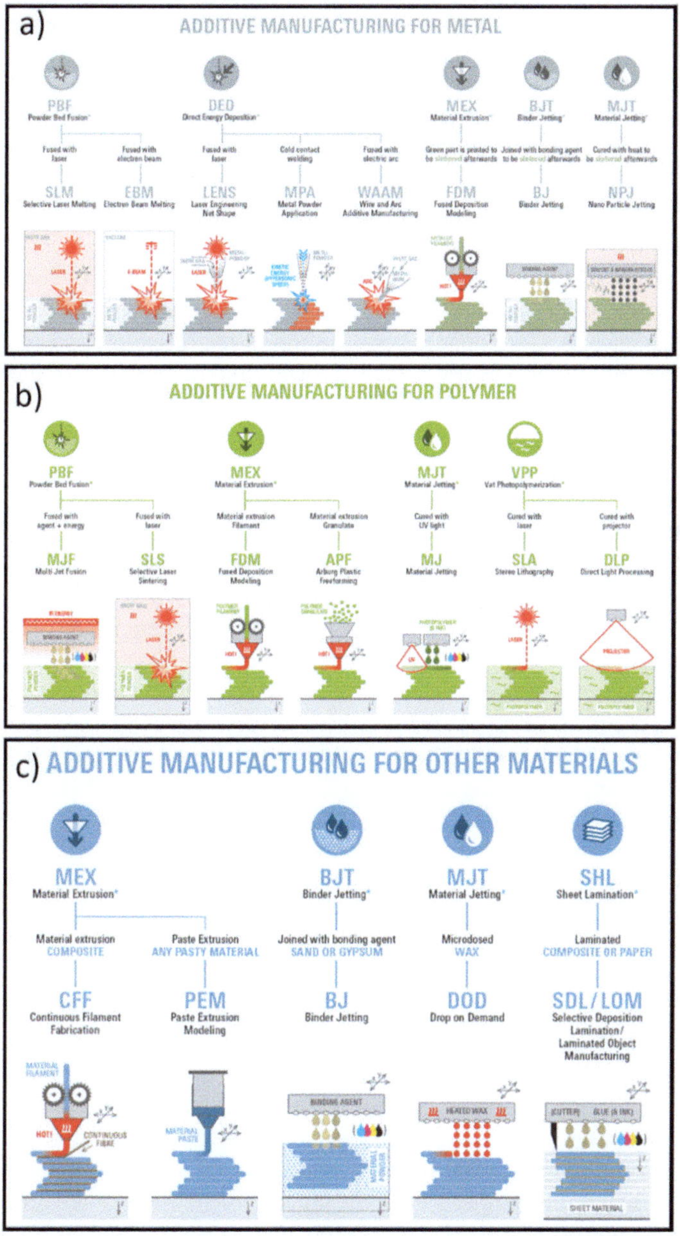

Fig. 1.5 Classification of AM technologies based on the method of fabrication and processable feedstock materials as **a** metal; **b** non-metals and **c** other materials. *Source* Ritter, Steffen (2020). Formnext AM field Guide compact 2021. Frankfurt Main: Mesago Messe Frankfurt GmbH ISBN 978-3-9,823,018-7-3 [8]

A vital characteristic of AM technologies is the need for adequate feedstock material in the required form such as powder, filament, pellets, granules, sheets, and resins. The range of feedstock materials for AM technologies is expanding exponentially and new materials are constantly coming into the market [4–6, 9]. The materials can be broadly classified as metals and non-metals and come in a variety of forms depending on AM technology ranging from powder to resins. The properties of feedstock material such as particle size, shape, density, composition, and purity are governed by the operational requirements of individual AM technologies and are specified by the manufacturer of the technology. Specific materials can and have been developed for an individual AM technology for dedicated applications with material properties. A more detailed description of the different feedstock material characteristics and properties for each AM technology is provided in the following chapters.

Another very important characteristic feature of AM is the surface roughness of manufactured parts and components and anisotropic mechanical properties arising from the layered nature of the build. The variation in mechanical properties for example can arise from the directional solidification of the melt pool in the case of metallic feedstocks and interlayer bonding deficiencies in the case of polymer feedstocks. In general, the quality of a printed part's surface is mainly determined by the thickness and overlap of each printed layer. Larger layer thickness and overlaps result in a rougher surface especially when building at an angle. Employing smaller layer thicknesses result into corresponding increase in object production time which can seriously impact the process economics. This adverse effect arising from the layer-by-layer manufacturing methodology in AM is also called the "stair-stepping" effect and governed by the underlying deposition technology. Extrusion-based AM systems typically have the largest layer thicknesses (0.2 mm) due to the large diameter of the deposition nozzle. Alternatively, powder bed fusion and vat photopolymerization AM technologies systems have much smaller layer thicknesses ($\sim$0.1 mm), and thus result in smoother surfaces because of the system ability to precisely focus the energy beam radius. Material jetting systems also offer fine layer thicknesses (as low as 0.02 mm) due to the ultrasmall size of jetted droplets. Surface quality of the manufactured objects also depend on the employed feedstock materials. AM processes employing metallic powders have poorer surface quality than others due to large and partially melted powder particles that adhere to the surface of manufactured object. The surface roughness of powder bed systems is also a function of the build orientation where the sides of the build have different surface roughness in comparison to the top and bottom of the build. A large majority of AM systems still employ only a single material at a time. New multi-material AM systems are now emerging that allowing manufacturing with functionally graded materials in both polymer and metal. However, the technology is very much in the research phase with limited data on material behavior at the interface and a lack of design software support [10, 11].

1.4 Additive Manufacturing in Chemical Sciences and Engineering

A chemical is a substance with a characteristic composition and distinct property. Chemicals are the building blocks of our modern technological society and it would be impossible to sustain our current lifestyle without them. Chemical sciences refer to the group of fundamental sciences which deal with the scientific study of all aspects of chemicals such as their properties, production, storage, and transportation. Thus, the advancement and practice of chemical sciences is essential for the production of medicines that allow us to recover from illness, health supplements that shape and improve our body, perfumes and air-fresheners for our sensory pleasures, soaps, and detergents to clean us and our clothes, artificial fabrics such as nylon, terylene, velvet and velcro for our fashion needs, paints, and varnishes for the beautification of objects all around us, plastics for storage, fuels for electricity, heating, cooking and vehicles, fertilizers, and pesticides for the production of the food we eat, the asphalt in our roads and so on. Due to the multidisciplinary nature of applications involved, chemical sciences operate at the interface of chemistry, physics, biology, and material science. Chemical engineering uses the fundamental knowledge from chemical sciences and deals with the design, manufacturing, safety, operation and maintenance of chemical processing instruments, equipment, and plants. Thus, chemical engineering operates at the interface of chemical sciences, mathematics, economics, mechanical and civil engineering. Continuous progress in the chemical sciences and engineering is clearly critical to:

1. Future advances in sustainable production, storage, and supply of energy
2. Conservation of the environment and mitigation of climate change
3. Increase in food production and its sustainability with less demand on arable land
4. Advanced diagnostics and novel disease prevention therapies based on exploitation of the human genome
5. Designing processes and products that preserve resources
6. Provision of accessible clean drinking water for all

As the leaders of process-based manufacturing, chemical scientists and engineers must face the modern challenges of chemical production such as process economics, competitiveness, supply chain and logistics, environment protection and sustainability. To meet these challenges a continuous evolution of chemical industries is required which relies on the fast and innovative development of modern and cutting-edge manufacturing processes, materials, and products. In addition to the customary demands of low price and best quality, the cut-throat market competition demands products that are intricate, possess shorter life cycles, exhibit shorter delivery times, involve customization, and require less skilled workers. In fact, the current breed of products is highly sophisticated and challenging to design. Accordingly, there is a strong incentive toward the design, development, and implementation of new and ingenious chemical manufacturing processes which can be realized by employing additive manufacturing technologies [12, 13].

By 2030 the world's population is expected to increase to over eight billion, with the majority of those people living in cities. Global energy requirements will continue to increase along with the pressure on Earth's natural resources to provide this rapidly expanding population with enough food, water, and shelter. It is estimated that more than one billion people now live-in poverty without sufficient food, water or adequate sanitation and healthcare provision. The newly industrialized nations of Asia and Latin America are undergoing rapid industrial expansion and experiencing large economic growth but also contributing to modern society's environmental problems such as waste disposal, air, water, noise, and light pollution. The human ambition to grow and expand has seriously impacted the earth's ecological system and climate change is getting wide recognition as the number one threat to humanity. Mitigation and permanent resolution of these challenges will rely on adoption of sustainable development principles i.e., development of capabilities to meet the needs of the present without compromising the ability of future generations to meet their own needs. The foundation of this sustainable society must be laid down with eco-efficient processes and products that the chemical sciences can provide. Advances in chemical sciences will be needed to develop functional materials to construct our homes and buildings and develop lightweight, safe and energy efficient vehicles which will help reduce greenhouse gas emissions. The chemical sciences, therefore, have a clear role in enabling sustainable development and providing technological solutions to face the challenges of society today and in the future. The technologies that the chemical sciences engender will improve the quality of daily life, underpin prosperity, and increase our readiness to face the challenges of the future.

The adoption of AM technologies in some industry sectors such the aerospace, defense and bioengineering has been significant over the last five years with new AM products or AM based processes being reported in the open press on a regular basis. However, the use of AM technologies within the chemical sciences has not seen the same level and scale of adoption. Nevertheless, the potential is significant in the context of health, energy and food production and laboratory scale demonstrations of the application of the AM technology have now started to emerge in the chemical sciences [5, 14]. A thorough and critical review summarizing the most frequently applied AM technologies in the context of design and functional prototypes, the printing materials and technologies is given in [15]. In the chemical sciences, polymeric materials have found a larger acceptance due to their favorable processing and post-processing options, range of product properties, low cost of printing, and accessibility of printers. Designed and manufactured objects have been successfully applied as lab-scale reaction ware, for optimization of reactors and reaction protocols and development of operation units for process scale-up. This union of AM from the mechanical sciences and chemical reaction engineering from chemical sciences opens doors for design of smart reactors that enable high conversion, superior product selectivity and favorable process economics. Thus, through the synergistic utilization of computational fluid dynamics and AM technologies, a new generation of batch and continuous reactors will emerge for chemical reactions. For example, Zhao et al. [16] designed a fractal type photobioreactor based on bifurcation tree algorithm and fabricated it using SLA technology as shown in Fig. 1.6. The novel design improved mass transfer, photosynthetic conversion and CO_2 capture efficiency enabling high growth rates for biomass.

The scope of AM in chemical reaction engineering is not just limited to manufacturing of sophisticated reactor designs but also allows multi-device integration, rapid prototyping and testing and improvisation of lab scale continuous and batch reactors which can be manufactured in polymers and metals. For instance, Fig. 1.7 depicts a state-of-the-art 3D printed split-and-recombine stainless-steel reactor with an integrated cooling shell and oxygen sensors for online reaction monitoring of the oxidation reaction of a Grignard reagent in homogeneous phase. The reactor

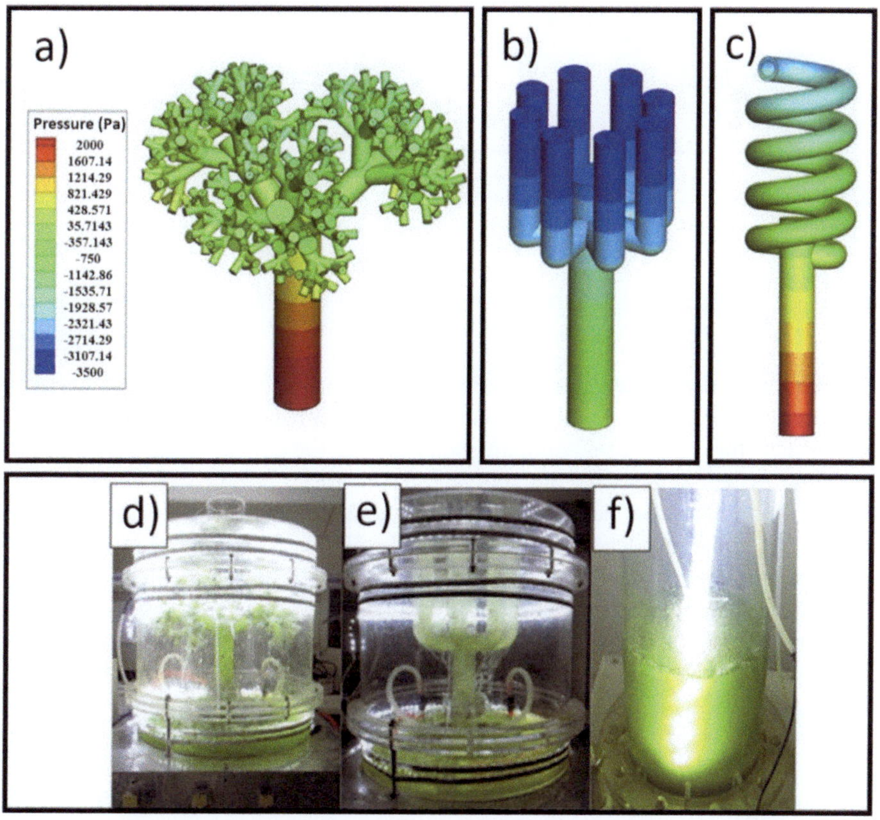

Fig. 1.6 A fractal type photobioreactor (PBR) designed based on bifurcation tree algorithm and manufactured using stereolithography; (Top) Time averaged relative pressure in three-types of PBRs modelled using CFD; **a** Fractal tree-like PBR; **b** multitubular PBR; **c** tubular reactor with a spiral column; (Bottom) Chlorella culture in the different PBRs on the 4th day in the **d** Fractal tree-like PBR; **e** multitubular PBR; and **f** tubular reactor with a spiral column. Reproduced from Ref. [15] with permission from Elsevier

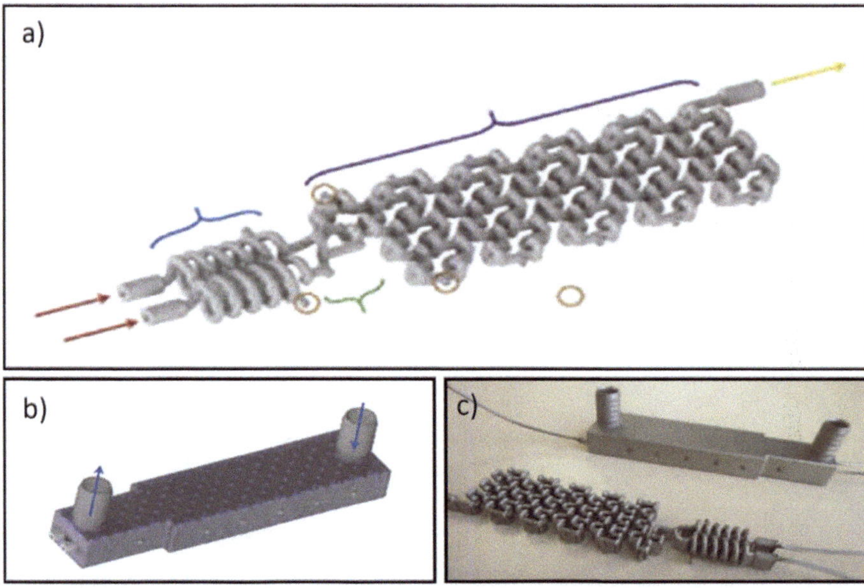

Fig. 1.7 A stainless-steel split and recombine reactor fabricated by additive manufacturing **a** Overview and CAD design of reactor section in the split and recombine reactor. Repetition of the structural elements in any direction allow arbitrary scaling of the reactor; **b** Design of the cooling jacket encapsulating the reactor section; and **c** Image of the reactor with and without the external cooling shell. Reproduced from Ref. [17] with permission from the Royal Society of Chemistry

comprises pre-cooling sections, a series of split-and-recombine mixing elements which can be increased in numbers or scaled for larger reactor lengths or volumes, a cooling shell for recirculating coolant encapsulating the entire reactor, and multiple fiber optic oxygen sensors for online reaction monitoring. Compared to a plug flow micro coiled reactor, the novel reactor exhibited comparable conversion levels but significantly higher product selectivities. The developed AM rector allowed product yields up to hundred's milliliter range per day [17].

Another vital component of chemical reaction engineering, heterogeneous catalysis is an integral part of our large-scale energy, fuel, chemical, and pharmaceutical industries. Implementation of alternative, small-scale AM technologies in heterogeneous catalysis can potentially change the entire technological landscape and anticipated to have an enormous and immediate impact. A detailed review of the different aspects and application of various AM technologies in heterogeneous catalysis such as functionalization of printed structures with catalysts, direct printing of catalytic materials or catalyst supports and rapid prototyping and printing of reactionware is available in the literature [18]. One of the major objectives in chemical manufacturing is the development of microreactors which enable superior process efficiencies and allow sustainable production of chemicals on demand and onsite. This objective has taken the center stage in the wake of COVID-19 pandemic where disruptions in global supply chains and international border closures have put extraneous pressure on the local manufacturing sector and provided impetus for localized production of pharmaceuticals, fine chemicals, food additives and other similar chemicals of importance. Thus, just a reduction in the scale of chemical operations achieved at comparable process efficiencies using AM technologies can address this major objective. Secondly, traditional large-scale chemical manufacturing units and plants are difficult to adapt once built as the cost for modifications at the large scale are quite steep. Thus, in the case of poor resource management, hazardous processing practices or poor environmental sustainability resulting from the final design, implementation of the necessary modifications can prove to be unviable and non-profitable. Therefore, introduction of AM in heterogeneous catalysis and the chemical industry can dramatically improve plant adaptability by employing modular design and smaller sized equipment and reactor components to reduce manufacturing and design costs. In addition, the relative cost efficiency of AM technologies for short production runs makes them ideal for late-term adjustments to chemical plants either due to unknowns in the process, or just the need for new process functionality. Images of some of these AM devices in metals and polymers which were developed at RMIT university for various catalytic, separation and sensing applications, are shown in Fig. 1.8.

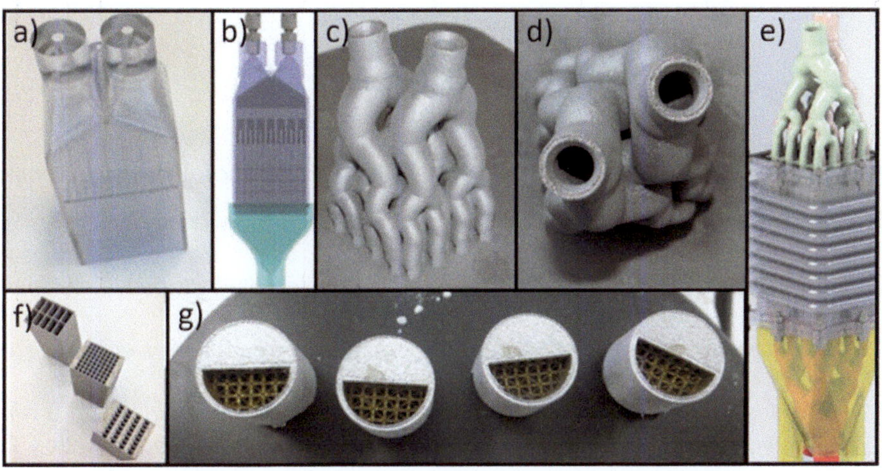

Fig. 1.8 Images of the additively manufactured flow devices, monoliths and reactionware developed at RMIT university; **a** Perspective view of the concept of a dual inlet fractal 3D micromixer with decrease in channel cross section with reactor depth; **b** Front view of the fractal micromixer; **c** Image of the 3D micromixer fabricated using selective laser melting; **d** Top view of the fractal 3D micromixer; **e** The concept of the complete reactor assembly with micromixer inlets and outlets; **f** rectangular monolith reactors; **g** cylindrical monolith reactors. (*Source* RMIT Centre for Additive Manufacturing)

1.5 Summary

Additive Manufacturing is truly an enabling technology and its enormous potential in generation of new products and processes with improved process economics has only started to emerge in the chemical manufacturing sector. Implementation of AM technologies in chemical reaction engineering has enabled the development of state-of-the-art automated reactor systems with online reaction monitoring and sophisticated reactor designs with functionalized surfaces. The design freedom allows control of reactant stoichiometry by modification of the reactor geometry and incorporation of desired features, such as proton exchange membranes and purification columns. AM also has enormous potential in catalyst exploration, material development and advanced catalytic systems with gradient catalytic activities as multi-material AM systems become commercialized and accessible. While the application of AM to chemical sciences and engineering is still in an early stage, the reported research and developments in the field demonstrate that AM will become an integral tool in chemical science and engineering and shape the manufacturing processes of the future. New innovations in materials, technology and software will ensure that such progress will only be accelerated in the future.

1.6 Test Your Knowledge

After reading the above chapter you should be able to answer the following questions:

1. *What are the three most common manufacturing philosophies used for fabrication of complex shaped objects?*
2. *What is the typical process flow involved in additive manufacturing of an object from a design concept?*
3. *What are the seven standard additive manufacturing processes according to ASTM F52900-15 committee on additive manufacturing?*
4. *What is stair-stepping effect in additive manufacturing and what is the consequence of this effect on the surface finish of the printed object?*

References

1. Wohler (2021) 3D Printing and additive manufacturing, global state of the industry
2. Brandt M (2017) The role of lasers in additive manufacturing. In: Brandt M (ed) laser additive manufacturing. Woodhead Publishing, pp 1–18. https://doi.org/10.1016/B978-0-08-100433-3.02001-7
3. ISO/ASTM52900-15, standard terminology for additive manufacturing—general principles—terminology ASTM international, West Conshohocken, PA (2015), www.astm.org

4. Gao W, Zhang Y, Ramanujan D, Ramani K, Chen Y, Williams CB, Wang CCL, Shin YC, Zhang S, Zavattieri PD (2015) The status, challenges, and future of additive manufacturing in engineering. Comput Aided Des 69:65–89. https://doi.org/10.1016/j.cad.2015.04.001
5. Zhakeyev A, Wang P, Zhang L, Shu W, Wang H, Xuan J (2017) Additive manufacturing: unlocking the evolution of energy materials. Adv Sci 4(10):1700187. https://doi.org/10.1002/advs.201700187
6. Molotnikov A, Kingsbury A, Brandt M (2021) 23—current state and future trends in laser powder bed fusion technology. In: Yadroitsev I, Yadroitsava I, du Plessis A, MacDonald E (eds) Fundamentals of laser powder bed fusion of metals. Elsevier, pp 621–634. https://doi.org/10.1016/B978-0-12-824090-8.00011-1
7. Kardys G (2017), https://insights.globalspec.com/article/7447/factors-to-consider-when-3d-printing-or-additive-manufacturing-metal-parts
8. Formnext AM Field guide compact, https://formnext.mesago.com/frankfurt/en/themes-events/am-field-guide.html. (2020)
9. Bourell D, Kruth JP, Leu M, Levy G, Rosen D, Beese AM, Clare A (2017) Materials for additive manufacturing. CIRP Ann 66(2):659–681. https://doi.org/10.1016/j.cirp.2017.05.009
10. Walker J, Middendorf JR, Lesko CCC, Gockel J (2021) Multi-material laser powder bed fusion additive manufacturing in 3-dimensions. Manuf Lett,https://doi.org/10.1016/j.mfglet.2021.07.011
11. Wei C, Sun Z, Chen Q, Liu Z, Li L (2019) Additive manufacturing of horizontal and 3D functionally graded 316L/Cu10Sn components via multiple material selective laser melting. J Manuf Sci Eng 141(8), https://doi.org/10.1115/1.4043983
12. Chang F, Zhang X, Zhan G, Duan Y, Zhang S (2020) Review of methods for sustainability assessment of chemical engineering processes. Ind Eng Chem Res 60(1):52–66. https://doi.org/10.1021/acs.iecr.0c04720
13. Boydston AJ, Cui J, Lee C-U, Lynde BE, Schilling CA (2020) 100th anniversary of macromolecular science viewpoint: integrating chemistry and engineering to enable additive manufacturing with high-performance polymers. ACS Macro Lett 9(8):1119–1129. https://doi.org/10.1021/acsmacrolett.0c00390
14. Femmer T, Flack I, Wessling M (2016) Additive manufacturing in fluid process engineering. Chem Ing Tec 88(5):535–552. https://doi.org/10.1002/cite.201500086
15. Zentel KM, Fassbender M, Pauer W, Luinstra GA (2020) Chapter four—3D printing as chemical reaction engineering booster. In: Moscatelli D, Sponchioni M (eds) advances in chemical engineering, vol 56. Academic Press, pp 97–137, https://doi.org/10.1016/bs.ache.2020.08.002
16. Zhao L, Zeng G, Gu Y, Tang Z, Wang G, Tang T, Shan Y, Sun Y (2019) Nature inspired fractal tree-like photobioreactor via 3D printing for CO_2 capture by microaglae. Chem Eng Sci 193:6–14. https://doi.org/10.1016/j.ces.2018.08.057
17. Maier MC, Lebl R, Sulzer P, Lechner J, Mayr T, Zadravec M, Slama E, Pfanner S, Schmölzer C, Pöchlauer P, Kappe CO, Gruber-Woelfler H (2019) Development of customized 3D printed stainless steel reactors with inline oxygen sensors for aerobic oxidation of Grignard reagents in continuous flow. React Chem Eng 4(2):393–401. https://doi.org/10.1039/c8re00278a
18. Hurt C, Brandt M, Priya SS, Bhatelia T, Patel J, Selvakannan PR, Bhargava S (2017) Combining additive manufacturing and catalysis: a review. Catal Sci Technol 7(16):3421–3439. https://doi.org/10.1039/c7cy00615b

An Introduction to the World of Additive Manufacturing

Background, Features, Standardization, AM in Chemical Science and Engineering

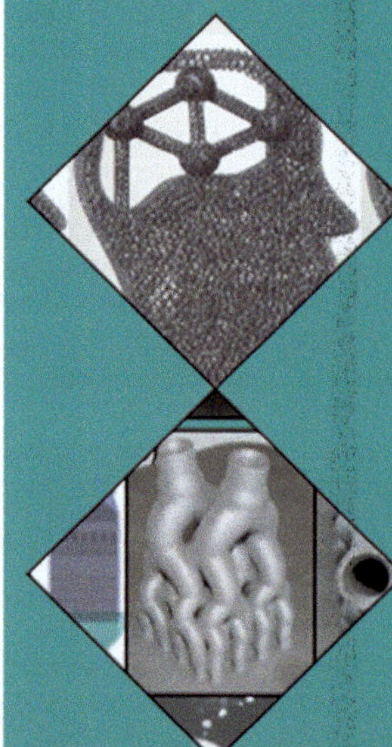

Milan Brandt, Suresh K. Bhargava

1

Chapter 2
History and Evolution of Additive Manufacturing

Sunpreet Singh, Sunil Mehla, Suresh K. Bhargava, and Seeram Ramakrishna

Contents

Abstract Since its first use to build sculptures, additive manufacturing has come a long way to become a prominent and prolific modern technology with wide acceptance due to significant benefits and advantages arising from its adoption and implementation in different industrial and consumer sectors. AM is rapidly paving a path towards an intensified technological society by facilitating unprecedented accomplishments ranging from digital storage of production data and on-demand manufacturing to repair and maintenance of malfunctioning machines, tools, and even human's anatomy. This chapter presents a state-of-the-art review on the history and evolution of additive manufacturing technologies along with the emerging paradigm for the incorporation of AM in major industrial and research sectors. Overall, the chapter provides all the necessary insights essentially required by young researchers to understand where AM came from and where it is going.

S. Singh · S. Ramakrishna
Centre for Nanofibers and Nanotechnology, National University of Singapore, Singapore, Singapore
e-mail: seeram@nus.edu.sg

S. Mehla · S. K. Bhargava (✉)
Centre for Advanced Materials and Industrial Chemistry, RMIT University, Melbourne, VIC, Australia
e-mail: suresh.bhargava@rmit.edu.au

S. K. Bhargava et al. (eds.), *Additive Manufacturing for Chemical Sciences and Engineering*, https://doi.org/10.1007/978-981-19-2293-0_2

Keywords History · Evolution · Additive manufacturing · Customization · Modularization · Merits · Principles

Abbreviations

2D	Two-dimensional
3D	Three-dimensional
4D	Four-dimensional
AM	Additive Manufacturing
ASTM	American Society for Testing and Materials
CAD	Computer Aided Design
CEN	European Committee for Standardization
CT	Computer Tomography
DED	Directed Energy Deposition
DMLS	Direct Metal Laser Sintering
EBM	Electron Beam Melting
FDM	Fused Deposition Modelling
HA	Hydroxyapatite
IOP	Integrated Organ Printing
IoT	Internet of Things
IP	Intellectual Property
ISO	International Organization for Standardization
LM	Layered Manufacturing
MRI	Magnetic Resonance Imaging
PEEK	Polyether ether ketone
RPD	Rapid Plasma Deposition
SLA	Stereolithography
SLS	Selective Laser Sintering
SMP	Shape Memory Polymers
UV	Ultraviolet

2.1 Introduction

Additive manufacturing (AM), also known as three-dimensional (3D) printing, is a manufacturing methodology in which material is added to fabricate a 3D object as per CAD design instead of removing material to shape the object as performed in conventional manufacturing methodologies such as substractive machining. Thus, using AM solid objects can be fabricated directly from polymeric, metallic, ceramic and a possible combination of these materials without minimal use of tooling and fixtures which improves ease of operation and reduces cost of fabrication [1]. AM also has great synergy with 3D scanning tools that can directly generate the CAD file needed for fabrication. For example, magnetic resonance imaging (MRI) and computer tomography (CT) scan data can be converted to CAD models using medical modelling software, such as MIMICS or 3D Doctor. Thus, with adequate feedstock materials, AM can in principle fabricate any organ, tissue, nerve, artery, and vein that can be scanned, mimicked, or designed. For instance, in 2013 AM of a bionic ear was demonstrated using biological tissues in a hydrogel matrix and electronic components which exhibited capability to listen stereophonic music and enhanced auditory sensing of radio frequency [2]. Likewise in 2019, 3D printing of a cellularized human heart with a natural biological architecture was successfully demonstrated as shown in Fig. 2.1 [3]. Using biopsy of omental tissues from patient, thick, vascularized, and perfusable cardiac patches that completely matched the immunological, cellular, biochemical, and anatomical response of the patient could be developed. Such technological feats demonstrate the limitless potential of AM in engineering of personalized tissues and organs and drug screening in patient specific biochemical microenvironment [3] Replication of objects by scanning is not limited to just bodies' internal organs and tissues. Today, desktop 3D scanners are easily accessible at low costs and can be used to generate a CAD model of natural objects like a flower, plant or a pet and high value artistic objects like sculptures, pottery, and handicraft. The CAD model can subsequently be used to fabricate a copy of the parent object in the material of choice.

Moshe Dayan, a prominent Israeli military leader and politician believed freedom to be the oxygen of the soul. Scientists and engineers have always craved true freeform in fabrication of objects since the rise of modern science. Today, freeform fabrication is well-understood to be a subset of additive manufacturing and refers to the collection of technologies such as rapid prototyping, rapid manufacturing, and solid freeform fabrication. Latest generation AM technologies not only serve the needs of rapid prototyping but also allow small to medium scale production of customized functional and non-functional end-user products. These technological advances are paving the way for replacement of conventional manufacturing processes in industries with new processes exhibiting much higher process flexibilities, minimal tooling requirements and optimized material consumptions. The integration of the Industry 4.0 model, the internet of things (IoT), with AM is anticipated to be the harbinger of the fourth industrial revolution which will unleash a new era of manufacturing centered around customer-manufacturer interface and feedback loop. AM also has enormous potential to break the barriers of localized engineering and enable true globalization. It is now possible for design engineers to develop and analyze a model anywhere in the world and subsequently manufacture the matured model in an AM facility located in a different corner of the world altogether. Thus, the overall and niche advantages of AM compared to conventional manufacturing processes are stark and clear [4]. Nevertheless, AM is not flawless and exhibits some characteristic advantages and intrinsic limitations which are summarized in Fig. 2.2. The advantages and limitations of each specific AM technology will be elaborated in subsequent chapters. While the cost of industrial and high-end AM systems, especially those using metallic feedstock makes them impractical for household and common use,

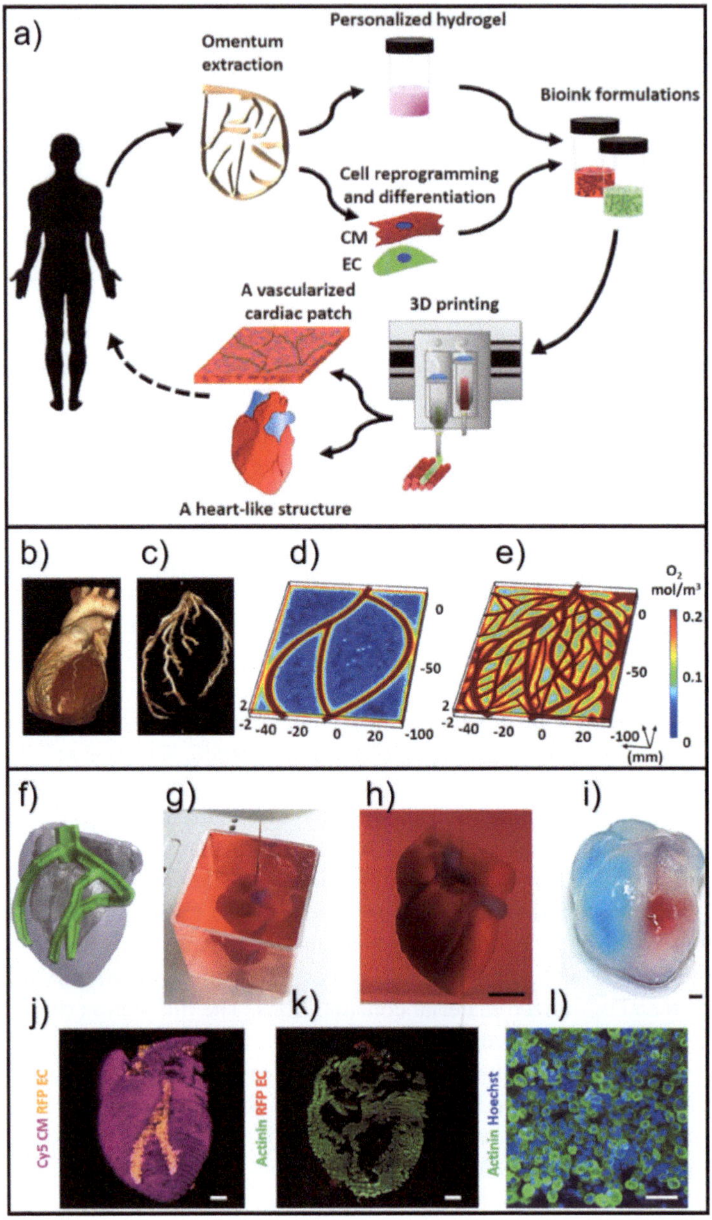
a)
Personalized hydrogel
Omentum extraction
Bioink formulations
Cell reprogramming and differentiation
CM
EC
A vascularized cardiac patch
3D printing
A heart-like structure
b)
c)
d)
e)
O2 mol/m3
0.2
0.1
0
(mm)
f)
g)
h)
i)
j)
Cy5 CM RFP EC
k)
Actinin RFP EC
l)
Actinin Hoechst

◄**Fig. 2.1** The concept and demonstration of additive manufacturing of a human heart; **a** Concept schematic illustrating the process involved in additive manufacturing of a human heart. An omentum tissue is extracted from the patient followed by separation of cells from the matrix. The matrix is extracted into a personalized thermoresponsive gel whereas the cells are reprogrammed to become pluripotent and encapsulated within the gel to obtain the bioink. The bioinks can be further used for additive manufacturing of vascularized patches and complex cellularized structures; **b** CT image of a human heart; **c** Left ventricle coronary arteries; **d** A model of oxygen concentration profile in an engineered patch; **e** Optimization and redesign of the model to obtain better oxygen diffusion in the patch enough to support cell viability; **f** CAD design of a human heart; **g** Printing of the heart in a support matrix; **h** Front view of the printed heart in support matrix; **i** Image of the printed heart after extraction from the support matrix where blue and red dyes were injected to exhibit hollow chambers and the interconnecting septum; **j** 3D confocal image of the printed heart; **k** Cross section of the heart immunostained against sarcomeric actinin (green); and k) High magnification image of the immunostained heart. Reproduced from Ref. [3] with permission from Wiley Online Library

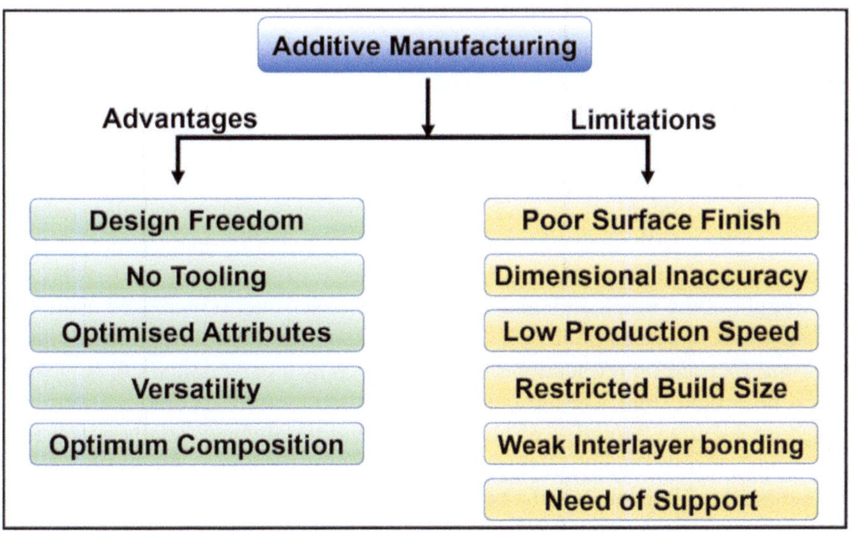

Fig. 2.2 Advantages and limitations of Additive Manufacturing

desktop 3D printers that can be mass produced and supplied at affordable prices have captured enormous public interest. Today, the applications of AM systems have expanded, beyond prototyping, into architectural modelling, production of obsolete automobile and mechanical parts, human tissues, photovoltaic panels, hearing aids, dental implants and braces, prosthetics, blades of jet engine, sports shoes, toys, jewelry, and even food items at your local bakery.

In order to guide global innovation and development efforts in the AM sector, independent bodies, published literature, international conferences and roadmap studies have laid out clear cut objectives and challenges that need to be surpassed. For instance, a team of 65 experts from academia, industry and government met at the US—Turkey workshop on rapid technologies in Washington DC in 2009 to develop a national roadmap study and define the research needs and possibilities for AM in the next 10–15 years [5]. The workshop discussed several vital aspects of AM such as its historic evolution, industry targets, technological goals and barriers, design and analysis, processes and machines, materials and processing, bio-additive manufacturing, and energy and sustainability [5]. Evolving from the modelling and carving technique of 'photosculpture' in 1860, industrial 3D printing for rapid prototyping, customization, sophistication, and design has gone mainstream in a massive way with greater than 20% adoption levels [6]. This chapter captures the main movements of this advancement and encapsulates the historic evolution of AM, its standardization, current trends, and future potential.

2.2 The Beginning

AM is also known as layered manufacturing (LM) and solid freeform fabrication (SFF) because it employs a layer-by-layer fabrication process and provides freedom to fabricate in any form and shape [7]. However, in the beginning, each of these two essential aspects of AM i.e., freeform fabrication and layered deposition belonged to entirely different schools of solid modelling. It is believed that AM emerged from the union of photosculpture which evolved in the nineteenth century to build sculptures and topography which evolved in twentieth century to build topographical maps [5, 8]. Each of these solid modelling techniques were experiencing their own evolution throughout the nineteenth and twentieth centuries due to increased availability of better tools, materials, and scientific understanding. Photosculpture techniques were more focused on replication of subjects such as a person, animal, or object to the best possible extent. On the other hand, the topography techniques were more focused on accurate depiction of contour data and topological details which found use in preparation of maps. These two schools merged through the works of Morioka between 1930 to 1940 and laid the foundation of the modern era of additive manufacturing indicated as the green line in Fig. 2.3. After the middle twentieth century most innovations in additive manufacturing were related to the use of new materials and manufacturing methods. By 1970s, new themes based on materials could be clearly identified in the practiced AM methods. These themes later evolved into different classes of AM technologies such as powder bed fusion, sheet lamination and vat photopolymerization as shown in Fig. 2.3. This historic evolution of AM and later progress is briefed in this section.

The photosculpture process which involved the reproduction of a 3D object based on synchronized photo projections to create a sculpture, can be regarded as the oldest AM technique. The technique was popular throughout the nineteenth century but successfully demonstrated by François Willème, a French artist, sculptor, and photographer who also patented the technique in 1860 [9]. As shown in Fig. 2.4a, a model would be placed on a circular stage and photographed by 24 cameras arranged in a circular fashion to cover the entire form of the model. An artisan will then carve 1/24th portion of the sculpture based on the silhouette of each photograph

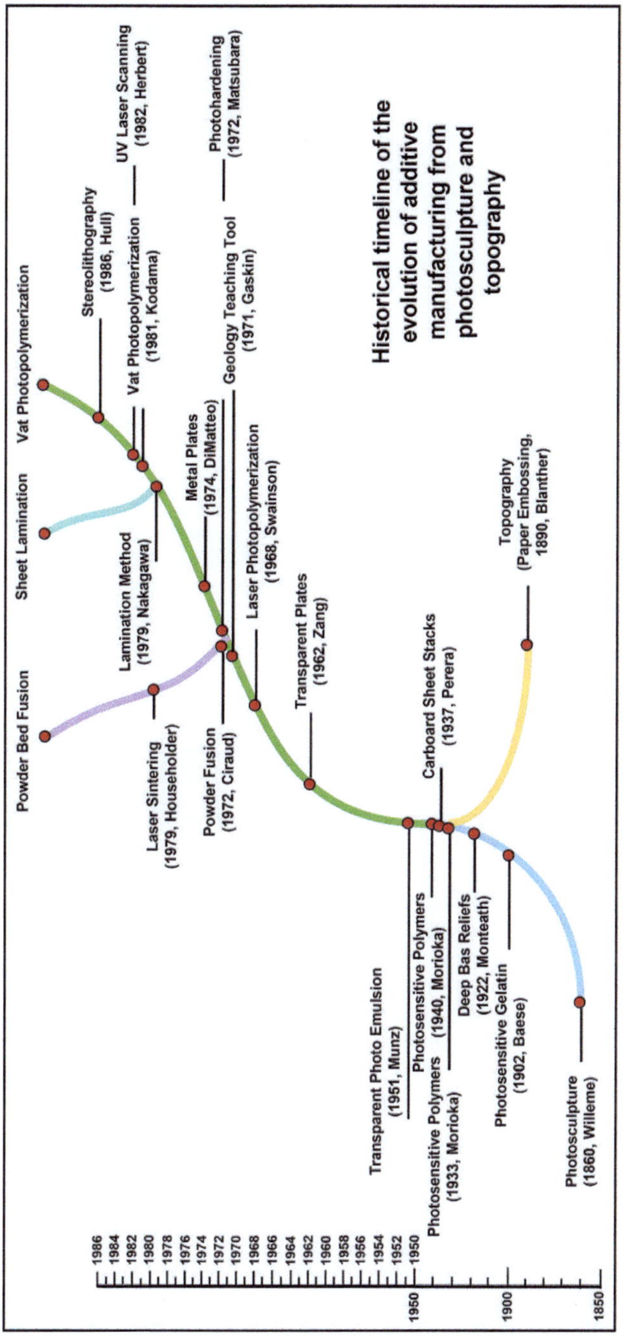

Fig. 2.3 Historical timeline depicting the evolution of modern additive manufacturing technologies from photosculpture and topography solid modelling techniques till the launch of the first commercial equipment at 3D systems in 1986. Timeline developed from patents and published literature

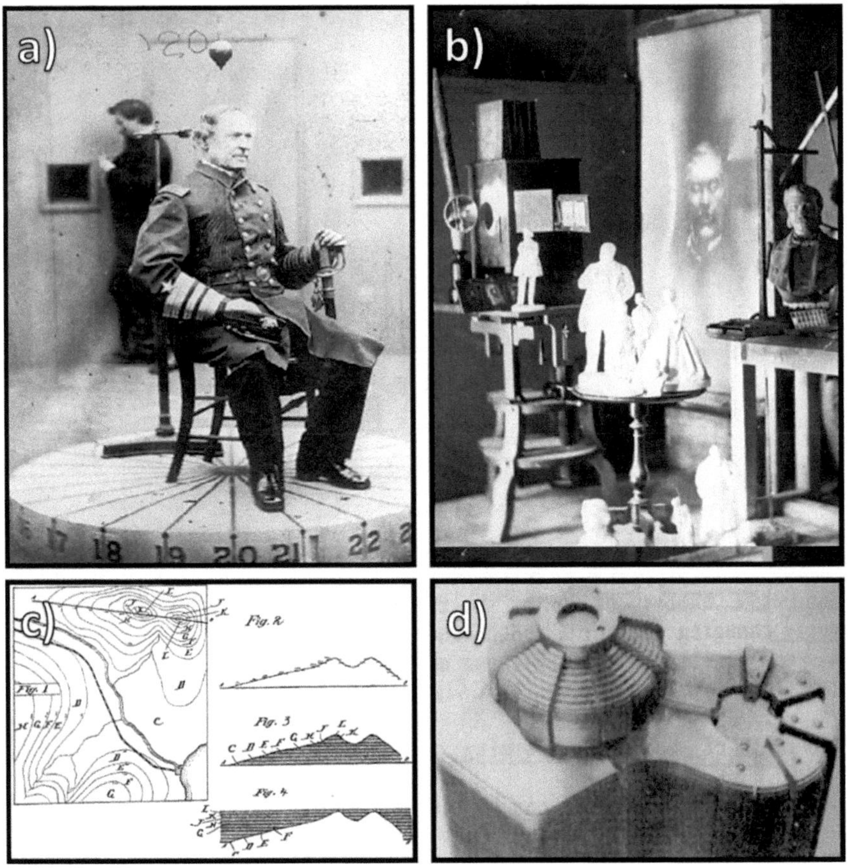

Fig. 2.4 Some examples of early solid modelling techniques; **a** Admiral Farragut posing for the photosculpture technique developed by Francois Willeme in 1860s which involved taking 360° photographs using 24 cameras arranged in a circular manner; **b** Sculptures developed in Willeme's studio from projected photographs; **c** Depiction of Blanther's topography method involving paper embossing between two protruding and engraved wax blocks which were carved and processed based on contour data; **d** Example of an unfinished punch tool manufactured by the lamination method developed by Prof. Nakagawa at Tokyo University. Reproduced from Ref. [8] with permission from Springer Nature

as shown in Fig. 2.4b. To eliminate the labor-intensive process, Baese developed a light-based technique relying on the expansion of a photosensitive gelatin in water and filed a patent in 1902 [10]. The technique used annular rings of exposed gelatin to develop a replica of desired object. By 1922, a British and Irish artist, F. H. Monteath, while functioning from Sydney expanded the photosculpture technique to fabricate bas reliefs which is a method of projecting sculpture elements out of a solid surface while remaining attached to the solid background [11]. Monteath identified several shortcomings in the known method of using gelatin such as limited depth of relief, repeatability, mechanical imperfections at the surface and the issue of bubble removal and developed an improvised photomechanical method capable of producing high-definition deep bas reliefs. By 1933, Morioka from Japan was already experimenting with cameras, projector, and photosensitive films to improvise the method and became successful in hybridization of the photography based sculpting technique with the earlier known topography techniques for contour mapping which he later improved in his 1940 patent [12, 13]. Thus, the decade of 1930–1940 can be considered as the period in time where topography and photosculpture had already merged and beginning to emerge as a new hybrid technique. The chronology of development of both the photosculpture and topography branches of AM is depicted in Fig. 2.3. The other school of solid modelling, topography, evolved independently from the need for topographical relief maps which had obvious usage in historic times for governance, security, expansion, and war. In 1890, Blanther filed a patent for an embossing method which described a technique for developing topographical relief maps by pressing of paper between two pieces of engraved and protruding wax blocks as depicted in Fig. 2.4c [14]. The engraved and protruding wax blocks were initially obtained from pressing of topographical contour lines on wax plates and subsequent cutting, stacking, and smoothing of the wax plates. By 1937, Perera was working to improve the method by cutting contour lines on cardboard sheets, stacking, and pasting to obtain 3D relief maps for which the patented was granted in 1940 [15]. Based on previous developments in combining photosculpture with topography, Munz filed a patent for a new stereolithography technique in 1951 which was very similar to the modern-day dynamic light projection stereolithography technique [16]. Munz exposed a transparent photo emulsion in layer wise manner where each layer was derived from the cross section of a scanned object. The lowering of the piston to expose consequent layers and continuation of the process resulted in a solid cylindrical object embedded with the image of the scanned object which could be carved, or chemical etched to obtain the 3D model. Zang developed a topographical solid modelling technique using transparent plates and patented the method in 1962 [17]. Gaskin further expanded the topographical method towards development of solid models for teaching geology and filed a patent in 1971 [18]. By 1968, Swainson was already employing polymerization of a photosensitive polymer at the intersection of two laser beams to develop a 3D pattern. A patent was filed for the developed technique in 1971 and the method termed photochemical machining which could either use photochemical crosslinking or degradation of a polymer [19]. Parallel work on the photochemical machining method was ongoing at Batelle's Columbus Laboratories and later published in 1984 [20]. A technological leap was made by Matsubara of Mitsubishi motors in 1972 when he used photo-hardening materials to shape a coherent solid sheet and patented the developed technology [21]. The photo hardening materials were developed by coating of photopolymer resins onto refractory materials like graphite powder or sand. When spread in the form of a layer and heated, the particles formed a coherent solid sheet. On exposure of light to selective sections of this sheet, significant hardening of exposed sections could be obtained. Unexposed sections were later removed by a suitable solvent. The obtained adequately shaped sheets were later stacked to form a casting mold. In-fact this original method of Matsubara laid the foundation for many matured freeform fabrication techniques available today such as lithography and stereolithography.

In 1972 Ciraud patented a method of fusing powdered materials which could be applied to a matrix by different mechanisms such as gravity, magnetostatics, electrostatics or flow from a nozzle and joined by partial melting induced by energy input using a laser, plasma, or electron beam [22]. Householder later introduced the concept of selective sintering of powders using laser in his patent filed in 1979 and successfully demonstrated solidification of a portion of each layer by controlled heat scanning process or heat applied through a mask [23]. Thus, in the historical timeline of the evolution of additive manufacturing, a new branch of AM employing powder binding emerged from the work of Ciraud and Householder during 1970–80 and can be regarded as the foundation of modern-day powder bed fusion AM methods as shown in Fig. 2.3.

DiMatteo continued the work of sheet binding and realized that the same stacking technique could be extended to metal plates. The method was patented in 1976 and DiMatteo used it for fabrication of solid components and tools that were difficult to fabricate by conventional manufacturing methods [24]. Prof. Nakagawa from Tokyo university further expanded the technique in 1979 by employing lamination techniques and demonstrated fabrication of different types of tools such as blanking, press forming, injection molding and punch tools [25–27]. An example of an unfinished punch tool fabricated by his lamination techniques is depicted in Fig. 2.3d. This point in the historical timeline of the evolution of additive manufacturing can be regarded as the birth of modern-day sheet lamination AM methods. Nevertheless, the use of photosensitive polymers remained popular and further developed by Hideo Kodama of Nagoya Municipal Industrial Research Institute. In 1981, Kodama published a vat photopolymerization based rapid prototyping system where he described different methods to expose the photosensitive polymer with UV light such as exposure from the top, exposure from a window in the bottom of the vat and use of an x–y plotter and optical fiber for exposure of the layer [28]. Herbert was independently developing a UV laser-based AM method using a x–y plotter and mirror system for photopolymer curing at 3M and published the work in 1982 [29]. By the 1980's, development of new AM techniques had become mainstream represented by a rapid surge in new discoveries and technological innovations. One of the most popular AM technology, stereolithography (SLA), and the corresponding CAD file format '.STL' was invented by

2.3 Standardization of AM Technologies

Beyond the initial phase of invention and commercialization of AM technologies the scientific community felt the need for their classification and standardization. This was a glorious phase for AM since it experienced enormous uptake in different engineering and scientific domains. The term AM was officially adopted in 2009 at the inaugural meeting of the ASTM F52-900-15 Technical Committee on AM at West Conshohocken, Pennsylvania where 59 attendees voted to form an ASTM technical committee and develop standards for AM. Numerous names, including 3D printing, were suggested because of the similarity with the process flow in 2D printing i.e. printing text or design on paper from a soft document. However, at the time the term 3D printing was already being used for the binder jetting process [32]. The committee decided not to brand all technologies under a single name and the term AM was preferred by the committee. Today, the terms AM and 3D printing are used interchangeably but refer to two different user groups. The technical community prefers AM whereas 3D printing has found large acceptance in the general public and widely used by the mass media, marketing campaigns, decision makers, as well as in political references. ISO/ASTM defined AM as a process of joining materials in layer-by-layer fashion to make parts from 3D model data [33]. ASTM has further defined five key standards for better utilization of AM systems [34]:

1. **Standard Terminology for Additive Manufacturing Technologies (F2792)**
   This standard defines different terminologies, terms, descriptions of terms, nomenclature, and acronyms associated with AM technologies for common use by producers, researchers, educators, end users and press/media.
2. **ISO/ASTM 52,900—General Principles—Terminology, ISO/ASTM 52,900**
   This standard jointly defines AM terminologies and principles with the European Committee for Standardization (CEN). The standard attempted to unite the world's standards bodies around AM.
3. **Standard Specification for AM with Titanium-6, Aluminum-4, and Vanadium with Powder Bed Fusion (F2924)**
   This standard incorporated AM of biomedical grade of titanium using EBM, SLS, and DMLS. The F2924 serves as a valuable educational document and industrial standard.
4. **ISO/ASTM 52,901—AM—General principles—Requirements for Purchased AM Parts 52,901**
   This standard covers the quality and other fundamental requirements of the AM equipment and tools and connects producers to the needs and demands of the buyer.
5. **ISO/ASTM 52,910—Guidelines for Design for AM (ISO/ASTM 52,910)**
   As a process document, this standard describes opportunities and design freedom, discusses general design considerations, and lists "red-flags" for designers that can lead to problems in many AM systems.

Further, ASTM International has released specifications on the new format support textures, functionally graded material information, colour, curved triangles, and other elements not supported by the STL file format.

2.4 Advance of AM into Modern Technological Applications

The advance of AM in biomedical applications, termed 'bioprinting', was fueled by new innovations in biological feedstock materials that could be employed for additive manufacturing. Bioprinting is a simple, precise, cost-effective, and rapidly evolving technology which allows enormous control over distribution of cells leading to significant advantages in tissue printing. Extrusion based bioprinting methods use bio-inks with viscosity in the range of 6–30 $\times$ 10^7 mPa s and find wide usage because of the simplicity, diversity, and reliability of the manufacturing method [35]. The wide range of printable biomaterials and their cost-effectiveness are amongst the top benefits of this technology. Recent research has also seen the commercialization of new AM systems developed by modification and upgradation of conventional systems [36, 37]. A schematic illustration of the principles of 3D bioprinting is shown in Fig. 2.5. [38, 39]. The advance of AM into biomedical applications is evident from the vast number of success stories. In 2010, a Spanish research center (IQS) announced the use of AM for biomedical and dental applications based on two new hydroxyapatite (HA) formulations. In 2011, Kelyniam Global (New Britain, Connecticut) innovated the use of modelling data obtained from CT or MRI medical scan for AM of skull implants composed of thermoplastic poly-ether-ether-ketone (PEEK) using SLA. The surgeons were allowed to analyze the obtained stereolithography models of the skull while the implants got machined. This incredible technological and process innovation proved instrumental in improved planning of surgical procedures and reduction in surgery durations by 85% [31]. 3D bioprinting has also been actively pursued for tissue engineering since it allows effective control over scaffold fabrication and cell distribution, wherein the resolution of printing can be ranged from 10–10,000 μm with significantly higher flexibility compared to conventional routes [40, 41]. The very first test for 3D printed tissues was made on Orgonovo's NovoGen MMX bioprinter. For specialized biomedical applications of tissue engineering and artificial organs, Orgonovo Holdings (San Diego, California) and Autodesk (San Rafael, California) announced collaboration in 2012 for creating the first 3D bioprinting software. The high demand for bioprinters acted as a driving force for the development of commercial bioprinters and improvements in printing quality and suitability for printing wider range of biomaterials [42, 43]. A bioprinting technology termed integrated organ printing (IOP) was developed to produce tissue-like structures, especially with local differences in cell types and mechanical properties [44]. Another future trend in organ printing is in-situ printing, where living organs can be printed directly in the human body during surgery [45], which has already been tested for repairing external organs such as skin [46] and living structures [47]. Apart from extrusion, inkjet-based and laser-based [48] AM technologies have also been developed for bioprinting. Inkjet bioprinting generates droplets of bio-ink at the print head assisted by heater/piezoelectric actuator as shown in Fig. 2.5a. Similarly, the laser-assisted bioprinting can be conducted either by laser guided or laser

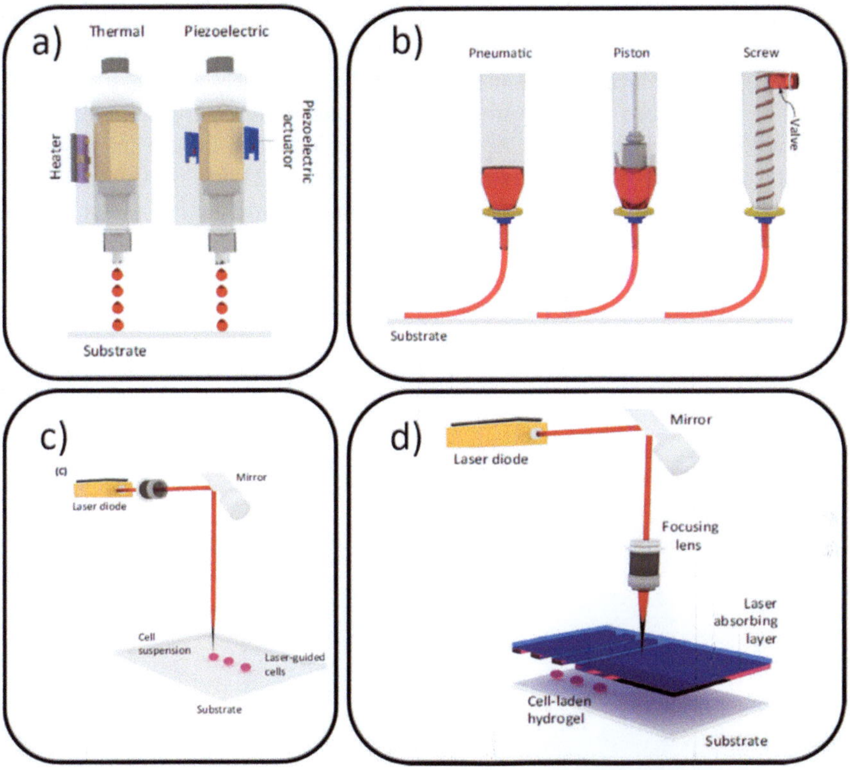

Fig. 2.5 Principles of 3D bioprinting; **a** Material jetting using thermal and piezoelectric actuators; **b** Material deposition using a nozzle driven by pneumatic, screw or piston driver; **c** Cell deposition and patterning using laser radiation in cell suspension; and **d** Cell deposition on a substrate from a cell-laden hydrogel through an absorption layer using a focused laser beam. Reproduced from Ref. [39] with permission from Elsevier

induced principles [49]. In the laser induced method as shown in Fig. 2.5d, a cell-laden hydrogel is deposited below a laser-absorbing layer, as a donor film, and placed parallel to the receiving substrate. The stereolithography approach of bioprinting as shown in Fig. 2.5c involves curing a photo-reactive material using light projected from a digital micro-mirror device in the desired spatial pattern [50, 51]. Thus, cells can be encapsulated in and seeded on 3D-patterned hydrogel scaffolds with a range of printable materials and control over microarchitecture and scaffold properties. Apart from the tissue engineering applications, these techniques have been widely used for printing 3D microfluidic devices and self-healing materials with embedded vasculature [52, 53]. Similarly, AM has also found applications in development of biological tissues, drug-screening, disease modelling and tissue regeneration [54, 55].

Parallel to the rise of AM in the biomedical sector, AM has made incremental but steady progress in the construction sector. Latest generation AM systems are versatile enough to build even a complete house or architectural models economically and in fast production times. Thus, civil engineers, construction managers and site supervisors can communicate more efficiently and clearly with their clients and satisfy the expectations of the customer beyond the rudimentary method of paper and pencil [56]. For instance, the Apis Cor Printed House in Russia and its contour crafting process as shown in Fig. 2.6 [57] and the Canal House in Amsterdam [56] are two demonstrated examples of 3D printing in the construction sector. The use of 3D printing could reduce the total cost of concrete construction by up to 35% to 60% simply by removing the need for formwork [58]. A schematic representation of the principles of 3D printing in the construction sector is shown in Fig. 2.6.

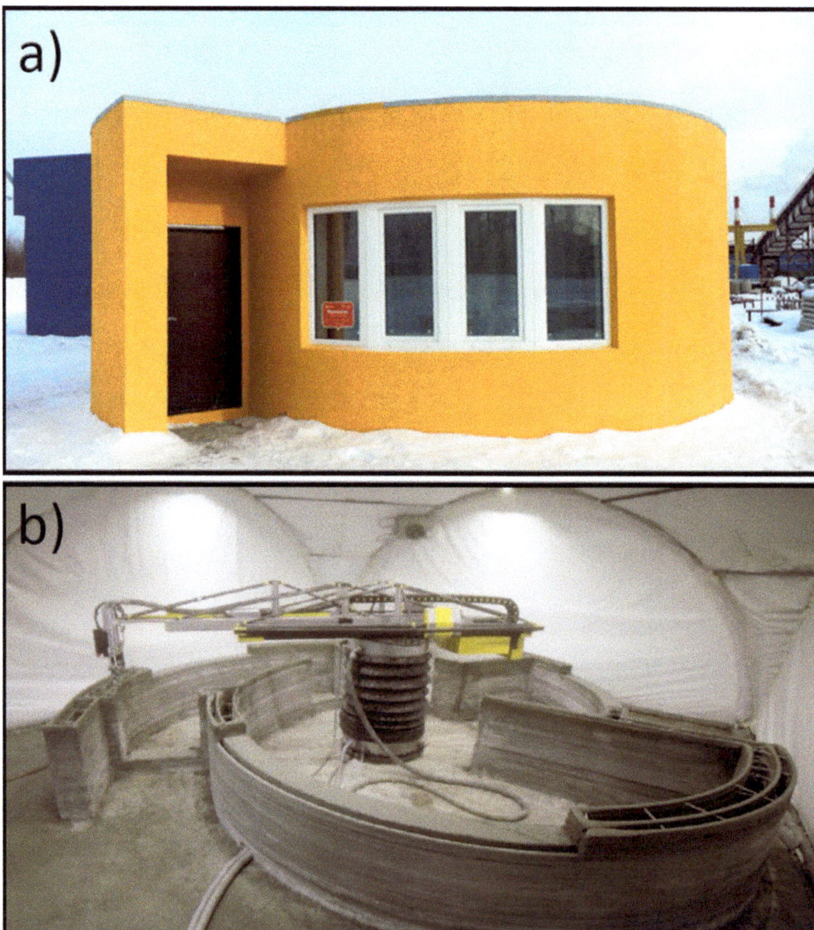

Fig. 2.6 Examples of 3D printing in construction; **a** The Apis Cor House in Russia; and **b** The contour crafting process used in the Apis Cor House. Reproduced from Ref. [57] with permission from Elsevier

and a detailed review of the application of 3D printing in construction is given by [57, 59, 60]. Some 3D printing technologies are capable of manufacturing solids for construction with strengths several times higher compared to conventional methods [60, 61]. New advances in software innovations have allowed modelling with standard building blocks and development of new fabrication strategies with improved production speeds [62]. Figure 2.7 depicts the image of a sophisticated large mono-

Fig. 2.7 A 3D printed large monolithic construction work. Reproduced from Ref. [60] with permission from Taylor & Francis

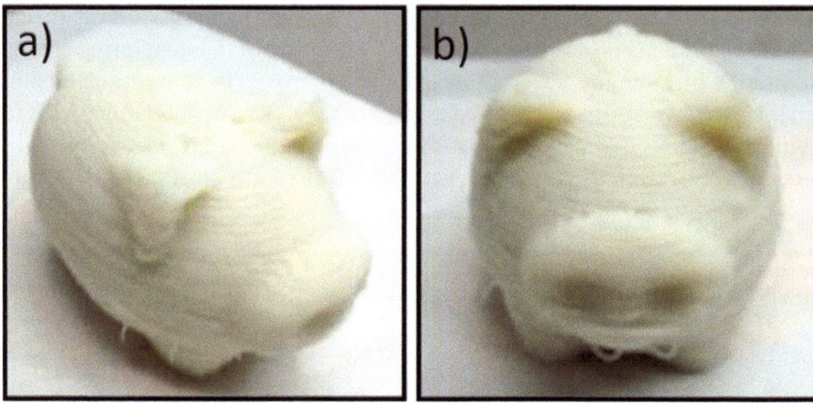

Fig. 2.8 Image of a 3D piggy printed using mashed potatoes. Reproduced from Ref. [66] with permission from Elsevier

lithic structure made using 3D printing [63]. Akin to the developments in the biomedical and construction sectors, AM is revolutionizing the food industry by digitizing and personalizing the nutrition and energy requirements of each according to their physical and nutritional needs [64, 65]. AM has the potential to transform the entire food supply chain by dramatic reductions in transport volume, packaging, distribution, and overriding costs [66–68]. The food printing technology can broaden the source of raw materials for processed food by employing non-traditional food materials such as insects, high fiber plant-based materials, and plant and animal-based by-products [67]. Figure 2.8 depicts an image of a 3D piggy printed using mashed potatoes [69]. A novel approach which uses plant cells to print foods resembling plant tissues in different ways has also been developed [70]. Different types of AM technologies that can be used for food printing and processing food materials are reviewed by [71] and [72], respectively. However, there are still some restrictions in food printing as processing conditions for different types of food feedstock vary extensively demanding high mechanical force, careful design of digital recipes, and appropriate feeding of ingredients [73]. In addition, different types of food formulations require optimization of feedstock temperature and feedstock flow through the nozzle [74]. Besides the application of AM in modern biomedical, construction and food sectors, the core field of engineering design has continued to reap the benefits of improved sophistication and rapid prototyping enabled by AM. In-fact some AM technologies were specifically developed for manufacturing of large footprint parts and components for use in aerospace and motor industries. For example, AM technologies like SLA, multi-jet modelling, fused deposition modelling (FDM), directed energy deposition (DED), rapid plasma deposition™ (RPD), SLS, etc., have been actively employed by Thales Alenia Space, Norsk Titanium AS, GKN Aerospace, Airbus A350 XWB, NASA, GE Aviation and Bell helicopter [59, 75]. Such AM technologies offered the following remarkable benefits

(1) manufacturing capabilities for complex shaped parts and components for integrated structures, heat dissipation, and airflow components [47].
(2) production of objects with difficult to machine materials and high buy-to-fly ratio [76].
(3) customized production for small batches of parts [77].
(4) high-performance to weight ratio to address the demand of light weight, high strength, and stiffness at reduced costs and emissions [78].

2.5 Four-Dimensional Printing

Four-dimensional (4D) printing is a newly coined term and refers to additive manufacturing of 4D objects i.e., objects that change their shape in the fourth dimension of time. 4D printing relies on the innovative use of functional feedstock materials that can be activated by environmental stimulus such as heat, moisture, UV lights, electricity, or magnetism for functional applications [63]. The 4D printing revolution is driven by innovations in functional materials such as hydrogels [80, 81] and shape memory polymers (SMPs) [82, 83]. 4D objects are currently being actively explored for applications in self-assembly, multi-functionality, and self-repair.

Self-assembly. Self-assembly is anticipated to be a great enabler of futuristic technological applications ranging from the nanoscale to the planetary scale. For instance, a 4D printed device can be transferred inside the human body, through a surgical hole, where it can self-assemble at the desired location for drug delivery, tissue recovery or tumor annihilation. Based on the 4D printing philosophy, individual parts printed with small 3D printers can be self-assembled into larger functional structures such as space antennae, satellites, and architectural components [84].

Multifunctionality. Using the 4D printing technology, it is possible to integrate sensors and actuators directly into the material of construction thereby reducing the number of parts in a structure, assembly time, material, and energy costs. Such philosophy of device fabrication also eliminates the impact of external electromechanical stimulus and reduces the number of failure-prone components [84, 85]. For example, self-adaptive 4D printed tissues, personalized medical devices and tracheal stents are some of the fascinating applications of 4D printing [86, 87].

Self-repair. The shape morphing abilities of 4D printed objects allows the structure to undergo self-corrections or self-repairs during their use such as failure prevention using self-healing implants and scaffolds [88].

4D printing is not only a disruptive technology with enormous potential to replace many conventional manufacturing processes but also an enabling technology which will allow new business models, new products, and new supply chains to flourish [90, 91]. For instance, a 4D flower which changes shape from a 2D flower-like sheet to a 3D structure on heating without any application of an external force is depicted in Fig. 2.9 [79]. A variety of constrained and unconstrained external stimuli can be employed to induce self-assembly, self-actuation, and self-repair based on different

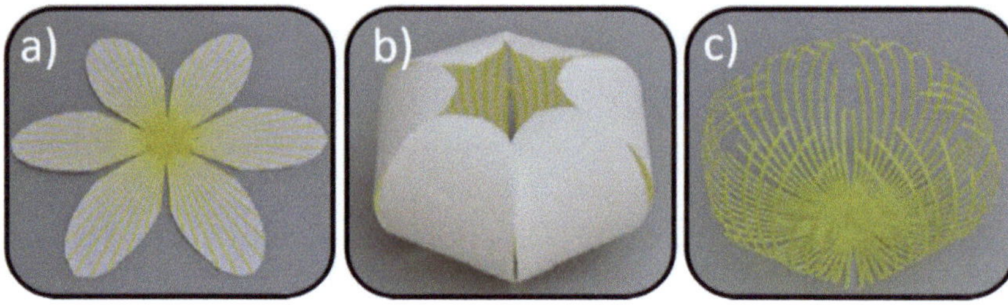

Fig. 2.9 Example of a 4D flower which changes shape from a 2D flower-shaped sheet into a 3D flower after heating; **a** Original flat sheet cut in the form of a flower; **b** Bending of the leaves in the 2D structure to form a 3D flower flower-like structure; and **c** A complex 3D flower-like structure formed by tearing the structure. Reproduced from Ref. [79] with permission from Springer Nature

mechanisms and driving forces such as hydromechanics, thermomechanics, hydro-thermo mechanics, pH mechanics, photothermal mechanics, osmosis, and dissolution mechanics [92]. In addition, composites of stimuli-responsive smart materials and functional materials can be readily employed in 4D printing to fine tune the product properties [89]. Figure 2.10 illustrates shape memory effect in 4D printed letters NTNU printed using the Polyjet method developed by NTU, Singapore. A pulling force was applied to the 3D printed letters after heating above glass transition temperature which leads to distortion of the letters and loss of information as shown in Fig. 2.10a. Cooling to room temperature while maintaining the pulling force allowed the return of printed letters back to the NTNU shape. Thus, 4D printing concept is already proven and has enormous potential for development of smart materials and devices with pre-programming, stimuli responsivity, self-adaptability, self- sensing, shape memory, multiple functionalities, and decision-making capabilities [93–95]. Information on list of feedstock materials, compatible AM technologies, potential applications, design concepts, role of external stimuli, and future prospects of 4D printing is available in the literature [92, 95, 96]. 4D printing also finds use in thin

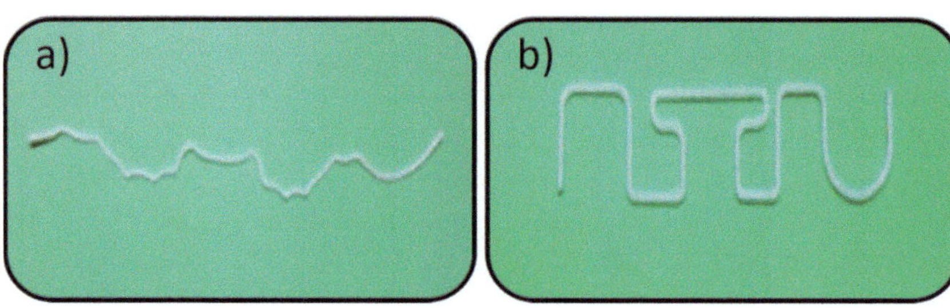

Fig. 2.10 Illustration of the shape memory effect in 4D printed letters 'NTNU' printed using the Polyjet method developed by NTU, Singapore. **a** Distortion of letters NTNU on application of a pulling force above glass transition temperatures **b** Return of the material to the original shape after cooling to ambient temperature while maintaining the pulling force. Reproduced from Ref. [89] with permission from Taylor & Francis

film applications where shape morphing capabilities can be imbibed by digital photopolymerization and gel swelling and wax uptake dynamics. Similarly, 4D printing is being enthusiastically pursued in biomedical applications such as tissue engineering and regenerative medicine. A micro-gripper or medical staple with self-tightening capabilities in response to stimulus has already been demonstrated [97]. The strength, reliability, and biocompatibility of 4D printed active composites have great potential for development of artificial muscles [98].

2.6 Impact of AM on Industries

The rise of AM in industrial manufacturing and supply chains has led to large growth spurts and development of a niche market worth billions of dollars. In 2013, the market for AM grew to \$3.07 billion with an impressive compound annual growth rate of 34.9%. By 2018, the global AM market was estimated to have reached \$9.3 billion, with 18% growth observed in the hardware, software, materials, and services market [63]. The growth in the AM sector is anticipated to continue strongly with predictions of the rise in market size to \$10.8 billion by 2021 and \$23.33 billion by 2026 [63, 100]. The penetration and impact of AM technologies in different industrial sectors is depicted in Fig. 2.11 [99]. Clearly the consumer electronics and goods sector has benefitted the most from AM technologies followed by the transportation and the medical sector. AM benefits these industries not only by facilitating favorable process economics, fast design iterations, product development, process intensification and inventory expansion but also by enabling environmental sustainability, improved process efficiencies and ease of operation.

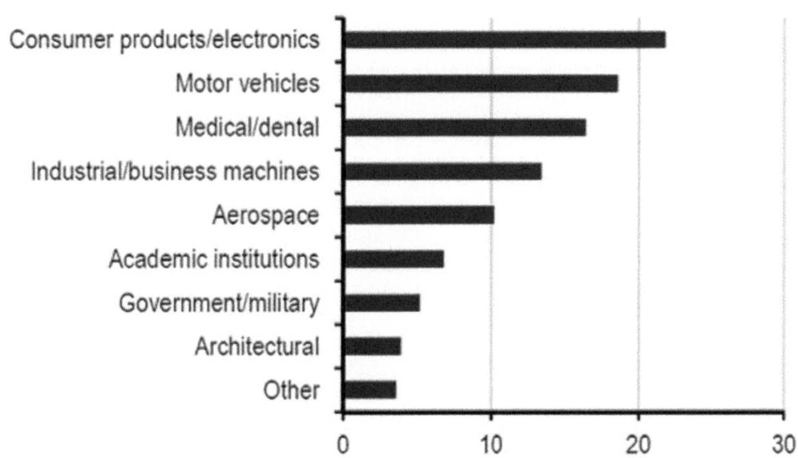

Fig. 2.11 Impact of AM on various industrial sectors in terms of profit. Reproduced from Ref. [99] with permission from Elsevier

AM technologies are also being pursued in other industrial sectors which employ AM directly or indirectly, with great trust and enthusiasm. The rapid prototyping of the past has evolved into AM of the present with the capability to directly manufacture products for the end user in a variety of extreme environments. Tool-less and waste-less production is unleashing new ways to tackle the sustainability challenges faced by the modern industry. AM technologies allow maximum design freedom with the capability to create sophisticated yet lightweight components. Since AM employs direct digital manufacturing from the 3D CAD model, design optimization and subsequent prototyping can be performed directly at the CAD modelling stage [74]. AM technologies also allow incorporation of a greater degree of variety in designed products and their finished form with potentially infinite possibilities and without incurring any additional costs. This allows production of highly customized products satisfying customer preferences [101]. AM technology also has the strength to offset the economy of scale and removing the barrier of high cost in production at low batch volumes [102]. Thus, design flexibility and catering to individual demand is one of the prime factors that allows improved profitability by the use of AM technologies [103].

The impact of AM technologies on industrial supply chains is well-documented and anticipated to rapidly change the supply chain management and logistics landscape by refurbishing the efficiencies of lean supply chains, pre-planned and timely manufacturing, and waste elimination. Direct digital manufacturing from 3D CAD data in AM reduces setup, change-over and assembly times and allows fulfilling of product demand and supply with reduced holdup in feedstock and storage inventory. Moreover, reduction in dependency of the shop and factory outlets on product logistics by manufacturing directly on the shop floor is a tremendous value addition for the commercial sector. This allows reduction of non-value-added activities such as material flow and inventory management to minimum possible levels and improves the responsiveness of an agile supply chain [104]. Thus, AM is indisputably emerging as a marvelous feat of modern engineering with applications ranging from rapid fulfillment of customer needs in the commercial sector to saving millions of lives in the biomedical sector by imbibing high responsiveness and customization capabilities [105].

The appearance of new open-source CAD modelling and process modelling software is also aiding the AM revolution and making it a true household phenomenon. Since the expiration of the IP rights and copyrights of the first-generation AM printers, numerous open-source printers have entered into the commercial market which are finding rapid uptake in the developing world such as South Africa, India, Ethiopia and Kenya for manufacturing and production of lightweight, replaceable, and flexible components [106, 107]. Thus, AM technology is dramatically reducing the life span of the process from CAD modelling to mass production, especially where the requirements of complex tooling, high process costs and large process lifetimes are involved [108]. The rapid progression and advancement of AM has truly revolutionized the manufacturing industry by providing numerous benefits at the industry and society levels such as customized healthcare products, reduced raw material and energy consumption, improved sustainability and on-demand manufacturing leading to reinforcement of supply chain. At the nano scale, new AM processes such as two-photon lithography have been developed that allow fabrication of a diverse range of functional microproducts with sophisticated architecture, overhanging parts, and high aspect ratios. Such AM technologies are anticipated to play an increasingly important role in various areas of modern technologies such as biochips, microfluidic devices, photonic crystals and so on. These nano-devices have enormous potential in delivering high-quality and economically efficient products for electronics, healthcare, and smart ornament sectors.

Another niche and superlative aspect of AM is the capability to manufacture functionally and compositionally graded components allowing a radical shift from contour modelling to performance modelling [109]. This can be achieved by employing different strategies such as variable densification within the same material, gradual density transitions in a heterogenous object composed of different materials and integration of variable densification and heterogeneous composition strategies. Thus, it is now possible to strategically control density and directionality of material deposition within a complex 3D or 4D physical component. In this way, seamless monolithic structures can be manufactured for functional technological applications [110]. AM is also reducing costs and production times in industrial rapid tooling applications such as metal casting, plastic molding, and other machine fixture and tool manufacturing industries [93]. Use of AM technology for production of thermoforming molds is opening new regimes in possible geometries and precision. For example, Junk et al. demonstrated rapid printing of mold within a few hours which is sufficiently heat-resistant for the molding processes [111]. The design flexibility and capability to manufacture sophisticated geometries using AM allows creation of uniquely shaped channels for gas bubble removal under vacuum and simplification of the manufacturing process for prototype production of thermoforming molds is opening new regimes in possible geometries and precision. Figure 2.12 depicts apt fitting of a carbon fiber reinforced composite mold produced by AM onto the original aircraft engine [112]. In some cases, issues of product deterioration in the manufacturing process or storage have also been resolved using AM [113, 114]. Other small-scale niche markets of AM include art, jewelry, museum displays, sculptures, architecture, and popular game avatars [115].

The rise of AM in to the chemical and pharmaceutical industries is impending subject to its uptake in various sectors of the industry. The uptake of AM into goods packaging, supply chains and logistics has already started in line with other industries as elaborated in this section. However, the application of AM for technological applications in the chemical and pharmaceutical industries largely depends on successful implementation and demonstration of pilot scale and large scale projects. The efficacy and scope of AM into simulation, design and testing of microscale reactors and microreactor assemblies is discussed in more details in Chap. 8 of the book. The design sophistication, customization, capability to produce components in a variety of materials and localized production are anticipated to dramatically change the technological landscape for existing chemical processes and upcoming chemical

Fig. 2.12 Fitting of a 3D printed carbon fiber reinforced composite mold onto a reciprocating engine of an aircraft. Reproduced from Ref. [112]

processes of global importance such as carbon dioxide utilization, green hydrogen production, production of batteries, biomass utilization and conversion of industrial and household waste [116–121].

2.7 Customization

The customization of a product refers to tailoring of goods and services as per the requirements of the customers [122]. Incorporation of large scale product customization in manufacturing processes requires the production enterprises to modify all stages of production cycles including design, manufacturing, and distribution. Thus, customization is largely circumvented in commercial production because of disruption of the economy of scale and associated losses due to decrease in production time and increase in material waste. The emergence of AM technologies is rapidly changing the customization landscape and allowing enormous process flexibility and agility to satisfy the diverse customization needs. However, it is important to balance the intensity of customization and process economics [123]. Manufacturing resources must be strategically managed for a high efficiency in point-of-delivery customization, services customization, quick responses, and modular production [124]. Many companies are experimenting with innovative strategies of mass customization to overcome process rigidity, customization intensity and resource scarcity [125]. Undoubtedly, high levels of customization can be achieved by AM when compared

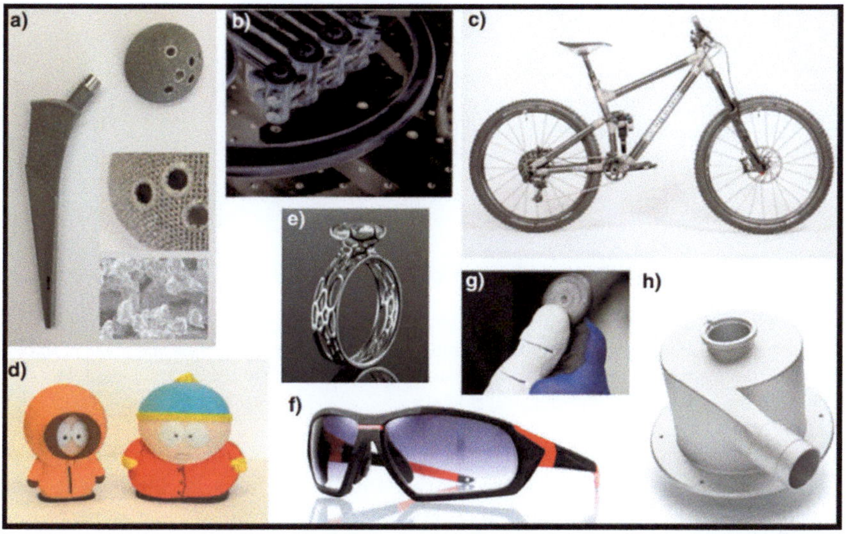

Fig. 2.13 A sample of different customized products fabricated using AM; **a** Hip-joint; **b** Fixture of a musical instrument; **c** Fiber-reinforced cycle; **d** Figurines of 'South Park' characters; **e** Diamond ring; **f** Sunglasses; **g** Ergonomic thumb augment for the automotive industry; and **h** Sophisticated plastic component. Adapted from Ref. [126] with permission from Wiley

with conventional subtractive manufacturing. Integration of AM with big data, cloud computing, and other digital tools is unleashing even finer approaches for improvement in productivity, optimization of supply chains, and reduction of unexpected delays. Products can now be designed anywhere in the world, delivered to manufacturing partners through the internet and manufacturing of end use products can be completed locally in regions of high demand [63]. Figure 2.13 depicts a collection of different products customized using AM [126]. costs. AM also enables the concept of 'global design, local production' wherein CAD design and modelling can be performed by product designer anywhere in the world. At the frontier of food technology, digital gastronomy and 3D printing of food is enabling mass customization in shape, colour, flavor, texture, and even nutritional value [71]. AM is a highly agile manufacturing philosophy with extremely low input requirements for conversion of matured CAD model into satisfactory products. In addition, a variety of common or sophisticated geometries can be built using the same machine platform. Thus, AM greatly enables product customization and modularization [127]. AM capability to hold stock as digital data and print on demand also has enormous potential to push customization towards the downstream sectors such as marketing, logistics, retail and even the consumer. This allows leaning and reduction in customization requirements for the upstream sourcing and production sectors where customization can lead to large increase in costs. Latest research in food technology has focused on quality optimization, behavior of edible printable feedstock during AM, and their influence on the food market and customer acceptability. However, to enhance the

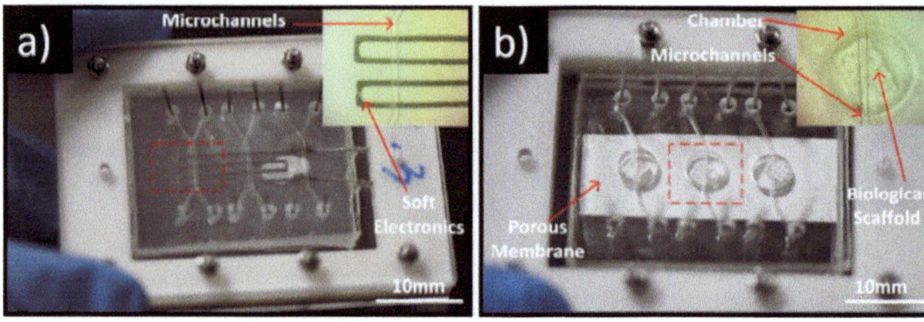

Fig. 2.14 Example of customization in additive manufacturing through modularization of components in a micro-fluidic device; **a** A microfluidic device with imbedded soft electronics; and **b** A microfluidic device with embedded membrane and biological scaffolding. Adapted from Ref. [132] with permission from American Chemical Society

nutritional value of food and ensure distribution of this value to the people at low cost, it is essential that AM focusses on economical printing of healthy protein and whole grain feedstocks instead of unhealthy sugary and fatty feedstocks.

Incorporation of customization in manufacturing cycle still remains a matter of large debate. Human-in-the-loop simulations have been used to investigate the reliability of customized products where individual designers use their freedom to completely personalize a product design under the limits of safety and manufacturing feasibility [128]. Such studies indicate that complexity of the customizable objects can be extremely high with involvement of dozens of parameters and multiple options leading to technological complexity and related issues [129]. Nevertheless, complete control over the digital design and physical [130] form of the manufactured object through AM is the right path to customized manufacturing [131]. Modularization has been often viewed as the harness which can tame the wild horse of customization. Software-based tools can be utilized for automatic generation of printable modules based on printing methodology, mechanism, and device hardware. For example, Qiu et al. described a new approach for fabrication of customized integrated microfluidic chips using a modular design to reduce time and cost [132] as shown in Fig. 2.14. The developed microfluidic chip was investigated for cell culture and biological analysis opening doors for the concept of organ-on-a-chip. This is just one example of a customized bioengineered device. Strategic utilization of surface functionalisation, coating and thin film deposition techniques with AM devices opens the doors for rapid customization of devices adapted to applications [133]. Thus, customized medicine and stimuli responsive materials and devices and are now emerging as the new frontier in AM [130]. Groeger et al. developed a new method termed 'HotFlex' that allowed a specific region or component of a printed object to transition from a solid into a deformable state on demand and back [134]. Thus, it can be concluded that AM enables new routes to mass customization for production of a high volume of products at low costs comparable to conventional manufacturing technologies. However, further studies are required to explore customization aspects in more details such production flexibility, reliability, performance characteristics, and cost-efficiency.

2.8 Summary

Additive manufacturing technology evolved from the photosculpture and topographic solid modelling techniques of the twentieth century. Since the establishment of the first commercial enterprise for supplying 3D printers in 1986, AM has seen tremendous growth and uptake in various industrial and research sectors. Recognizing the need of the global industry, ASTM defined standards to unify different nomenclatures, classifications, and channel global growth. AM has drastically changed the technological landscape in the manufacturing, logistics, services, food, biomedical, construction, consumer goods and electronics sector by generating overwhelming benefits. In the manufacturing sector, AM is enabling creation of sophisticated geometries, tailoring of material properties, design of functionally graded materials and mass customization for versatile and customer-oriented production. In the logistics and services sector AM has created new pathways for industries to supply goods and products while cutting down costs and timespans. In the construction sector AM technology has seen its use for development of entire homes and buildings with reduction in production costs, time and manpower and improvement in strength, visual appeal, and better customer satisfaction. In the food sector AM has enabled digitization and personalization with dramatic reductions in transport volume, packaging, distribution, and overriding costs. In the biomedical sector AM has facilitated tissue engineering, 3D microfluidic devices, self-healing materials with embedded vasculature, drug-screening, disease modelling and tissue regeneration. Thus, AM is a marvelous feat of modern technology and being hailed globally for enabling high standards of sustainability, resource utilization, and environmental protection. Seeking inspiration from the rapid expansion of AM, research on conventional manufacturing technologies is also showing an upwards trend. However, as of now, the interdisciplinary AM technology is bridging the gaps and paving a path towards unified universal engineering.

2.9 Test Your Knowledge

1. *Besides solid object modelling in CAD software what other methods can be used to obtain a CAD file usable for additive manufacturing?*
2. *What is the biggest advantage and limitation of additive manufacturing methodology?*
3. *Which historical figure can be regarded as the father of additive manufacturing and which historic modelling techniques gave birth to the modern form of additive manufacturing?*
4. *What are the different AM standards defined by ASTM and with what objectives?*
5. *Who established the first commercial venture to supply 3D printers and also developed the STL file format?*

6. *The advance of additive manufacturing into modern applications is evident from significant development in which technological sectors?*
7. *What does the term 3D bioprinting refers to and how does it differ from 3D printing?*
8. *What are the two examples of buildings constructed by 3D printing?*
9. *Give some examples AM technologies which have been specifically developed for the engineering design sector, derived benefits and some important clients.*
10. *What is 4D printing and why is it being regarded as a disruptive technology?*
11. *Which industries have benefitted the most from AM in terms of profit.*
12. *AM driven customization incentives are rapidly impacting which industrial sectors?*

References

1. Chua CK, Leong KF (2014) 3d printing and additive manufacturing: principles and applications (with companion media pack)—, 4th edn. World Scientific Publishing Company, Of Rapid Prototyping
2. Mannoor MS, Jiang Z, James T, Kong YL, Malatesta KA, Soboyejo WO, Verma N, Gracias DH, McAlpine MC (2013) 3D printed bionic ears. Nano Lett 13(6):2634–2639. https://doi.org/10.1021/nl4007744
3. Noor N, Shapira A, Edri R, Gal I, Wertheim L, Dvir T (2019) 3D printing of personalized thick and Perfusable cardiac patches and hearts. Adv Sci (Weinh) 6(11):1900344. https://doi.org/10.1002/advs.201900344
4. The history and application of additive manufacturing for design personalisation (2016). In: Kuksa I, Fisher T (ed) Design for Personalisation (1st ed.). Routledge. https://doi.org/10.4324/9781315576633
5. Bourell DL, Joseph J. Beaman J, Leu MC, Rosen DW (2009) A brief history of additive manufacturing and the 2009 roadmap for additive manufacturing: looking back and lookin ahead. US—Turkey workshop on rapid technologies, 5–11
6. D'Aveni R (2015) The 3-D printing revolution. harvard business review. https://hbr.org/2015/05/the-3-d-printing-revolution
7. Singh S, Ramakrishna S, Singh R (2017) Material issues in additive manufacturing: a review. J Manuf Process 25:185–200. https://doi.org/10.1016/j.jmapro.2016.11.006
8. Beaman JJ, Barlow JW, Bourell DL, Crawford RH, Marcus HL, McAlea KP (1997) Solid freeform fabrication: a new direction in manufacturing. Springer Science, Business Media, LLC.https://doi.org/10.1007/978-1-4615-6327-3
9. Sobieszek RA (2014) Sculpture as the sum of its profiles: François Willème and Photosculpture in France, 1859–1868. The Art Bulletin 62(4):617–630. https://doi.org/10.1080/00043079.1980.10787818
10. Baese C (1904) Photographic process for the production of plastic objects. US774549
11. Monteath FH (1924) Photomechanical process for producing bas reliefs. US1516199
12. Morioka I (1935) Process for manufacturing a relief by the aid of photography. US2015457
13. Morioka I (1944) Process for plastically reproducing objects. US2350796
14. Blanther JE (1892) Manufacture of contour relief maps. US473901
15. Perera BV (1940) Process of making relief maps. US2189592
16. Munz OJ (1956) Photo-glyph recording. US2775758
17. Zang EE (1964) Vitavue relief model technique. US3137080
18. Gaskin TA (1973) Earth science teaching device. US3751827

19. Swainson WK (1977) Method, medium and apparatus for producing three-dimensional figure product. US4041476
20. Woodin RL, Schwerzel RE, Kaldor A, Wood VE, McGinniss VD, Verber CM (1984) Three-dimensional photochemical machining with lasers. Paper presented at the applications of lasers to industrial chemistry
21. Matsubara K (1974) Molding method of casting using Photocurable substance. Jpn Kokai Patent Appl Patent Sho 51[1976]-10813
22. Ciraud PA (1972) Process and device for the manufacture of any objects desired from any meltable material. FRG Disclosure Publication 2263777
23. Householder RF (1981) Molding process. US4247508
24. DiMatteo PL (1976) Method of generating and constructing three-dimensional bodies. US3932923
25. Nakagawa T (1979) Blanking tool by stacked bainite steel plates. Press Technique, pp 93–101
26. Kunieda M, Nakagawa T (1984) Development of laminated drawing dies by laser cutting. Bull of JSPE, pp 353–354
27. Nakagawa T (1985) Laser cut sheet laminated forming dies by diffusion bonding. Proc 25th MTDR Conf, pp 505–510
28. Zheng X, Deotte J, Alonso MP, Farquar GR, Weisgraber TH, Gemberling S, Lee H, Fang N, Spadaccini CM (2012) Design and optimization of a light-emitting diode projection micro-stereolithography three-dimensional manufacturing system. Rev Sci Instrum 83(12):125001. https://doi.org/10.1063/1.4769050
29. Herbert AJ (1982) Solid Object Generation Jour Appl Photo Eng 8(4):185–188
30. Hull CW (1986) Apparatus for production of three-dimensional objects by stereolithography US4575330
31. Wohlers T, Gornet T (2016) History of additive manufacturing
32. Sachs EM, Haggerty JS, Cima MJ, Williams PA (1993) Three-dimensional printing techniques. US5204055
33. ASTM (2015) Standard terminology for additive manufacturing—general principles. Part 1: terminology ISO/ASTM Stand. 52792
34. Sprinkle T The 5 most important standards in additive manufacturing. https://sn.astm.org/?q=features/5-most-important-standards-additive-manufacturing-.html
35. Holzl K, Lin S, Tytgat L, Van Vlierberghe S, Gu L, Ovsianikov A (2016) Bioink properties before, during and after 3D bioprinting. Biofabrication 8(3):032002. https://doi.org/10.1088/1758-5090/8/3/032002
36. Yu Y (2014) Ozbolat IT Tissue strands as "bioink" for scale-up organ printing. In: 2014 36th annual international conference of the IEEE engineering in medicine and biology society, 26–30 Aug. 2014 2014, pp 1428–1431. doi:https://doi.org/10.1109/EMBC.2014.6943868
37. Zhao Y, Li Y, Mao S, Sun W, Yao R (2015) The influence of printing parameters on cell survival rate and printability in microextrusion-based 3D cell printing technology. Biofabrication 7(4):045002. https://doi.org/10.1088/1758-5090/7/4/045002
38. Derakhshanfar S, Mbeleck R, Xu K, Zhang X, Zhong W, Xing M (2018) 3D bioprinting for biomedical devices and tissue engineering: a review of recent trends and advances. Bioact Mater 3(2):144–156. https://doi.org/10.1016/j.bioactmat.2017.11.008
39. Knowlton S, Onal S, Yu CH, Zhao JJ, Tasoglu S (2015) Bioprinting for cancer research. Trends Biotechnol 33(9):504–513. https://doi.org/10.1016/j.tibtech.2015.06.007
40. Daly AC, Cunniffe GM, Sathy BN, Jeon O, Alsberg E, Kelly DJ (2016) 3D Bioprinting of developmentally inspired templates for whole bone organ engineering. Adv Healthc Mater 5(18):2353–2362. https://doi.org/10.1002/adhm.201600182
41. Kesti M, Eberhardt C, Pagliccia G, Kenkel D, Grande D, Boss A, Zenobi-Wong M (2015) Bioprinting complex cartilaginous structures with clinically compliant biomaterials. Adv Func Mater 25(48):7406–7417. https://doi.org/10.1002/adfm.201503423
42. Pati F, Jang J, Ha DH, Won Kim S, Rhie JW, Shim JH, Kim DH, Cho DW (2014) Printing three-dimensional tissue analogues with decellularized extracellular matrix bioink. Nat Commun 5:3935. https://doi.org/10.1038/ncomms4935

43. Schmolka IR (1972) Artificial skin 1. Preparation and properties of pluronic F-127 gels for treatment of burns. J Biomed Mater Res 6:571–582
44. Merceron TK, Burt M, Seol YJ, Kang HW, Lee SJ, Yoo JJ, Atala A (2015) A 3D bioprinted complex structure for engineering the muscle-tendon unit. Biofabrication 7(3):035003. https://doi.org/10.1088/1758-5090/7/3/035003
45. Binder KW (2011) In situ bioprinting of the skin
46. Zhang YS, Yue K, Aleman J, Moghaddam KM, Bakht SM, Yang J, Jia W, Dell'Erba V, Assawes P, Shin SR, Dokmeci MR, Oklu R, Khademhosseini A (2017) 3D Bioprinting for tissue and organ fabrication. Ann Biomed Eng 45(1):148–163. https://doi.org/10.1007/s10439-016-1612-8
47. Mironov V, Prestwich G, Forgacs G (2007) Bioprinting living structures. J Mater Chem 17(20). https://doi.org/10.1039/b617903g
48. Gaebel R, Ma N, Liu J, Guan J, Koch L, Klopsch C, Gruene M, Toelk A, Wang W, Mark P, Wang F, Chichkov B, Li W, Steinhoff G (2011) Patterning human stem cells and endothelial cells with laser printing for cardiac regeneration. Biomaterials 32(35):9218–9230. https://doi.org/10.1016/j.biomaterials.2011.08.071
49. Nahmias Y, Odde DJ (2006) Micropatterning of living cells by laser-guided direct writing: application to fabrication of hepatic–endothelial sinusoid-like structures. Nat Protoc 1(5):2288–2296. https://doi.org/10.1038/nprot.2006.386
50. Lu Y, Mapili G, Suhali G, Chen S, Roy K (2006) A digital micro-mirror device-based system for the microfabrication of complex, spatially patterned tissue engineering scaffolds. J Biomed Mater Res A 77(2):396–405. https://doi.org/10.1002/jbm.a.30601
51. Mehla S, Kandjani A, Coyle V, Harrison CJ, Low MX, Kaner RB, Sabri Y, Bhargava SK (2022) Gold sunflower microelectrode arrays with dendritic nanostructures on the lateral surfaces for antireflection and surface-enhanced Raman scattering. ACS Appl. Nano Mater. 5(2):1873–1890.https://doi.org/10.1021/acsanm.1c03501
52. Therriault D, White SR, Lewis JA (2003) Chaotic mixing in three-dimensional microvascular networks fabricated by direct-write assembly. Nat Mater 2(4):265–271. https://doi.org/10.1038/nmat863
53. Wu W, DeConinck A, Lewis JA (2011) Omnidirectional printing of 3D microvascular networks. Adv Mater 23(24):H178-183. https://doi.org/10.1002/adma.201004625
54. Kolesky DB, Homan KA, Skylar-Scott MA, Lewis JA (2016) Three-dimensional bioprinting of thick vascularized tissues. Proc Natl Acad Sci U S A 113(12):3179–3184. https://doi.org/10.1073/pnas.1521342113
55. Yan Q, Dong H, Su J, Han J, Song B, Wei Q, Shi Y (2018) A Review of 3D printing technology for medical applications. Engineering 4(5):729–742. https://doi.org/10.1016/j.eng.2018.07.021
56. Hager I, Golonka A, Putanowicz R (2016) 3D printing of buildings and building components as the future of sustainable construction? Procedia Engineering 151:292–299. https://doi.org/10.1016/j.proeng.2016.07.357
57. Sakin M, Kiroglu YC (2017) 3D printing of buildings: construction of the sustainable houses of the future by BIM. Energy Procedia 134:702–711
58. Lloret E, Shahab AR, Linus M, Flatt RJ, Gramazio F, Kohler M, Langenberg S (2015) Complex concrete structures. Comput Aided Des 60:40–49. https://doi.org/10.1016/j.cad.2014.02.011
59. Ngo TD, Kashani A, Imbalzano G, Nguyen KTQ, Hui D (2018) Additive manufacturing (3D printing): a review of materials, methods, applications and challenges. Compos B Eng 143:172–196. https://doi.org/10.1016/j.compositesb.2018.02.012
60. Tay YWD, Panda B, Paul SC, Noor Mohamed NA, Tan MJ, Leong KF (2017) 3D printing trends in building and construction industry: a review. Virt Phys Prototyping 12(3):261–276. https://doi.org/10.1080/17452759.2017.1326724
61. Xia M, Sanjayan J (2016) Method of formulating geopolymer for 3D printing for construction applications. Mater Des 110:382–390. https://doi.org/10.1016/j.matdes.2016.07.136
62. Mueller S, Mohr T, Guenther K, Frohnhofen J, Baudisch P (2014) faBrickation. Paper presented at the proceedings of the SIGCHI conference on human factors in computing systems

63. http://s671151578.initial-website.com/
64. Severini C. http://printinthemix.cad.rit.edu/Research/Show/127
65. Sun J, Zhou W, Huang D, Fuh JYH, Hong GS (2015) An overview of 3D printing technologies for food fabrication. Food Bioprocess Technol 8(8):1605–1615. https://doi.org/10.1007/s11947-015-1528-6
66. Liu Z, Zhang M, Bhandari B, Wang Y (2017) 3D printing: printing precision and application in food sector. Trends Food Sci Technol 69:83–94. https://doi.org/10.1016/j.tifs.2017.08.018
67. Chen Z (2016) Research on the impact of 3D printing on the international supply Chain. Adv Mater Sci Eng 2016:1–16. https://doi.org/10.1155/2016/4173873
68. Jia F, Wang X, Mustafee N, Hao L (2016) Investigating the feasibility of supply chain-centric business models in 3D chocolate printing: a simulation study. Technol Forecast Soc Chang 102:202–213. https://doi.org/10.1016/j.techfore.2015.07.026
69. Payne CLR, Dobermann D, Forkes A, House J, Josephs J, McBride A, Müller A, Quilliam RS, Soares S (2016) Insects as food and feed: European perspectives on recent research and future priorities. J Insects Food Feed 2(4):269–276. https://doi.org/10.3920/JIFF2016.0011
70. Vancauwenberghe V, Baiye Mfortaw Mbong V, Vanstreels E, Verboven P, Lammertyn J, Nicolai B (2019) 3D printing of plant tissue for innovative food manufacturing: encapsulation of alive plant cells into pectin based bio-ink. J Food Eng 263:454–464. https://doi.org/10.1016/j.jfoodeng.2017.12.003
71. Sun J, Peng Z, Zhou W, Fuh JYH, Hong GS, Chiu A (2015) A review on 3D printing for customized food fabrication. Procedia Manuf 1:308–319. https://doi.org/10.1016/j.promfg.2015.09.057
72. Pallottino F, Hakola L, Costa C, Antonucci F, Figorilli S, Seisto A, Menesatti P (2016) Printing on food or food printing: a review. Food Bioprocess Technol 9(5):725–733. https://doi.org/10.1007/s11947-016-1692-3
73. Yang F, Zhang M, Bhandari B (2017) Recent development in 3D food printing. Crit Rev Food Sci Nutr 57(14):3145–3153. https://doi.org/10.1080/10408398.2015.1094732
74. Lipson H, Kurman M (2013) Fabricated: the new world of 3D printing. Wiley
75. Masuch T A world first: additively manufactured titanium components now onboard the Airbus A350 XWB. https://www.etmm-online.com/a-world-first-additively-manufactured-titanium-components-now-onboard-the-airbus-a350-xwb-a-486310/
76. Wimpenny DI, Pandey PM, Kumar LJ (2017) Advances in 3D printing & additive manufacturing technologies
77. Khajavi SH, Partanen J, Holmström J (2014) Additive manufacturing in the spare parts supply chain. Comput Ind 65(1):50–63. https://doi.org/10.1016/j.compind.2013.07.008
78. SpaceX—Capabilities & Services. https://www.spacex.com/media/Capabilities&Services.pdf
79. Zhang Q, Zhang K, Hu G (2016) Smart three-dimensional lightweight structure triggered from a thin composite sheet via 3D printing technique. Sci Rep 6:22431. https://doi.org/10.1038/srep22431
80. Hong S, Sycks D, Chan HF, Lin S, Lopez GP, Guilak F, Leong KW, Zhao X (2015) 3D printing of highly stretchable and tough hydrogels into complex. Cellularized Structures. Adv Mater 27(27):4035–4040. https://doi.org/10.1002/adma.201501099
81. Wehner M, Truby RL, Fitzgerald DJ, Mosadegh B, Whitesides GM, Lewis JA, Wood RJ (2016) An integrated design and fabrication strategy for entirely soft, autonomous robots. Nature 536(7617):451–455. https://doi.org/10.1038/nature19100
82. Yu K, Ge Q, Qi HJ (2014) Reduced time as a unified parameter determining fixity and free recovery of shape memory polymers. Nat Commun 5:3066. https://doi.org/10.1038/ncomms4066
83. Zhao Q, Qi HJ, Xie T (2015) Recent progress in shape memory polymer: nw behavior, enabling materials, and mechanistic understanding. Prog Polym Sci 49–50:79–120. https://doi.org/10.1016/j.progpolymsci.2015.04.001
84. Choi J, Kwon OC, Jo W, Lee HJ, Moon M-W (2015) 4D printing technology: a review. 3D Printing Addit Manuf 2(4):159–167. https://doi.org/10.1089/3dp.2015.0039

85. Tibbits S, McKnelly C, Olguin C, Dikovsky D, Hirsch S 4D printing and univerrsal transformation. In, 2014
86. Jung JP, Bhuiyan DB, Ogle BM (2016) Solid organ fabrication: comparison of decellularization to 3D bioprinting. Biomater Res 20(1):27. https://doi.org/10.1186/s40824-016-0074-2
87. Zarek M, Mansour N, Shapira S, Cohn D (2017) 4D printing of shape memory-based personalized Endoluminal medical devices. Macromol Rapid Commun 38 (2). https://doi.org/10.1002/marc.201600628
88. Taylor DL, In Het Panhuis M (2016) Self-healing hydrogels. Adv Mater 28(41):9060–9093. https://doi.org/10.1002/adma.201601613
89. Khoo ZX, Teoh JEM, Liu Y, Chua CK, Yang S, An J, Leong KF, Yeong WY (2015) 3D printing of smart materials: a review on recent progresses in 4D printing. Virt Phys Prototyping 10(3):103–122. https://doi.org/10.1080/17452759.2015.1097054
90. Jiang R, Kleer R, Piller FT (2017) Predicting the future of additive manufacturing: a Delphi study on economic and societal implications of 3D printing for 2030. Technol Forecast Soc Chang 117:84–97. https://doi.org/10.1016/j.techfore.2017.01.006
91. Agarwal T, Hann SY, Chiesa I, Cui H, Celikkin N, Micalizzi S, Barbetta A, Costantini M, Esworthy T, Zhang LG, De Maria C, Maiti TK (2021) 4D printing in biomedical applications: emerging trends and technologies. J Mater Chem B 9(37):7608–7632. https://doi.org/10.1039/d1tb01335a
92. Momeni F, Mehdi Hassani MNS, Liu X, Ni J (2017) A review of 4D printing. Mater Design 122:42–79.https://doi.org/10.1016/j.matdes.2017.02.068
93. Kamila, (2013) Introduction, classification and applications of smart materials: an overview. Am J Appl Sci 10(8):876–880. https://doi.org/10.3844/ajassp.2013.876.880
94. Vaezi M, Chianrabutra S, Mellor B, Yang S (2013) Multiple material additive manufacturing—Part 1: a review. Virt Phys Prototyping 8(1):19–50. https://doi.org/10.1080/17452759.2013.778175
95. Mitchell A, Lafont U, Hołyńska M, Semprimoschnig C (2018) Additive manufacturing—a review of 4D printing and future applications. Addit Manuf 24:606–626. https://doi.org/10.1016/j.addma.2018.10.038
96. Shin D-G, Kim T-H, Kim D-E (2017) Review of 4D printing materials and their properties. Int J Precis Eng Manuf Green Technol 4(3):349–357. https://doi.org/10.1007/s40684-017-0040-z
97. Yang W, Lu H, Huang W, Qi H, Wu X, Sun K (2014) Advanced shape memory technology to reshape product design. Manuf Recycling. Polymers 6(8):2287–2308. https://doi.org/10.3390/polym6082287
98. Ramakrishna S, Mayer J, Wintermantel E, Leong KW (2001) Bomedical applications of polymer-composite materials: a review. Compos Sci Technol 61:1189
99. Stavropoulos P, Foteinopoulos P, Papacharalampopoulos A, Bikas H (2018) Addressing the challenges for the industrial application of additive manufacturing: towards a hybrid solution. Int J Lightweight Mater Manuf 1(3):157–168. https://doi.org/10.1016/j.ijlmm.2018.07.002
100. Wohlers Report 2012
101. Franke N, Piller F (2004) Value creation by toolkits for user innovation and design: the case of the watch market. J Product Innov Manag 21:401–415
102. Lott P, Schleifenbaum H, Meiners W, Wissenbach K, Hinke C, Bültmann J (2011) Design of an optical system for the in situ process monitoring of selective laser melting (SLM). Phys Procedia 12:683–690. https://doi.org/10.1016/j.phpro.2011.03.085
103. P R (2009) Additive manufacturing–a supply chain wide response to economic uncertainty and en-vironmental sustainability. In: Mass customization engineering and managing global operations
104. Huang SH, Liu P, Mokasdar A, Hou L (2012) Additive manufacturing and its societal impact: a literature review. Int J Adv Manuf Technol 67(5–8):1191–1203. https://doi.org/10.1007/s00170-012-4558-5

105. du Plessis A, Broeckhoven C, Yadroitsava I, Yadroitsev I, Hands CH, Kunju R, Bhate D (2019) Beautiful and functional: a review of biomimetic design in additive manufacturing. Addit Manuf 27:408–427. https://doi.org/10.1016/j.addma.2019.03.033
106. Beyers RN, Blignaut AS, Mophuti L (2012) Mobile fablabs: local and rural innovation in South Africa. In: EdMedia+ innovate learning association for the advancement of computing in education (AACE), pp 112–122
107. Ishengoma FR, Mtaho AB (2014) 3D Printing: developing counties perspectives. Int J Comput Appl 104(11)
108. Berman B (2012) 3-D printing: the new industrial revolution. Bus Horiz 55(2):155–162. https://doi.org/10.1016/j.bushor.2011.11.003
109. Loh GH, Pei E, Harrison D, Monzón MD (2018) An overview of functionally graded additive manufacturing. Addit Manuf 23:34–44. https://doi.org/10.1016/j.addma.2018.06.023
110. Oxman N (2011) Variable property rapid prototyping. Virt Phys Prototyping 6(1):3–31. https://doi.org/10.1080/17452759.2011.558588
111. Junk S, Sämann-Sun J, Niederhofer M (2010) Application of 3D printing for the rapid tooling of thermoforming moulds. In: Proceedings of the 36th international MATADOR conference Springer, London, pp 369–372
112. Wang PH, Kim G, Sterkenburg R (2019) Investigating the effectiveness of a 3D printed composite mold. Int Schol Scient Res Innov 13(11):684–688
113. Kumar K, Zindani D, Davim JP (2020) Rapid prototyping, rapid tooling and reverse engineering: from biological models to 3D Bioprinters. De Gruyter. https://doi.org/10.1515/9783110664904
114. D HP (2000) Rapid tooling: technologies and industrial applications
115. T C (2015) Wohlers Report 2015: additive manufacturing and 3D printing state of the industry annual worldwide progress report
116. Mehla S, Kandjani AE, Babarao R, Lee AF, Periasamy S, Wilson K, Ramakrishna S, Bhargava SK (2021) Porous crystalline frameworks for thermocatalytic CO_2 reduction: an emerging paradigm. Energy Environ Sci 14(1):320–352. https://doi.org/10.1039/d0ee01882a
117. Zhakeyev A, Wang P, Zhang L, Shu W, Wang H, Xuan J (2017) Additive manufacturing: unlocking the evolution of energy materials. Adv Sci 4(10):1700187. https://doi.org/10.1002/advs.201700187
118. Mehla S, Kukade S, Kumar P, Rao PVC, Sriganesh G, Ravishankar R (2019) Fine tuning H-transfer and β-scission reactions in VGO FCC using metal promoted dual functional ZSM-5. Fuel 242:487–495. https://doi.org/10.1016/j.fuel.2019.01.065
119. Nishiyama H, Yamada T, Nakabayashi M, Maehara Y, Yamaguchi M, Kuromiya Y, Nagatsuma Y, Tokudome H, Akiyama S, Watanabe T, Narushima R, Okunaka S, Shibata N, Takata T, Hisatomi T, Domen K (2021) Photocatalytic solar hydrogen production from water on a 100-m(2) scale. Nature 598(7880):304–307. https://doi.org/10.1038/s41586-021-03907-3
120. Mehla S, Krishsna V, Sriganesh G, Ravishankar R (2018) Mesoporous superacid catalysts for valorisation of refinery naphtha stream. RSC Adv 8(59):33702–33709. https://doi.org/10.1039/C8RA07024E
121. Path to hydrogen competitiveness: a cost perspective (2020). Hydrogen Council
122. Silveria GD, Borenstein D, Fogliatto FS (2001) Mass customization: literature review and research direction. Int J Product Econ 72:1–13
123. Mai J, Zhang L, Tao F, Ren L (2015) Customized production based on distributed 3D printing services in cloud manufacturing. Int J Adv Manuf Technol 84(1–4):71–83. https://doi.org/10.1007/s00170-015-7871-y
124. Liu N, Choi T-M, Yuen C-WM, Ng F (2012) Optimal pricing, modularity, and return policy under mass customization. IEEE Trans Syst Man Cybern Part A: Syst Hum 42(3):604–614. https://doi.org/10.1109/tsmca.2011.2170063
125. Hart CWL (1994) Mass customization: conceptual underpinnings, opportunities and limits. Int J Serv Ind Manag 6(2):36–45
126. Quinlan HE, Hasan T, Jaddou J, Hart AJ (2017) Industrial and consumer uses of additive manufacturing: a discussion of capabilities, trajectories, and challenges. J Ind Ecol 21(S1):S15–S20. https://doi.org/10.1111/jiec.12609

127. Salvador F, Forza C, Rungtusanatham M (2002) Modularity, product variety, production volume, and component sourcing: theorizing beyond generic prescriptions. J Oper Manag 20(5):549–575. https://doi.org/10.1016/S0272-6963(02)00027-X
128. Hu SJ (2013) Evolving paradigms of manufacturing: from mass production to mass customization and personalization. Procedia CIRP 7:3–8. https://doi.org/10.1016/j.procir.2013.05.002
129. Acher M, Baudry B, Barais O, Jézéquel J-M (2014) Customization and 3D printing. Paper presented at the proceedings of the 18th international software product line conference, Vol 1
130. Banks J (2013) Adding value in additive manufacturing: researchers in the United Kingdom and Europe look to 3D printing for customization. IEEE Pulse 4(6):22–26. https://doi.org/10.1109/MPUL.2013.2279617
131. Kyriakou H, Nickerson JV, Sabnis G (2017) Knowledge reuse for customization: Metamodels in an open design community for 3D printing. MIS Q 41(1):315–332
132. Qiu J, Gao Q, Zhao H, Fu J, He Y (2017) Rapid Customization of 3D integrated microfluidic chips via modular structure-based design. ACS Biomater Sci Eng 3(10):2606–2616. https://doi.org/10.1021/acsbiomaterials.7b00401
133. Mehla S, Das J, Jampaiah D, Periasamy S, Nafady A, Bhargava SK (2019) Recent advances in preparation methods for catalytic thin films and coatings. Catal Sci Technol 9:3582–3602. https://doi.org/10.1039/C9CY00518H
134. Groeger D, Chong Loo E, Steimle J (2016) HotFlex. Paper presented at the Proceedings of the 2016 CHI conference on human factors in computing systems

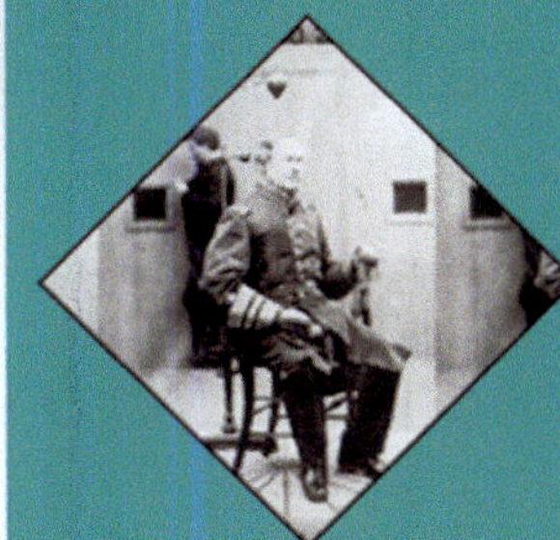

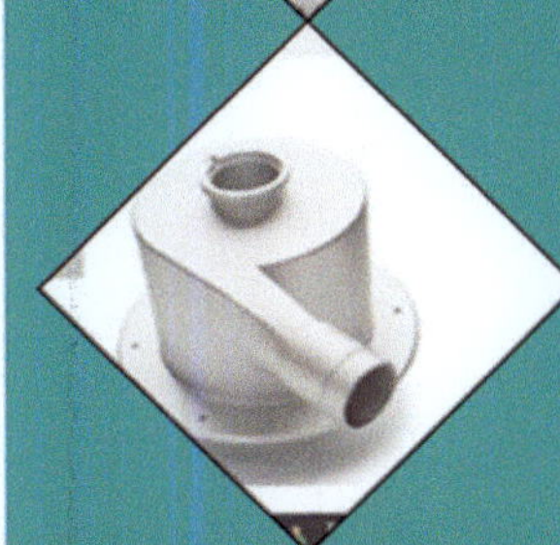

History and Evolution of Additive Manufacturing

Photosculpture, Topography, Sheet Lamination, Powder Bed Fusion, Vat Photopolymerization, Standardization, 4D Printing, AM in industries and construction, Mass Customization,

Sunpreet Singh, Sunil Mehla, Suresh K. Bhargava, Seeram Ramakrishna

2

Chapter 3
Material Extrusion and Vat Photopolymerization—Principles, Opportunities and Challenges

PR. Selvakannan, Maciej Mazur, and Xiaochen Sun

Contents

Abstract The additive manufacture (AM) of plastic components can be accomplished by a variety of methods which are classified according to how material layers are consolidated layer-by-layer to create physical objects from digital data. Two of the most widely used methods include Material Extrusion (MEX) and Vat Photopolymerization (VP). These methods include a range of commercialized AM processes often referred to by various trademarked terms such as Fused Deposition Modelling (FDM) and Stereolithography (SLA). Compared to conventional subtractive or formative manufacturing process, MEX and VP are able to manufacture complex parts with high ability for customization, as they impose few constraints on part geometry, and require low setup effort with no custom tooling. Furthermore,

---

PR. Selvakannan (✉)
Centre for Advanced Materials and Industrial Chemistry, RMIT University, Melbourne, VIC, Australia
e-mail: selvakannan.periasamy@rmit.edu.au

M. Mazur · X. Sun
Centre for Additive Manufacturing, RMIT University, Melbourne, VIC, Australia
e-mail: maciej.mazur@rmit.edu.au

S. K. Bhargava et al. (eds.), *Additive Manufacturing for Chemical Sciences and Engineering*, https://doi.org/10.1007/978-981-19-2293-0_3

MEX and VP are both well-established additive manufacturing processes and through ongoing refinement have achieved compatibility with a broad range of materials and part geometries, and comparatively low operating costs. Due to their high versatility, MEX and VP have been widely used in a broad range of industries including applications in chemical sciences, biotechnology, aerospace, defense, and automotive engineering. However, despite their high versatility, AM processes such as MEX and VP are subject to unique technological characteristics associated with the manufacturing process, the material properties, part design and suitable application areas. This chapter provides an overview of MEX and VP processes characteristics critical to the effective application of these additive manufacturing technologies to high performance products.

Keywords Additive manufacturing · Design for additive manufacture · Material extrusion · Photopolymerization · Polymers · Thermoplastics · Vat photopolymerization

Abbreviations

AM	Additive Manufacturing
ABS	Acrylonitrile Butadiene Styrene
ASTM	American Society for Testing and Materials
BAAM	Big Area Additive Manufacturing
CLIP	Continuous Liquid Interface Production
DIW	Direct Ink Writing
DLP	Digital Light Processing
FDM	Fused Deposition Modelling
ISO	International Organization for Standardization
MEAM	Material Extrusion-based Additive Manufacturing
MIM	Metal Injection Moulding
PBT	Polybutylene terephthalate
PC	Polycarbonate
PCL	Polycaprolactone
PGA	Polyglycolic acid
PIM	Powder Injection Moulding
PLA	Polylactic acid
PS	Polystyrene
SDS	Shaping Debinding and Sintering

3.1 Introduction

Additive manufacturing (AM) is an advanced manufacturing technology contributing to the transformative trend of the fourth industrial revolution (Industry 4.0) critical to future economic competitiveness [1]. The application of AM can be found in a broad range of industries including chemical engineering, biotechnology, healthcare, transport and defence [2]. AM enables new production and supply chain opportunities, leading to disruptive business models. This is largely facilitated by the unique ability of AM processes to build highly complex and customizable parts with short lead times. This contrasts with conventional manufacturing processes, which are often subject to stricter manufacturability limits, and require custom tooling resulting in long-lead times, and limited product customization. To exploit the benefits of additive manufacturing, a range of AM process for polymeric materials have been developed which offer unique levels of flexibility and capability in building functional structures where conventional manufacturing reaches its limits [3]. Two of the most widely used approaches to AM of polymeric materials include Material Extrusion (MEX), which mainly uses thermoplastic-based feedstock, and Vat Photopolymerization (VP) which builds parts using light mediated curing of photo-sensitive reins. These approaches include a range of commercialised AM processes referred to by trademarked terms, the most common of which being Fused Deposition Modelling (FDM) or Stereolithography (SLA). This chapter provides an overview of the operating principles, advantages, and limitation of MEX and VP additive manufacturing technologies.

3.2 Material Extrusion Additive Manufacturing

Material extrusion (MEX) is an 'additive manufacturing process in which material is selectively dispensed through a nozzle or orifice' [4]. It is one of the seven categories of additive manufacturing technologies defined by the American Society for Testing and Materials (ASTM) and the International Organization for Standardization (ISO). In a typical MEX process, the material utilized is either gel, paste or liquefied solid materials. The materials are extruded through a nozzle or orifice by a plunger or screw mechanism to form predefined material tracks in a layered fashion (Fig. 3.1). Fused Deposition Modelling (FDM), which build parts from thermoplastic polymer feedstock, was the first MEX process to be commercialized and remains one of the

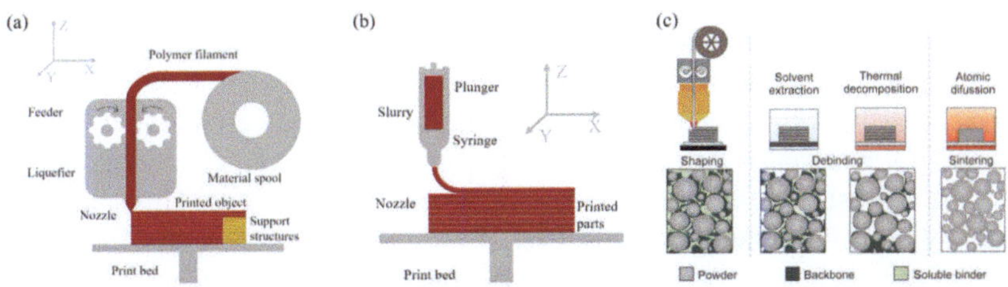

Fig. 3.1 Schematics of three MEX technologies **a** Fused deposition modelling (FDM) process **b** Robocasting **c** typical metal and ceramic MEX process [9]

most popular AM technologies today. Although the term FDM is a trademarked term, it is often used to synonymously refer to MEX processes due to its ubiquity and historical impact.

Over the last few years, MEX systems have been extended to be compatible with metal and ceramic materials developed from corresponding injection moulding technologies and commercialized under various trademark terms (Sect. 3.2.1). These systems use polymeric binders mixed with fine metal or ceramic powders as feedstock, and require printed parts to go through post-processing operations involving de-binding and sintering prior to use. MEX systems have also been adapted to enable multi-material printing using either shared or additional extrusion nozzles, further broadening possible application areas. Composite materials using fiber reinforcement are a particularly exciting example of this development. With the addition of continuous or chopped fibers extruded from one nozzle combined with thermoplastic from another, the mechanical properties of the additively manufactured parts can be significantly enhanced compared to purely thermoplastic counterparts [5].

Biological material have also been used in MEX systems in the form of bioinks, which are a combination of biologicals cells and biopolymer gels, to build tissue and organ constructs [6]. These are subject to research and development in applications involving regenerative medicine, transplantation, tissue engineering, drug screening, and cancer research [7]. Edible materials have also been used in MEX processes where typically purred food feedstock are mixed with hydrocolloids (e.g. starch) to modify the extrudate's rheological properties for improved printability [8]. Chocolate has also been as used in MEX food printing as feedstock to manufacture confectionary. The specific operating principles of various MEX process technologies are discussed in the following section.

3.2.1 MEX Process Variants

There are two major approaches to material fusion which differentiate MEX process implementations, namely solidification and curing. The solidification material fusion approach typically involves solid-to-liquid-to-solid phase changes and is predominantly used with thermoplastics (either pure or mixed with fine metal or ceramic powders). Feedstock materials used in this approach are commonly in filament, rod or pellet form and are softened and liquefied in the printing head through heating (Fig. 3.1a). The solidification of material occurs after deposition as the extruded material cools back to ambient temperature. In the curing material fusion approach, the feedstock material is typically in slurry form and is cured following extrusion by drying, exposure to UV light, or further chemical treatment. This approach is commonly used for ceramics, foodstuffs, or bio feedstocks. Several MEX process technologies based on the two major material consolidation approaches are discussed below.

FDM–Filament-based MEX process

Fused Deposition Modelling (FDM) was the first commercialized material extrusion-based additive manufacturing process and continues to remain highly utilized (Fig. 3.1a). According to Sculpteo's Survey Report, FDM is the most commonly used type of AM technology, with more than 70% of users being consumers [10]. FDM uses a continuous filament of thermoplastic material extruded from a heated nozzle which is deposited track-by-track and layer-by-layer to build up solid 3D parts. Detailed working principles of FDM are discussed in detail in Sect. 3.3.1.

Robocasting–Slurry-based MEX process

Robocasting, also known as Direct Ink Writing (DIW) [11], is one of the most versatile AM technologies from a material feedstock compatibility perspective (Fig. 3.1b). In a typical robocasting process, a paste or gel-like material with desired rheological properties is selectively deposited in layer-wise manner by a syringe onto a substrate to form a specific shape. After deposition, the material stabilises and becomes pseudo-solid due to its shear thinning behaviour. A second step is applied to dry, solidify or cure the print. The multi-material capacity of robocasting enables manufacturing with different classes of materials [12]. Initially developed for manufacturing ceramic products, it is now also compatible with metal, foodstuffs and bioink feedstocks.

Metal and ceramic MEX

The metal and ceramic MEX is usually a multi-step process, starting with an extrusion process analogous to the typical FDM or robocasting approach, after which several additional steps are applied to densify the components. The process steps can be summarised as Shaping, Debinding and Sintering (SDS), and are similar to those used in Powder Injection Moulding (PIM) [13] or Metal Injection Moulding (MIM) from which corresponding MEX processes are derived. These processes typically use metal or ceramic powders blended with thermoplastic or water-based binders to form extrudable feedstock. The process may be referred to by various trademark terms such as Bound Metal Deposition, Automated Multi-material Deposition, Composite Extrusion Modelling, and Pellet Additive Technology (Table 3.1).

3.3 Fused Deposition Modelling (FDM)

Fused Deposition Modelling (FDM) is a trade name registered by Stratasys Ltd. corporation which originally developed and commercialized this AM process in the early 1990s. Since then FDM has become the most commonly used AM process due to its simplicity, robust and low-cost thermoplastic material feedstock, and simple post-processing requirements [16]. Due to the expiration of the original Stratasys FDM patents in 2009, many FDM 3D printers have become available in the market ranging in cost, resolution, and build volume. Furthermore, the process is still under continuing development with active research focusing on expanding material and

Table 3.1 Comparison of three common forms of MEX technologies

Process	Feedstock	Compatible materials	Typical minimum layer thickness	Process principles
FDM	Filament-based	Mainly thermoplastic polymers and composites	50–100 μm	Single-step process
Robocasting	Slurry-based	Ceramics, metals, polymers, composites, bioink and foodstuffs [12]	5 μm [14]	Single or multi-step process
Metal and ceramics MEX	Slurry-based and filament-based	Metal or ceramics with binder mixture	50 μm [15]	Multi-step process

manufacturing capability in various application areas including chemical sciences, biotechnology, aerospace, defense, and automotive engineering.

3.3.1 FDM Operating Principles

FDM builds three-dimensional structures from thermoplastic materials using a layer-by-layer approach, as shown in Fig. 3.1a. In the initial stage of the process, the geometry of a 3D model of the part is algorithmically sliced into individually processable layers for which material deposition toolpath are generated and saved into machine instructions. The support structures are also generated at this stage if needed. The machine instructions are then executed in the printing process to deposit material according to the predefined toolpaths, layer-by-layer. A thermoplastic material is fed into the printer extrusion nozzle in filament form (or in some cases as pellets) where resistive heaters soften the material and enable extrusion and deposition into a single layer through the planar movement of the nozzle. Next, the build platform is lowered by one layer thickness, and a subsequent layer is deposited. The process repeats until all layers of the part model are consolidated into a complete part. Following completion of the build, the parts are removed from the build platform and any support structures necessary for overhanging surfaces are removed (support structures in additive manufacturing processes are discussed in detail in Chap. 4). Figure 3.2 shows some example components manufactured using FDM.

3.3.2 FDM Materials

The FDM process is compatible with a broad range of thermoplastics materials. Figure 3.3 shows the classification of typical thermoplastics used in FDM, based on

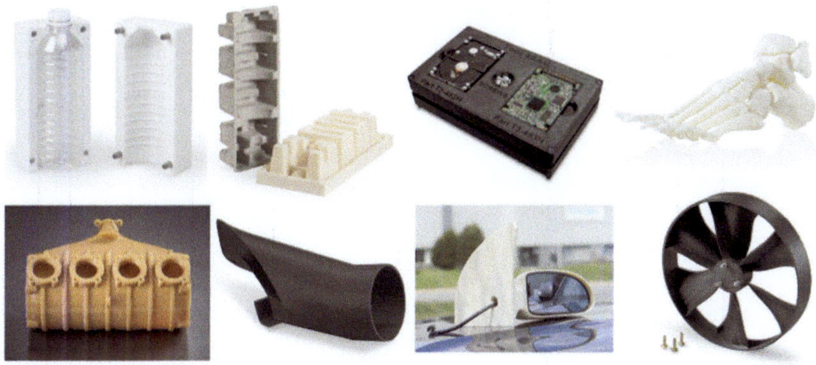

Fig. 3.2 Example FDM manufactured components including moulding tools, electronics housing, medical models, prototype automotive components (Image courtesy: Stratasys Ltd.)

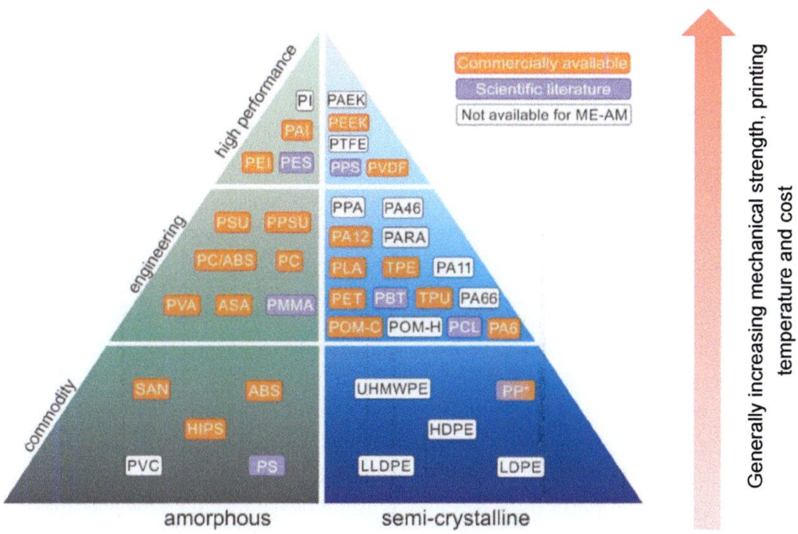

Fig. 3.3 Common thermoplastic materials used in the FDM process [17]

their semi-crystalline or amorphous nature, availability, and physical performance. These materials exhibit a range of physical and mechanical properties which should be selected in accordance with the desired application requirements. For general engineering applications, acrylonitrile butadiene styrene (ABS), polystyrene (PS), polylactic acid (PLA) and polycarbonate (PC) are the preferred material candidates due to low cost, good mechanical properties, and good compatibility with FDM processing. Recently, FDM materials have been expanded to biological applications, where biocompatible polymers such as polyglycolic acid (PGA), polycaprolactone (PCL) and polybutylene terephthalate (PBT) may be used.

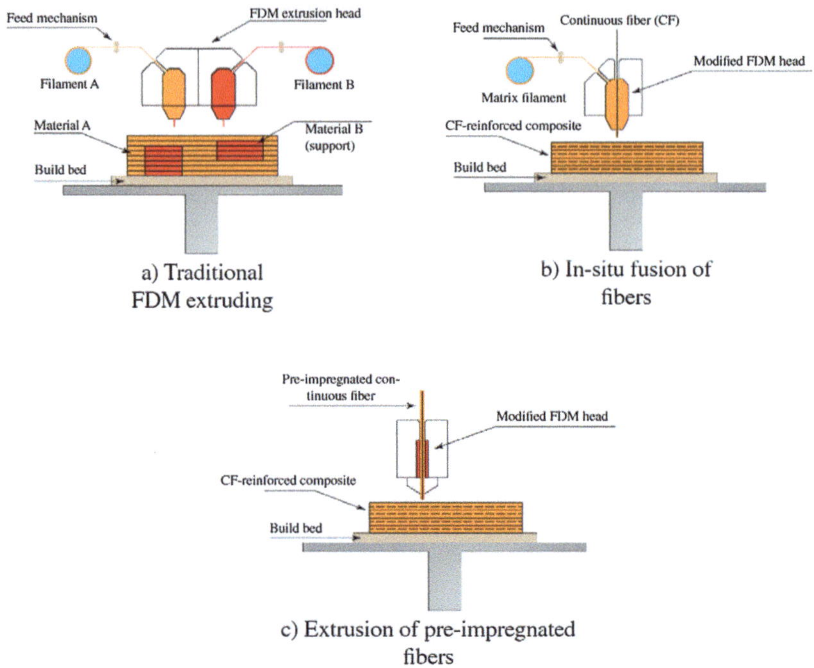

Fig. 3.4 FDM processes variants **a** dual extruder FDM **b** composite material FDM with in-situ fusion of fibres **c** composite material FDM with pre-impregnated composite fibres [18]

Furthermore, FDM has also been applied to composite materials in addition to neat polymers by either extrusion of pre-impregnated filament (Fig. 3.4b) or through in-situ composite fusion in a modified extrusion system (Fig. 3.4c).

There is ongoing research into developing further material alternatives for FDM using inorganic fillers and ceramic particles in the filament to achieve unique properties. For example, iron fillers have been added to Nylon-6 filament to achieve ferromagnetic properties in parts built directly with a conventional FDM machine [19].

3.3.3 Process Parameters

A range of process and part parameters influence the manufacturability, quality, and cost of FDM manufactured components. These can be broadly classified into process, part, deposition, and material parameters (Table 3.2). A careful consideration of these parameters is required in order to optimize the part properties for a given application. For example, studies have shown that the mechanical strength of FDM parts depends on the build orientation [20], infill pattern [21, 22], infill density [23], layer thickness, temperature, and build speed [24, 25].

The layer thickness in FDM processes typically ranges between 0.05 and 0.3 mm and can have a significant influence on several part attributes such as resolution,

Table 3.2 Key parameters which influence the cost, quality and capability of the FDM process and manufactured components

Process and part parameters	Deposition parameters	Material parameters
Layer thickness Build time Build volume Batch size Surface finish Overhanging surfaces and support structures Build orientation Platform and build chamber temperature Porosity Residual stress	Nozzle diameter Number of extrusion nozzles Extrusion speed Track width Deposition toolpath strategy Infill geometry Extrusion temperature	Material composition and process compatibility Mechanical properties Thermal properties Density Rheological properties Shrinkage factor Anisotropy Humidity

surface roughness, and build time. Due to the layered FDM construction process, the accuracy of curved or inclined surfaces is limited by stair-stepping effects which depend on layer thickness and surface orientation relative to the layer plane. Reduced layer thickness results in a more accurate curve approximation, however the build times are increased as more deposition steps are required for a given part height. Vertical or horizontal surfaces typically exhibit lowest surface roughness values due to the absence of stair-stepping. Downfacing surfaces with acute angles generally show the roughest surface finish. Typical FDM part build times range in the order of hours to days depending on part volumes, layer thickness, deposition rates, and part infill density.

The build orientation of the part relative to the build platform is a particularly critical parameter in FDM that affects a range of part and process attributes, including build time, amount of support structures required, part strength and surface roughness. The choice of build orientation is usually a trade-off between competing objectives prioritized by the designer; Chap. 4 provides detailed discussion of these objectives. The influence of build orientation is particularly prominent on the mechanical properties of FDM parts, where the tensile strength of parts loaded along to the printing directions (Z axis) is typically lower than when parts are loaded in the perpendicular direction (X & Y axes) with the differences being up to 50% or more [26]. The orientational strength differences are due to the layer-by-layer construction approach resulting in lower inter-layer material strength compared to intra-layer material strength. The anisotropy in FDM components can become a significant design challenge when FDM parts are expected to have isotropic mechanical properties. Some of the strength differences are attributable to residual voids/porosity introduced between layers and tracks formed during material deposition [26].

Figure 3.5 shows an example Computed Tomography (CT) X-ray cross section of a typical FDM component where the dark regions within the part are voids. The voids reduce part strength and increases stress concentrations which may lead to accelerated fatigue failure under cyclic loading conditions [27]. Research efforts are

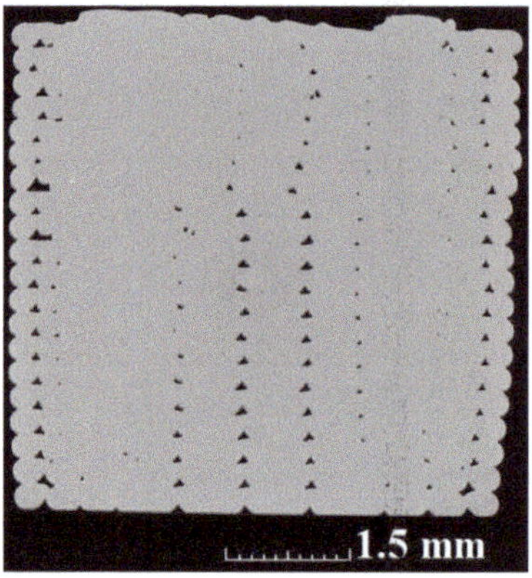

Fig. 3.5 FDM part cross-section indicating commonly seen inter-layer and inter-track voids

focused on void reduction strategies through optimisation of material composition and toolpath planning, ultrasonic vibration or laser heating during layer deposition, or post-processing with thermo-mechanical or chemical treatments. The research is ongoing and the elimination of porosity in FDM components is an unsolved problem.

Also critical to the strength of FDM parts is the geometry and density of the part infill structure. Typically, a low density (sparse) infill structure is used with a basic periodic cellular geometry such as rectangular, diamond or honeycomb pattern. Several infill geometry examples are shown in (Fig. 3.6). Sparse infill reduces material usage and build times, and basic infill geometry simplifies the generation of material deposition toolpaths. The infill pattern parameters such as topology, cell size and wall thickness may need to be modified if increased part mechanical strength is required, however this typically increases infill density and build time [22]. Alternatively, instead of using homogenous periodic infill patterns, there is growing interest

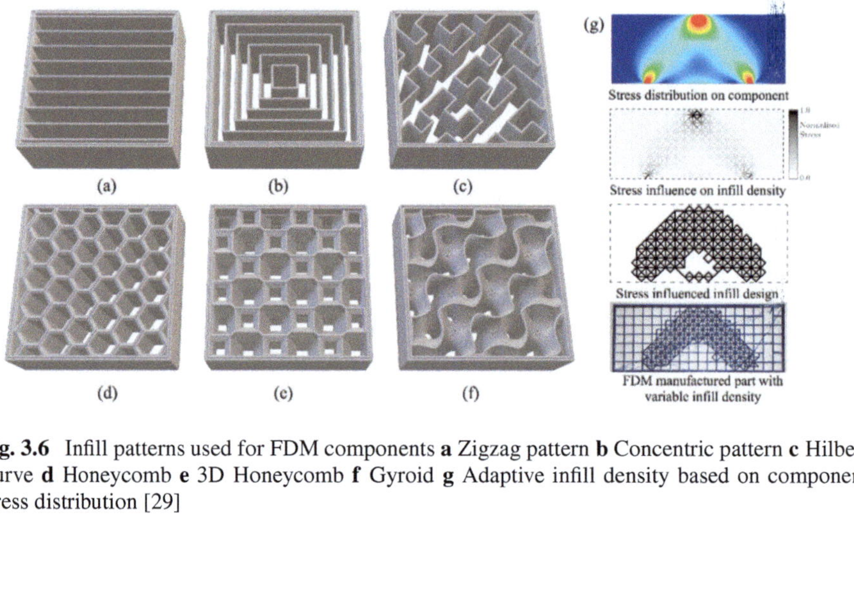

Fig. 3.6 Infill patterns used for FDM components **a** Zigzag pattern **b** Concentric pattern **c** Hilbert Curve **d** Honeycomb **e** 3D Honeycomb **f** Gyroid **g** Adaptive infill density based on component stress distribution [29]

in using heterogenous infill geometry locally optimised in proportion to the imposed loads (for example Fig. 3.6g). The approach is analogous to variable cell density in natural systems such as bone and can result in increased material use efficiency [28]. However, the generation and manufacture of such optimised infill patterns is challenging and the subject of ongoing research.

3.3.4 Challenges and Outlook

MEX is a well-established additive manufacturing process with proven utility across a range of applications. However, further development of the technology is ongoing to address outstanding challenges and exploit further possibilities. One the main challenges with MEX process is the relatively slow material deposition rates which can result in long part build times, particularly compared to rapid VP processes such CLIP (see Sect. 3.4). The slow built rates are further exaggerated in MEX processes utilizing bioinks, ceramics and metal materials which generally require time-consuming post-processing operations involving part curing, drying and sintering. Attempt to address low build rates are focused on machine hardware changes targeting increased material deposition rates. Recent work has demonstrated a high-speed polymer extrusion FDM system capable of approximately sevenfold higher build rates when compared to typical commercial desktop FDM systems [30]. The system uses a combination of a laser-heated polymer liquefier, rapid filament extrusion and high-speed servo gantry system to achieve high extrusion force, fast filament heating, and rapid motion. Other efforts addressing increased material deposition rates have focused on scaling up the FDM system concept with larger machines and extrusion nozzles such as in the Big Area Additive Manufacturing (BAAM) system developed for industries requiring large-scale parts such as architecture and aerospace [31].

Another challenge with MEX processes is the high level of anisotropy in mechanical properties of parts due to the presence of porosity. The porosity is resulting from voids which form between material tracks and layers and is inherent to the nature of MEX material deposition process. Ongoing research effort is focused on reducing the presence of voids either by preventing their formation during the build (in-situ) or by reducing their presence following the build using post-processing methods (ex-situ). These approaches include: toolpath optimisation, use of non-uniform material track widths, post-build hot isostatic pressing, or use of expanding feedstock material [32, 33].

The rough surface finish of MEX parts compared to VP components is another ongoing challenge which can limit potential application areas for MEX parts. The roughness may be reduced with the use of finer layers however this results in increased build times. Research efforts aimed at improving surface roughness of MEX parts are focused on the use of non-planar layer deposition to reduce stair stepping effects [34, 35], post-build chemical surface treatment [36], or post-build mechanical treatment with mechanical abrasion machining processes [37]. However, these methods are

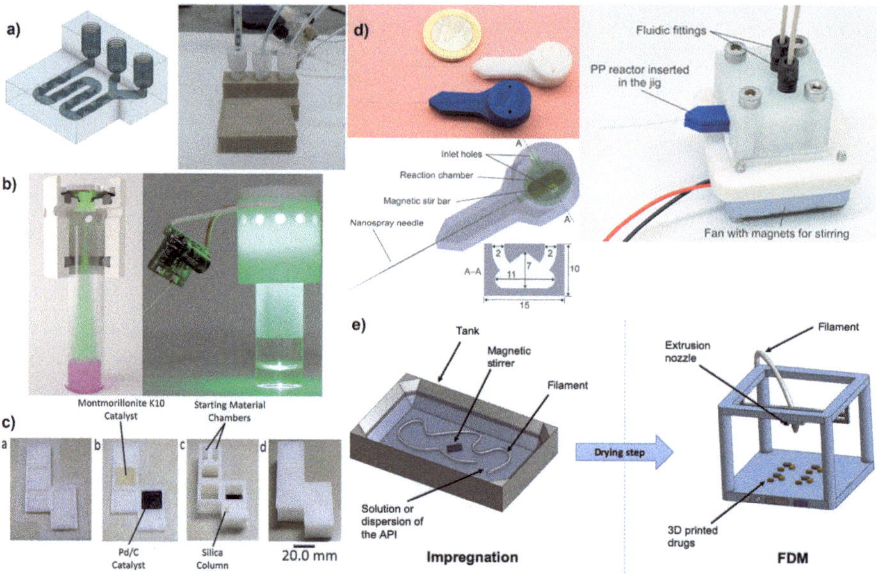

Fig. 3.7 Examples of FDM manufactured components in chemical engineering applications **a** Flow mixers [38] **b** Photo-catalytic reactor [39] **c** Catalytic reaction-ware [40] **d** reaction ware for online mass spectroscopic analysis[41] **e** personalized medicine production [42]

associated with drawbacks such as limits on machine and material compatibility and increased processing time and further research is needed.

Despite some of the challenges with MEX processes, there are exciting prospects for further development in the future. Particular research focus is on expanding the range of application areas utilizing MEX components, with the technology being applied to novel applications such as chemical reaction mixtures, photocatalytic reactors, reaction wares and tailored medicine (Fig. 3.7).

Although a broad range of MEX compatible feedstock materials are already available, further development in feedstock materials is expected to drive future innovation with the incorporation of additives such as pharmaceuticals, surfactants, nanoparticles, blowing agents and elastomers to obtain unique properties for new applications. In particular, advanced applications in pharmaceutical sciences such as personalized medicines are receiving significant research attention and driving further advances in MEX additive manufacturing technology.

3.4 Vat Photopolymerization (VP)

Vat Photopolymerization (VP) is an additive manufacturing method in which liquid photopolymer resin contained in a vat is selectively cured by light-activated polymerization using a layer-by-layer approach to build three-dimensional parts. A range of VP based additive manufacturing processes have been developed which are commonly referred to by the trademark terms of stereolithography (SLA) (Fig. 3.8a), Digital Light Processing (DLP) (Fig. 3.8b), Continuous Liquid Interface Production (CLIP) (Fig. 3.8c) as well as Two-Photon Polymerization (TPP). The oldest of these processes is SLA which was commercialised in the late 1980's [43]. Since then, the processes have been refined to improve accuracy, printing speed, and material properties. Concurrently the costs of VP systems have declined, and low-cost desktop consumer machines are now available from a range of suppliers. In comparison to material extrusion additive manufacturing processes such as FDM, VP systems offer superior build resolution with a layer thickness typically between 25 and 100 μm in contrast to FDM layer thicknesses typically in the range of 50 to 250 μm. A two-photon stereolithography process variant can also extend this resolution to a sub-micron level and offers the highest build resolution of all AM techniques [44–46]. Compared to material extrusion AM processes, VP imposes greater restrictions on the range of compatible materials as these are constrained to a relatively small class of photopolymeric resins with fine-tuned viscosity and rapid photoinitiated curing times. These characteristics are required to facilitate fast-coating of the build surface with smooth and stable liquid surfaces, as well as rapid layer solidification for timely and robust layer-by-layer manufacture. Consequently, VP materials are predominantly limited to resins based on acrylate, epoxide and vinylether monomers. Furthermore, the stability of these materials over time may be adversely affected by prolonged light exposure. However, VP has many practical benefits and has been demonstrated to be an effective manufacturing process for various chemical and biological applications [47–51]. The following sections provide detailed descriptions of the underlying principles of vat photopolymerization processes.

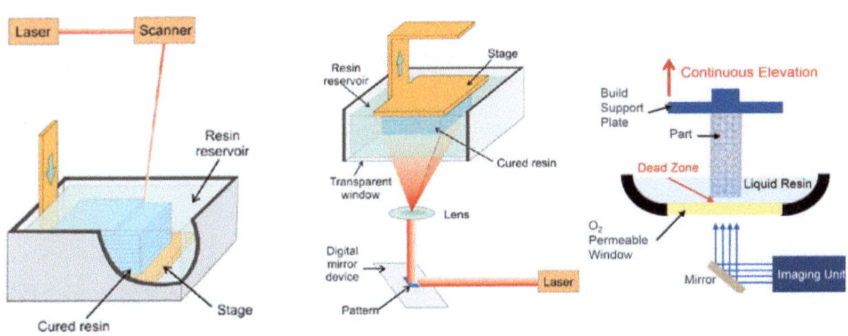

Fig. 3.8 Schematics of Vat Photopolymerization (VP) processes including **a** Stereolithography **b** Digital Light Processing (DLP) [53] **c** Continuous liquid interface production (CLIP) [52]

3.4.1 Vat Photopolymerization—Operating Principles

All VP processes work on fundamentally similar principles and comprise of the key system components of a vat of liquid photopolymer resin, a spatially controllable light source, and a movable build platform/stage (Fig. 3.8). During the VP process, the movable build platform is submerged into a vat of resin and a light source is exposed onto the platform surface according to the part layer cross-section where material requires solidification. The light exposure causes the liquid polymer resin to undergo cross-linking and solidification through photo-polymerization. The light source can be ultraviolet (UV) or visible light, focused as a beam of laser energy (in the case of SLA) or projected using an imaging unit (DLP or CLIP). Depending on whether the VP process is being carried out using a top-down or bottom-up layering approach, the light is projected on top of the resin vat (Fig. 3.8a), or through a transparent window in the vat base (Fig. 3.8b), respectively. After each layer projection, the build platform position is moved vertically by one layer thickness (typically 25–100 μm) and a new layer of resin is allowed to flow onto the solidified layer. In the top-down build approach, a wiper blade may be moved across the top of the vat surface prior to light exposure to level the resin layer thickness on the build platform to a controlled amount. In the bottom-up approach, the solidified part is detached and lifted from the bottom of the transparent vat and unfused resin flows into the void in proportion to the layer thickness. In the Continuous Liquid Interface Production (CLIP) vat photopolymerization process, an oxygen-permeable membrane inhibits photopolymerization at the base of the vat preventing the resin from attaching to the vat wall. Although system complexity is increased, the need to detach the solidified part from the base at each layer is avoided, allowing the CLIP approach to build in a continuous rather than layered manner, thereby reducing part stair-stepping artefacts and significantly increasing build speed compared to SLA and DLP techniques [52]. Photopolymerization is usually not complete following light exposure during the part build process, and most parts require post-build UV light curing, typically for 60–90 min, following a cleanup operation to removed unfused resin residue.

3.4.2 Photopolymers

The VP process builds 3D objects by curing of a liquid photopolymer resin through light-mediated solidification. The resin consists of various chemical components including monomers, photoinitiator, diluents, and chain transfer agents. Resin curing is an exothermic photopolymerization process, wherein chemical cross-linking of monomers leads to the formation a framework of rigid polymers (Fig. 3.9). These chemical components and reactions are critical to the dimensional and mechanical properties of VP manufactured parts and are discussed in detail below.

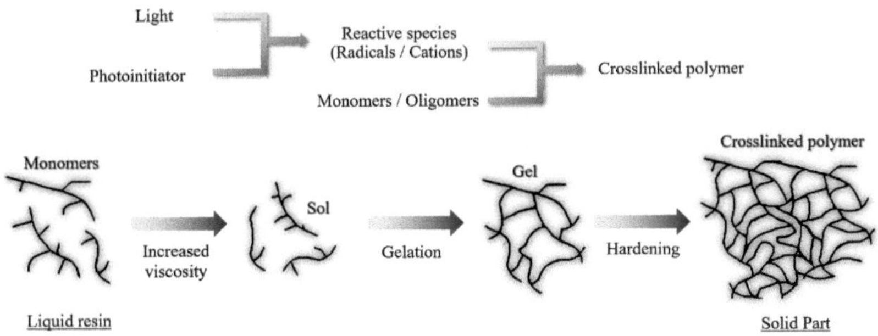

Fig. 3.9 Photocuring of resins used in vat photopolymerization

Functional monomers/oligomers

Acrylate and epoxide monomers are the major classes of monomers used in vat photopolymerization processes (Fig. 3.10c,d). Historically acrylates were used in early VP systems as they are known to exhibit high curing reaction speed under light irradiation thereby enabling increased build speeds. However, acrylates also undergo significant shrinkage during curing (between 5 and 20%) which can lead to layer warping resulting in dimensional instability during the VP printing process. Consequently, the use of acrylate monomers has been displaced in modern VP systems with resins primarily based on epoxy monomers. Epoxy resins undergo minor volume changes during curing (between 1 and 2%), produce parts with superior strength and

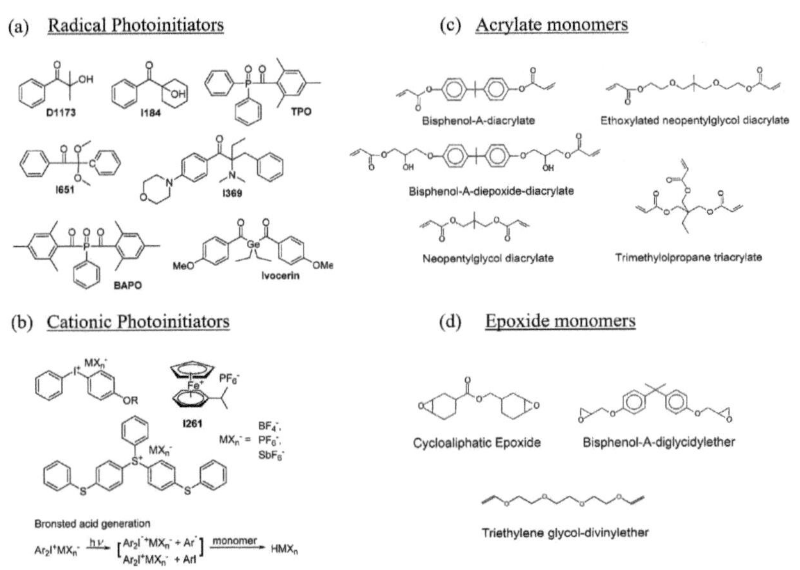

Fig. 3.10 **a** Radical photoinitiators **b** Cationic photoinitiators **c** Acrylate monomers **d** and Epoxide monomers [47, 57]

hardness, and are not subject to polymerization inhibition by exposure to oxygen (which can be problematic for top-down VP systems using acrylates that are exposed to the atmosphere) [54, 55]. However, epoxy resins require longer cure times and can result in parts with low ductility. To mitigate these drawbacks, contemporary VP resin feedstocks use a combination of epoxy monomers with minor addition of acrylates to accelerate cure times and reduce part brittleness.

Photo-initiators

The VP process requires the chemical transformation of acrylate or epoxide-based monomers into cross-linked polymers through a chain reaction triggered by free radicals or cations formed by light irradiation. Thereby in order to promote the formation of radicals or cations it is necessary to add organic photoinitiator molecules into the resin system [56]. The photo-initiators generate radicals through two major mechanism; the first involving the intramolecular cleavage of the initiator into two radicals, and the second involving the intermolecular hydrogen abstraction from a co-initiator molecule [54, 55]. High solubility within the monomer and high absorption cross-section are the two important requirements that dictate the selection of a photoinitiator. Figure 3.11a, b show commercially available radical initiators for acrylate monomers and cation initiators for epoxide monomers [47, 57].

Photopolymerization

The key mechanism enabling additive manufacture of VP parts with high accuracy is a rapid and spatially controllable photopolymerization reaction. The reaction involves conversion of monomers in a colloidal solution, known as a sol phase, through a liquid-to-rubber gelation transition into a viscous gel phase, with subsequent cross-linking of gel into a hard network of polymer. This reaction is initiated by localised interaction of light with a photoinitiator that forms radicals (in the case of acrylate monomers) or cations (with epoxide monomers). Figure 3.11a shows the mechanism of radical polymerization when acrylate monomers are used for the VP process. Upon light exposure, free radicals are formed from the photoinitiator which react with the acrylate monomer and triggers a chain reaction. Every chain reaction produces a radical required for the next step allowing the chain reaction to proceed continuously until there is no monomer left for cross-linking. The rate of this reaction decreases with increased viscosity due to the trapping of radicals into the gel. Radical recombination leads to chain termination and arrests the polymerization process. Free radical formation is a rapid process, therefore acrylate-based structures can be built in a shorter time period, however as discussed earlier, acrylates exhibit high shrinkage during curing, resulting in dimensional instability of VP components and as such, epoxy monomers are more commonly used. Figure 3.11b shows the mechanism of cationic polymerization wherein epoxy monomers are used for the VP process. Photoinitiators used for the epoxy resin have the tendency to produce Bronsted acid sites upon UV irradiation.

These acid sites are formed by abstraction of a proton from another molecule through the formation of radical cation intermediate. This proton reacts with the epoxide monomer to give an oxonium cation, which react with another monomer

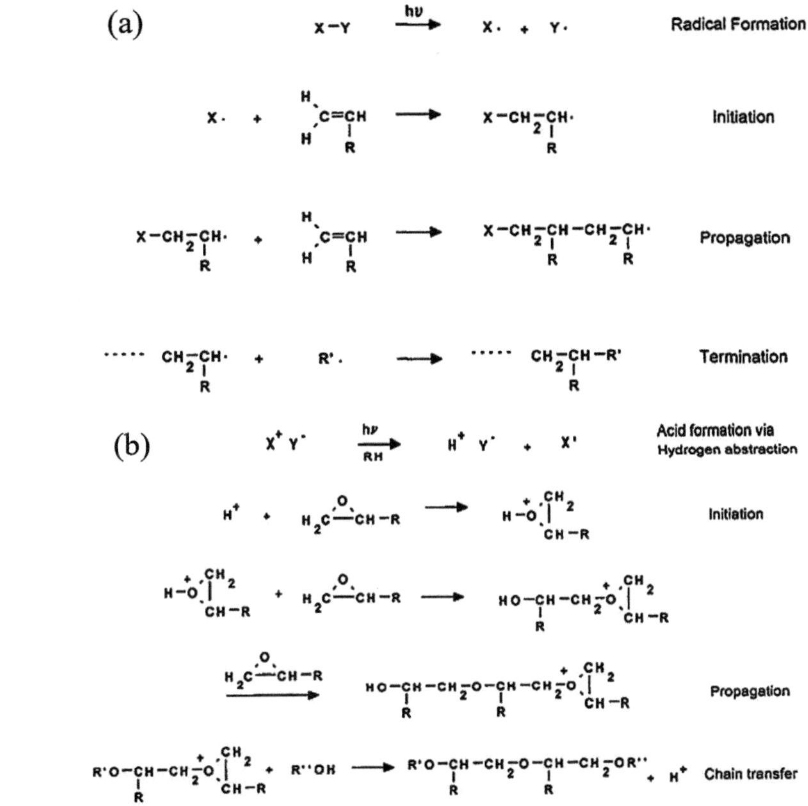

Fig. 3.11 **a** Mechanism of Radical polymerization **b** Mechanism of cationic polymerization (reprinted from [47, 57]).

to produce a new oxonium ion. Cross-linking of the epoxy propagates via these cation formations until the addition of a chain transfer agents such as an alcohol is introduced in order is to terminate this cationic polymerization.

Diluents

The viscosity of the monomer affects settling times, wetting and impregnation characteristics which directly impact the build speed and layer coating uniformity. The high molecular weight resins used in VP can exhibit high viscosity which increases settling times and may compromise layer recoating. Additives or fillers, which may be added to the resin to alter its visual and mechanical properties can also increase the resin viscosity. To control viscosity to desired levels, diluents are added to the resin. The diluents are generally expected to be active during the photopolymerization process but should remain unreactive after curing. Common diluents include low viscosity monomers or reactive diluents such as dipropyleneglycol diacrylate (DPGDA) and isobornyl acrylate (IBOA) [58].

3.4.3 *Vat Polymerization Parameters and Applications*

The VP process can produce parts with small feature sizes, high resolution, and unprecedented geometric complexity if critical process parameters are carefully selected. Key parameters which affect the manufacturability, quality, and cost of VP parts can be grouped into the categories of process, part, light source, and material parameters, as summarised in Table 3.3. Many of these parameters are common to other AM process and are discussed in detail in Sect. 3.3.3 and Chap. 4. Influential parameters unique to the VP process mainly relate to the chemistry of resin photopolymerization. These include fundamental aspects the photoinitiation, propagation and termination of polymerization, which directly influence the required light wavelength, exposure time, dwell time, and depth of focus for a given layer thickness.

VP parts are limited to a single build material mostly based on acrylate and epoxy monomer resins, which can be limiting compared to the larger range of materials compatible with material extrusion processes such as FDM. Therefore, the functionality of VP parts can be mainly derived by combining geometry and the material properties including porosity and elasticity. Figure 3.12 shows example VP manufactured parts using single build materials for a variety of applications.

Due to the capabilities of VP systems, there is an interest in increasing the range of compatible materials. Recently, it was demonstrated that various alkyne carbonate based compounds are manufacturable with VP and exhibit enhanced mechanical properties [59]. To significantly improve the strength of the VP manufactured parts, there is growing interest in introducing inorganic and organic additives into the resin, which enhance the strength and add new functionality to manufactured parts. A recent study demonstrated the DLP printing of silica and UV curable resin, followed by the removal of the polymer to produce silica structures as the end product [60]. In

Table 3.3 Key parameters which influence the cost, quality, and capability of the vat photopolymerization process and manufactured components

Process and part parameters	Light source parameters	Material parameters
Layer thickness	Wavelength	Resin composition
Build time	Power	Resin viscosity
Dwell time	Exposure time	Density
Build volume	Dwell time	Light energy absorption
Batch size	Spot size and depth of focus	Curing speed
Surface finish	Scan strategy	Mechanical properties
Overhanging surfaces and support structures		Post build curing
Build orientation		Additives
Temperature		
Post-processing		

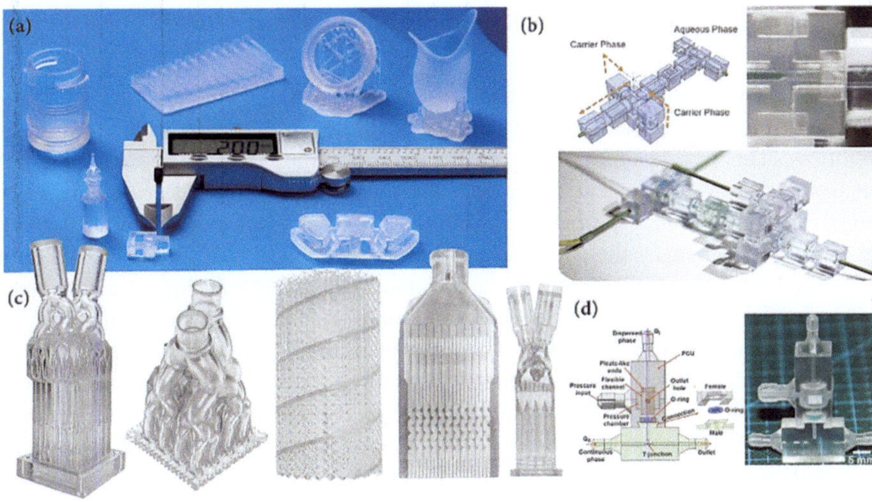

Fig. 3.12 VP manufactured example components **a** general components for prototyping applications [. (Image courtesy: Formlabs Inc.)] **b** modular microfluidic system [61] **c** fluid mixing and separation devices and cellular material scaffolds **d** pneumatic device for droplet generation [62]

this approach, the polymer functions as a sacrificial template to prepare the final silica structure. Similarly, high strength graphene oxide has been blended with polymer resin and used to manufacture parts with enhanced strength (Fig. 3.13). The approach of manufacturing such nanocomposite using VP methods is promising area of research.

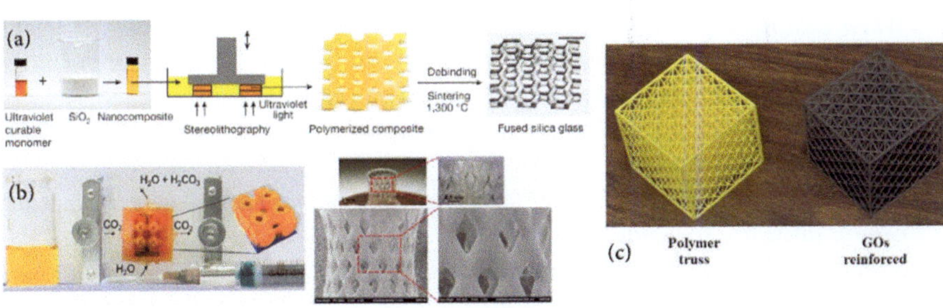

Fig. 3.13 VP manufacture of nanocomposites **a** Silica modified resin [60] **b** composite for gas separation **c** Graphene oxide reinforced composite[63]

3.4.4 Vat Photopolymerization–Challenges and Outlook

Vat photopolymerization exhibits unique capability in the manufacture of high-resolution components with unprecedented levels of complexity from a range of polymers. The process relies on specific photopolymer properties such as rapid reactive response to incident light, high thermal stability, low reaction shrinkage and robust mechanical properties. Although several photopolymers have been developed which meet these demands, the range of material options in VP can be limited compared to other AM technique such as MEX. Specifically, there is a strong demand in producing biocompatible materials for biomedical application. It is therefore desirable to develop biocompatible, water-soluble, high-performance, and high-migration-stable photoinitiators which can promote applicability in bio-printing. Moreover, there are significant opportunities in the development of monomers/oligomers with specific properties (e.g., high reactivity, fluorescence, electrical conductivity, tunable mechanical properties, stimuli-responsive properties, degradability, etc.) for new applications. The development of photocurable biocompatible synthetic and natural polymers for application in 3D bio-printing is one of the main frontiers of exciting future research in VP.

3.5 Test Your Knowledge

After reading the above chapter you should be able to answer the following questions:

1. *Explain the fundamental principle by which additive manufacturing enables the manufacture of parts with higher geometric complexity then traditional methods*
2. *When additively manufacturing a VP part with an inclined plane, at which plane angle relative to the horizontal build platform would you expect the plane surface roughness to be generally the smallest?*
3. *List 3 advantages and 3 challenges associated with FDM and SLA additive manufacturing processes.*
4. *Why do parts manufactured by fused deposition modelling (FDM) typically exhibit anisotropic behavior?*
5. *Suggest 3 components relevant to chemical engineering application that could be good candidates for redesign for additive manufacture using MEX or VP processes? What would be the advantages and disadvantages of making those parts using additive manufacturing?*

References

1. Dilberoglu UM, Gharehpapagh B, Yaman U, Dolen M (2017) The role of additive manufacturing in the era of industry 4.0. Proc Manuf 11:545–554. https://doi.org/10.1016/j.promfg.

2017.07.148
2. Shahrubudin N, Lee TC, Ramlan R (2019) An overview on 3D printing technology: technological, materials, and applications. Proc Manuf 35:1286–1296. https://doi.org/10.1016/j.promfg.2019.06.089
3. Daminabo SC, Goel S, Grammatikos SA, Nezhad HY, Thakur VK (2020) Fused deposition modeling-based additive manufacturing (3D printing): techniques for polymer material systems. Mater Today Chem 16:100248. https://doi.org/10.1016/j.mtchem.2020.100248
4. Astm I (2015) ASTM52900-15 standard terminology for additive manufacturing—general principles—terminology. ASTM International, West Conshohocken, PA 3(4):5
5. Goh GD, Yap YL, Agarwala S, Yeong WY (2019) Recent progress in additive manufacturing of fiber reinforced polymer composite. Adv Mater Technol 4(1):1800271
6. Hospodiuk M, Dey M, Sosnoski D, Ozbolat IT (2017) The bioink: a comprehensive review on bioprintable materials. Biotechnol Adv 35(2):217–239
7. Ozbolat IT, Peng W, Ozbolat V (2016) Application areas of 3D bioprinting. Drug Disc Today 21(8):1257–1271
8. Tan C, Toh WY, Wong G, Li L (2018) Extrusion-based 3D food printing–materials and machines. Int J Bioprint 4 (2)
9. Gonzalez-Gutierrez J, Cano S, Schuschnigg S, Kukla C, Sapkota J, Holzer C (2018) Additive manufacturing of metallic and ceramic components by the material extrusion of highly-filled polymers: a review and future perspectives. Materials 11(5):840
10. Sculpteo (2021) The state of 3D printing. Retrieved July 27, 2021, from https://www.sculpteo.com/en/ebooks/state-of-3d-printing-report-2021/
11. Lewis JA, Gratson GM (2004) Direct writing in three dimensions. Mater Today 7(7–8):32–39
12. Rocha VG, Saiz E, Tirichenko IS, García-Tuñón E (2020) Direct ink writing advances in multi-material structures for a sustainable future. J Mater Chem A 8(31):15646–15657
13. Gonzalez-Gutierrez J, Godec D, Kukla C, Schlauf T, Burkhardt C, Holzer C (2017) Shaping, debinding and sintering of steel components via fused filament fabrication. In: Proceedings of the 16th International scientific conference on production engineering, pp 99–104
14. Wang X, Jiang M, Zhou Z, Gou J, Hui D (2017) 3D printing of polymer matrix composites: a review and prospective. Compos B Eng 110:442–458. https://doi.org/10.1016/j.compositesb.2016.11.034
15. Rane K, Strano M (2019) A comprehensive review of extrusion-based additive manufacturing processes for rapid production of metallic and ceramic parts. Adv Manuf 7(2):155–173
16. Dhinakaran V, Manoj Kumar KP, Bupathi Ram PM, Ravichandran M, Vinayagamoorthy M (2020) A review on recent advancements in fused deposition modeling. Mater Today: Proc 27(Part_2):752–756. https://doi.org/10.1016/j.matpr.2019.12.036
17. Spoerk M, Holzer C, Gonzalez-Gutierrez J (2020) Material extrusion-based additive manufacturing of polypropylene: a review on how to improve dimensional inaccuracy and warpage. J Appl Polym Sci 137(12):48545
18. Rafiee M, Farahani RD, Therriault D (2020) Multi-material 3D and 4D printing: a survey. Adv Sci 7(12):1902307. https://doi.org/10.1002/advs.201902307
19. Garg H, Singh R (2016) Investigations for melt flow index of Nylon6-Fe composite based hybrid FDM filament. Rapid Prototyping J
20. Son J, Choi S (2015) Orientation selection for printing 3D models. In: 2015 International Conference on 3D Imaging (IC3D). IEEE, pp 1–6
21. Chynybekova K, Kim D-H, Choi S-M (2016) Weight dependency of 3D prints from their interior structure. In: Proceedings of HCI Korea. pp 151–156
22. Baich L, Manogharan G, Marie H (2015) Study of infill print design on production cost-time of 3D printed ABS parts. Int J Rapid Manuf 5(3–4):308–319
23. Webbe Kerekes T, Lim H, Joe WY, Yun GJ (2019) Characterization of process–deformation/damage property relationship of fused deposition modeling (FDM) 3D-printed specimens. Addit Manuf 25:532–544. https://doi.org/10.1016/j.addma.2018.11.008
24. Rahim TNAT, Abdullah AM, Md Akil H (2019) Recent developments in fused deposition modeling-based 3D printing of polymers and their composites. Polym Rev 59(4):589–624

25. Chacón JM, Caminero MA, García-Plaza E, Núñez PJ (2017) Additive manufacturing of PLA structures using fused deposition modelling: effect of process parameters on mechanical properties and their optimal selection. Mater Des 124:143–157. https://doi.org/10.1016/j.matdes.2017.03.065
26. Ziemian C, Sharma M, Ziemian S (2012) Anisotropic mechanical properties of ABS parts fabricated by fused deposition modelling. Mech Eng 23(10.5772):2397
27. Shanmugam V, Das O, Babu K, Marimuthu U, Veerasimman A, Johnson DJ, Neisiany RE, Hedenqvist MS, Ramakrishna S, Berto F (2021) Fatigue behaviour of FDM-3D printed polymers, polymeric composites and architected cellular materials. Int J Fatigue 143:106007
28. Kuipers T, Doubrovski EL, Wu J, Wang CC (2020) A framework for adaptive width control of dense contour-parallel toolpaths in fused deposition modeling. arXiv preprint arXiv:200413497
29. Gopsill JA, Shindler J, Hicks BJ (2018) Using finite element analysis to influence the infill design of fused deposition modelled parts. Prog Addit Manuf 3(3):145–163
30. Go J, Hart AJ (2017) Fast desktop-scale extrusion additive manufacturing. Addit Manuf 18:276–284
31. Duty CE, Kunc V, Compton B, Post B, Erdman D, Smith R, Lind R, Lloyd P, Love L (2017) Structure and mechanical behavior of Big Area Additive Manufacturing (BAAM) materials. Rapid Prototyping J
32. Mohamed OA, Masood SH, Bhowmik JL (2015) Optimization of fused deposition modeling process parameters: a review of current research and future prospects. Adv Manuf 3(1):42–53
33. Jiang J, Ma Y (2020) Path planning strategies to optimize accuracy, quality, build time and material use in additive manufacturing: a review. Micromachines 11(7):633
34. Ahlers D, Wasserfall F, Hendrich N, Zhang J (2019) 3D printing of nonplanar layers for smooth surface generation. In: 2019 IEEE 15th International Conference on Automation Science and Engineering (CASE). IEEE, pp 1737–1743
35. Nisja GA, Cao A, Gao C (2021) Short review of nonplanar fused deposition modeling printing. Mater Design Proc Commun 3(4):e221
36. Garg A, Bhattacharya A, Batish A (2017) Chemical vapor treatment of ABS parts built by FDM: analysis of surface finish and mechanical strength. Int J Adv Manuf Technol 89(5–8):2175–2191
37. Nguyen TK, Lee B-K (2018) Post-processing of FDM parts to improve surface and thermal properties. Rapid Prototyping J
38. Harding MJ, Brady S, O'Connor H, Lopez-Rodriguez R, Edwards MD, Tracy S, Dowling D, Gibson G, Girard KP, Ferguson S (2020) 3D printing of PEEK reactors for flow chemistry and continuous chemical processing. React Chem Eng 5(4):728–735. https://doi.org/10.1039/C9RE00408D
39. Zalesskiy SS, Shlapakov NS, Ananikov VP (2016) Visible light mediated metal-free thiol–yne click reaction. Chem Sci 7(11):6740–6745. https://doi.org/10.1039/C6SC02132H
40. Kitson PJ, Symes MD, Dragone V, Cronin L (2013) Combining 3D printing and liquid handling to produce user-friendly reactionware for chemical synthesis and purification. Chem Sci 4(8):3099–3103. https://doi.org/10.1039/C3SC51253C
41. Scotti G, Nilsson SME, Haapala M, Pöhö P, Boije af Gennäs G, Yli-Kauhaluoma J, Kotiaho T, (2017) A miniaturised 3D printed polypropylene reactor for online reaction analysis by mass spectrometry. React Chem Eng 2(3):299–303. https://doi.org/10.1039/C7RE00015D
42. Cailleaux S, Sanchez-Ballester NM, Gueche YA, Bataille B, Soulairol I (2020) Fused Deposition Modeling (FDM), the new asset for the production of tailored medicines. J Control Release. https://doi.org/10.1016/j.jconrel.2020.10.056
43. Hull CW (1986) Apparatus for production of three-dimensional objects by stereolithography. 4575330
44. Goncalves FAMM, Fonseca AC, Domingos M, Gloria A, Serra AC, Coelho JFJ (2017) The potential of unsaturated polyesters in biomedicine and tissue engineering: synthesis, structure-properties relationships and additive manufacturing. Prog Polym Sci 68:1–34. https://doi.org/10.1016/j.progpolymsci.2016.12.008

45. Li J, Fejes P, Lorenser D, Quirk BC, Noble PB, Kirk RW, Orth A, Wood FM, Gibson BC, Sampson DD, McLaughlin RA (2018) Two-photon polymerisation 3D printed freeform micro-optics for optical coherence tomography fibre probes. Sci Rep 8(1):1–9. https://doi.org/10.1038/s41598-018-32407-0
46. Miwa M, Juodkazis S, Kawakami T, Matsuo S, Misawa H (2001) Femtosecond two-photon stereolithography. Appl Phys A: Mater Sci Proc 73(5):561–566. https://doi.org/10.1007/s003390100934
47. Ligon SC, Liska R, Stampfl J, Gurr M, Mülhaupt R (2017) Polymers for 3D printing and customized additive manufacturing. Chem Rev 117(15):10212–10290. https://doi.org/10.1021/acs.chemrev.7b00074
48. Yu C, Schimelman J, Wang P, Miller KL, Ma X, You S, Guan J, Sun B, Zhu W, Chen S (2020) Photopolymerizable biomaterials and light-based 3D printing strategies for biomedical applications. Chem Rev 120(19):10695–10743. https://doi.org/10.1021/acs.chemrev.9b00810
49. Zanchetta E, Cattaldo M, Franchin G, Schwentenwein M, Homa J, Brusatin G, Colombo P (2016) Stereolithography of SiOC ceramic microcomponents. Adv Mater 28(2):370–376. https://doi.org/10.1002/adma.201503470
50. Manapat JZ, Chen Q, Ye P, Advincula RC (2017) 3D printing of polymer nanocomposites via stereolithography. Macromol Mater Eng 302(9):1600553. https://doi.org/10.1002/mame.201600553
51. Ambrosi A, Pumera M (2016) 3D-printing technologies for electrochemical applications. Chem Soc Rev 45(10):2740–2755. https://doi.org/10.1039/C5CS00714C
52. Tumbleston JR, Shirvanyants D, Ermoshkin N, Janusziewicz R, Johnson AR, Kelly D, Chen K, Pinschmidt R, Rolland JP, Ermoshkin A, Samulski ET, DeSimone JM (2015) Continuous liquid interface production of 3D objects. Science 347(6228):1349. https://doi.org/10.1126/science.aaa2397
53. Gross BC, Erkal JL, Lockwood SY, Chen C, Spence DM (2014) Evaluation of 3D printing and its potential impact on biotechnology and the chemical sciences. Anal Chem 86(7):3240–3253. https://doi.org/10.1021/ac403397r
54. Jakubiak J, Rabek JF (2000) Three-dimensional (3D) photopolymerization in stereolithography. Part I: fundamentals of 3D photopolymerization. Polimery 45(11–12)
55. Jakubiak J, Rabek JF (2001) Three-dimensional (3D) photopolymerization in the stereolithography. Part II: technologies of the 3D photopolymerization. Polimery 46(3)
56. Bernhard P, Hofmann M, Schulthess A, Steinmann B (1994) Taking lithography to the third dimension. CHIMIA Int J Chem 48(9):427–430
57. Bernhard P, Hofmann M, Schulthess A, Steinmann B (1994) Taking lithography to the third dimension. Chimia 48(9):427–430
58. Guo Y, Ji Z, Zhang Y, Wang X, Zhou F (2017) Solvent-free and photocurable polyimide inks for 3D printing. J Mater Chem A 5(31):16307–16314
59. Oesterreicher A, Wiener J, Roth M, Moser A, Gmeiner R, Edler M, Pinter G, Griesser T (2016) Tough and degradable photopolymers derived from alkyne monomers for 3D printing of biomedical materials. Polym Chem 7(32):5169–5180. https://doi.org/10.1039/C6PY01132B
60. Kotz F, Arnold K, Bauer W, Schild D, Keller N, Sachsenheimer K, Nargang T, Richter C, Helmer D, Rapp B (2017) Three-dimensional printing of transparent fused silica glass. Nature 544:337–339. https://doi.org/10.1038/nature22061
61. Bhargava KC, Thompson B, Malmstadt N (2014) Discrete elements for 3D microfluidics. Proc Natl Acad Sci 111(42):15013–15018
62. Ji Q, Zhang JM, Liu Y, Li X, Lv P, Jin D, Duan H (2018) A modular microfluidic device via multimaterial 3D printing for emulsion generation. Sci Rep 8(1):4791. https://doi.org/10.1038/s41598-018-22756-1
63. Lin D, Jin S, Zhang F, Wang C, Wang Y, Zhou C, Cheng G (2015) 3D stereolithography printing of graphene oxide reinforced complex architectures. Nanotechnology 26(43):434003

Material Extrusion and Vat Photopolymerization—Principles, Opportunities and Challenges

Fused deposition modelling, stereolithography, material extrusion, photopolymerization, thermoplastics

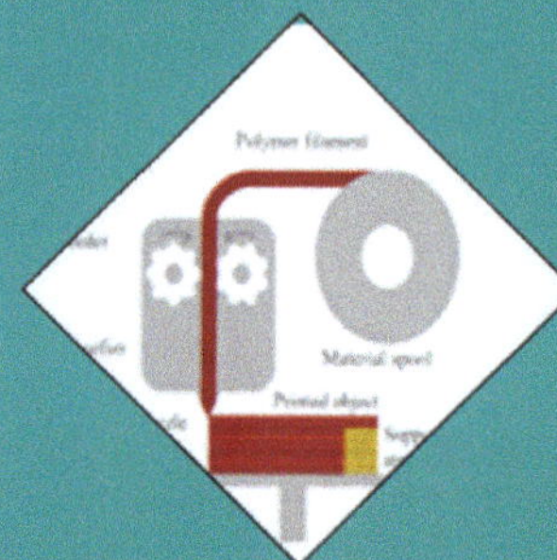

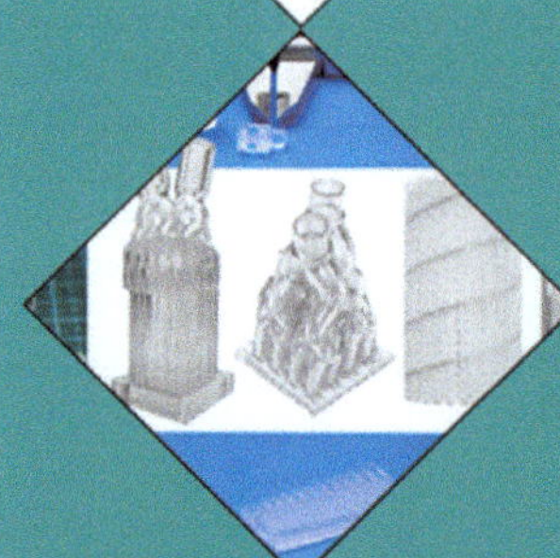

PR. Selvakannan, Maciej Mazur, Xiaochen Sun

3

Chapter 4
Laser Powder Bed Fusion—Principles, Challenges, and Opportunities

Maciej Mazur and PR. Selvakannan

Contents

Abstract Laser Powder Bed Fusion (L-PBF) is an additive manufacturing process which uses a scanning laser beam to selectively melt metal powder in a layer-wise manner to produce solid metal parts. In comparison to conventional subtractive or formative manufacturing processes, L-PBF imposes few design constraints on part geometry and requires low setup effort without the requirement for custom tooling. These unique capabilities enable the manufacture of complex parts that are highly customizable. A broad range of metals can be processed provided they are in a specific powdered form, and that they meet the process compatibility requirements associated with laser weldability. L-PBF provides an exciting opportunity for the manufacture of novel devices for chemical engineering applications, previously difficult or infeasible with conventional manufacturing processes including, cellular materials, high-efficiency heat exchangers, and reactors with highly optimised geometry. To utilize this technology to its maximum potential, it is necessary to understand the practical technological benefits as well as the operational constraints associated with L-PBF materials, process characteristics, part design, and suitable application areas. Particularly, as the final part quality is determined by many influential parameters. The following chapter provides an overview of L-PBF to assist designers in

M. Mazur
School of Engineering, RMIT University, Melbourne, VIC 3001, Australia
e-mail: maciej.mazur@rmit.edu.au

PR. Selvakannan (✉)
Centre for Advanced Materials and Industrial Chemistry, RMIT University, Melbourne, VIC, Australia
e-mail: selvakannan.periasamy@rmit.edu.au

S. K. Bhargava et al. (eds.), *Additive Manufacturing for Chemical Sciences and Engineering*, https://doi.org/10.1007/978-981-19-2293-0_4

optimizing the functional performance, manufacturability, reliability, and cost of additively manufactured products.

Keywords Additive manufacturing · Selective laser melting · Laser powder bed fusion · Metal components · Design for additive manufacturing · Material properties · Direct metal laser sintering

Abbreviations

3D	Three-dimensional
AM	Additive Manufacturing
Ag	Silver
Al	Aluminium
Au	Gold
DMLS	Direct Metal Laser Sintering
WEDM	Wire Electro Discharge Machining
HEA	High Entropy Alloys
L-PBF	Laser Powder Bed Fusion
Mn	Manganese
Mo	Molybdenum
Ni	Nickel
PSD	Particle Size Distribution
SLM	Selective Laser Melting
STL	Stereolithography file format
Ti	Titanium
TM	Traditional Manufacturing
W	Tungsten

4.1 Additive Manufacturing—General Benefits

Additive Manufacture (AM) is a "process of joining materials to make objects from 3D model data, usually layer upon layer, as opposed to subtractive manufacturing methodologies, such as traditional machining" (ASTM International, 2015). The AM process workflow begins with a digital representation of the part geometry, typically formulated as surface mesh (such as a stereolithography (STL) file), which is then segmented into individually processable layers for which material deposition or fusion tool paths are generated (commonly referred to as slicing). The tool paths are then sequentially consolidated in a layer-wise manner to produce parts without the need for any custom tooling [1]. The resultant layer-by-layer approach to part manufacture effectively decomposes a complex and difficult to manufacture 3D part geometry, into a series of comparatively simple to manufacture 2D layers. This key factor enables AM to be applied to the production of parts with unprecedented levels of geometric complexity and design freedom.

AM has been identified as a disruptive technology due to the unique relationships it offers between production cost, level of customization and geometric complexity of manufactured parts when compared to traditional subtractive or formative manufacturing methods [2, 3]. Traditional Manufacturing (TM) processes typically incur high initial costs due to the outlay for tooling, fixtures, and process setup. As these costs need to be amortized per component, small batch sizes result in higher part costs compared to larger batch sizes. Furthermore, the cost of traditionally manufactured components typically increases with higher part complexity. This is a direct result of the increased number of processes, equipment and specialized tooling required, at a higher level of intricacy, to achieve the desired product [1]. The part complexity can also exceed the capability of traditional manufacturing processes due to constraints associated with tool access in subtractive machining, or material flow in formative processes. Conversely, as AM processes do not require special tooling or machinery to produce varying geometric features, the associated set-up costs are substantially reduced. Similarly, the number and intricacy of manufacturing processes and equipment required for additive manufacture is generally independent of part complexity, and the constraints on manufacturable part geometry are low [4]. As such, the key advantages of AM over conventional methods are in the economic manufacture of components: in low batch sizes; with complex geometry where conventional manufacture is expensive; or where the geometric complexity is beyond the capabilities of traditional methods. However, traditional manufacture has significant economic advantages at higher production volumes and lower part complexities, primarily due to the substantially higher speed of manufacture compared to AM methods [5]. The three unique areas of opportunity for AM, as well as the areas of advantage of traditional methods, are highlighted in Fig. 4.1, where typical cost relationships of both manufacturing processes are shown in relation to batch size (Fig. 4.1a) and product complexity (Fig. 4.1b). Generally, the costs of AM are more independent of batch size and complexity than traditional methods. Albeit, a degree of dependence is associated with AM regarding set-up costs (e.g., design customization, model pre-processing and machine setup) and the additional engineering design effort associated with increasing part complexity [6].

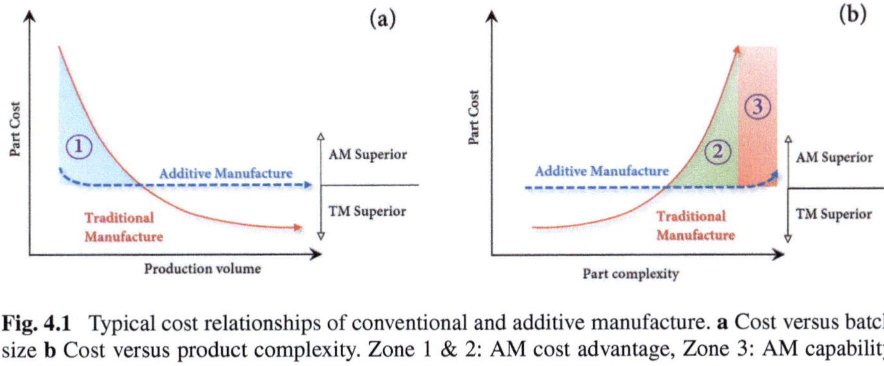

Fig. 4.1 Typical cost relationships of conventional and additive manufacture. **a** Cost versus batch size **b** Cost versus product complexity. Zone 1 & 2: AM cost advantage, Zone 3: AM capability advantage

In addition to the above key benefits, AM enables further opportunities including: part consolidation for reduced assembly part counts; rapid design changes for reduced lead times; the ability to process materials difficult to form conventionally or multiple/functionally graded materials; as well as the ability to achieve savings in material use during manufacture and in inventory costs through just-in-time manufacturing. With these opportunities also come challenges, such as limited achievable build volumes and production rates, comparatively greater feedstock material costs, and underdeveloped methods for process monitoring and part quality control [1, 7–9]. These general opportunities and challenges of additive manufacturing are summarized in Table 4.1.

4.1.1 Laser Powder Bed Fusion—Process Workflow

Laser Powder Bed Fusion (L-PBF) builds solid metal parts using a layer-by-layer approach common to all additive manufacturing processes (Fig. 4.2). The process involves the key steps of machine setup, layer recoating, layer scanning, and part removal. Machine setup requires loading the L-PBF system with feedstock material in the form of fine metal powder, as well as a build plate which acts as a foundational substrate for any manufactured parts. The build plate is typically made from the same or similar material as the powder. The build chamber is subsequently sealed and purged of air with an inert gas such as Argon that is continuously pumped throughout the build operation. An inert gas is required during the build to limit oxidation which can have a detrimental effect on material properties. The build plate is then preheated (typically to 200 °C) to reduce thermal gradients within the part as

Table 4.1 General opportunities and challenges associated with additive manufacturing

Opportunities	Challenges
Economic manufacture of customized components in low batch sizes Ability to manufacture parts with very high geometric complexity High ability for part consolidation Capacity for rapid design changes throughout the design cycle with low-cost penalties Low material wastage Just-in-time manufacturing for inventory saving Manufacture of rapid tooling for augmenting traditional manufacturing process capability and decreasing lead times	Not well suited for large production quantities as machine build volumes and production rates are limited compared to conventional manufacturing processes Feedstock material costs are generally higher than for traditional manufacturing processes May require high expertise to achieve robust results Part surface finish may not be as good as that achieved by conventional processes such as machining Material properties can differ from traditionally processed materials and in some cases result in lower performance Post treatment may be required to achieve desired properties (e.g., machining, stress relieving, heat treatment)

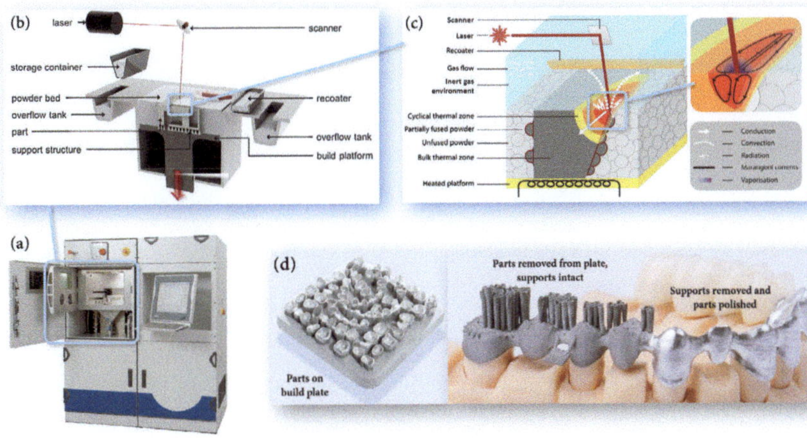

Fig. 4.2 L-PBF process **a** Machine **b** Process build chamber schematic [10] **c** Laser and powder bed interaction schematic including the melt pool [11] **d** Example completed dental components on the build plate, after removal from the build plate with intact support structures, and following support structure removal and polishing (EOS GmbH, 2021)

these can induce high thermomechanical stresses causing part distortion. The build plate is also typically thick (25 mm or greater) to further limit distortion. Following machine setup, the build process commences with a layer recoating operation in which a powder recoating system, comprising of a roller or wiper device, spreads a thin layer of powder (typically 0.03–0.09 mm) onto the build plate.

During the subsequent layer scanning phase, one or more high-powered laser beams are irradiated over the surface of the metal powder bed using actuated mirrors to scan out one layer cross-section of the part. The irradiating laser power heat source (typically 100–1000 W) is sufficient for fusing the powder particles and the solid substrate through melting. Following each cross-sectional laser scan, the build plate is lowered by one layer thickness and a new layer of material is applied by the powder recoating system. The process repeats until all cross-sectional layers of the part geometry are consolidated into a complete three-dimensional part. Upon build completion, the L-PBF build chamber is left to cool down before unfused powder surrounding the solidified part is removed for reuse in subsequent builds. The part and build substrate are then removed for further clean-up and processing. The ability to rapidly reuse unfused material in L-PBF provides a significant advantage in material use efficiency compared to conventional subtractive machining processes which may generate substantial quantities of swarf.

The layer thickness, the number of layers and lasers, the laser scan speed, and overall part volume are the primary factors which determine the build time of the part, which is typically in the order of hours to days. The process is also commonly referred to by trademarked terms such as Selective Laser Melting (SLM) and Direct Metal Laser Sintering (DMLS), popularized by machine manufacturers. However,

Table 4.2 List of process parameters applicable to the L-PBF process

Process and part parameters	Laser scanning parameters	Material and powder parameters
Layer thickness Build time Number of lasers Build volume Batch size Surface finish Overhanging surfaces and support structures Build orientation Temperature Residual stress Post-processing	Wavelength Laser absorption Power Scan speed Spot size Beam profile Scan strategy	Material composition and process compatibility Powder particle, shape, size, and distribution Mechanical and thermal properties Density Post-processing heat treatment

the preferred terminology standardized by ASTM International is Laser Powder Bed Fusion (L-PBF) (ASTM International, 2015).

4.1.2 Laser Powder Bed Fusion—Influential Parameters

L-PBF can build parts from a broad range of alloys with small features sizes, high resolution, and unprecedented geometric complexity, thereby facilitating the manufacture of parts previously not possible with conventional technologies. However, to fully realise the novel capabilities of this technology, it is necessary to consider the influence of key process, laser, material and part parameters which affect the manufacturability, quality, and cost of L-PBF parts. The process is subject to complex multi-physics phenomena such as powder rheology, powder bed laser absorption [12], rapid melting and solidification [13, 14], melt pool dynamics [15], transient heat transfer[16], microstructure development [17], and thermomechanical stress. As such, the number of influential parameters can be vast, with upper estimates suggesting 130 parameters of relevance [18].

The comprehensive understanding of all influential parameters is presently limited and, as such, the process is still subject to ongoing research [19]. However, many key process, material, and part parameters as defined in Table 4.2 are well understood and will be addressed in this section. Common values for a range of influential L-PBF parameters are displayed in Table 4.3.

4.1.2.1 Process and Part Parameters

A range of process and part parameters influence the manufacturability, quality, and cost of L-PBF parts (Table 4.2). The layer thickness in L-PBF typically ranges between 0.02 and 0.09 mm and can have a substantial effect on several part attributes such as resolution, surface roughness, and build time. Due to layer-based construction, the accuracy of curved or inclined surfaces is limited by stair-stepping effects which depend on layer thickness and surface orientation (Fig. 4.3).

Reduced layer thickness results in a finer curve approximation, however this can increase processing times due to the need for additional layer recoating and scanning steps for a given part height. Vertical or horizontal surfaces typically exhibit the lowest surface roughness values due to the absence of stair-stepping. Downfacing surfaces with acute angles generally show the roughest surface finish. Typical surface roughness values achievable with L-PBF are shown in Fig. 4.4.

The L-PBF process build time can be roughly approximated by Eq. (4.1) (excluding setup and post processing, and assuming all parts are identical), and depends on the number of layers, the time to scan and recoat each layer, and the number parts built concurrently. Typical L-PBF build times range in the order of hours to days depending on part volumes. The build time can be reduced by shortening layer scanning times with faster laser scan speeds (within limits which do not compromising material fusion, see Sect. 4.3) or the use of multiple lasers working

Table 4.3 Typical process parameters in L-PBF

Process parameters	Typical parameter range
Heat source power	50–1000 W
Scanning speed	10–1000 (mm s^{-1})
Deposition rate	25–180 (cm^3h^{-1})
Particle size distribution	15–45 μm
Cooling rate during solidification	10^5–10^7 (K s^{-1})
Temperature gradient	10^6–10^7 (K m^{-1})
Solidification growth rate	10^{-1}–10^0 (m s^{-1})

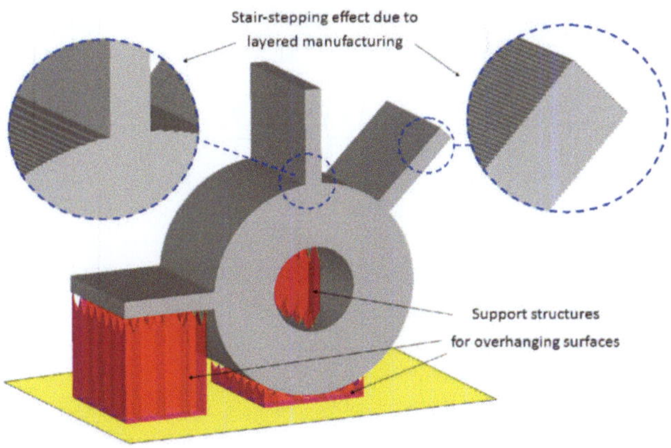

Fig. 4.3 Part model showing stair-stepping effects on curved and inclined surfaces and support structures for overhanging geometry [20]

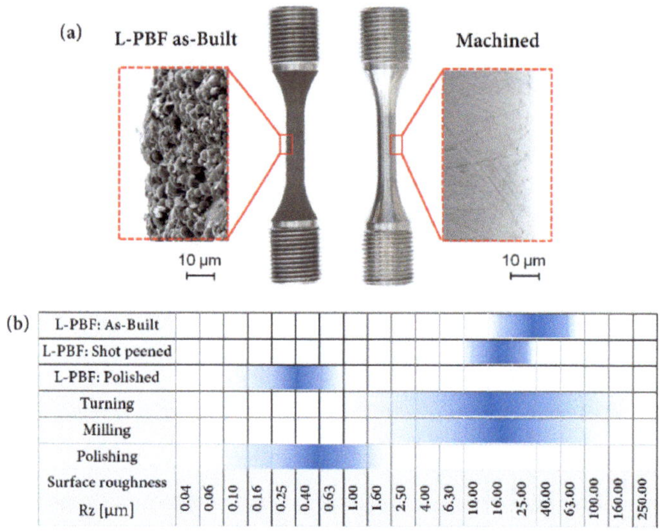

Fig. 4.4 **a** Example L-PBF surface finish before and after machining process for a Ti-6Al-4 V part [21] **b** Typical surface roughness achievable by L-PBF compared to conventional machining

in parallel. Due to these characteristics and the desire for increased productivity, L-PBF machine manufacturers have been increasing in laser power (and thereby the possible scan speed, see Eq. (4.2)) and the number of multi-laser machines in the market. However, multi-laser machines can significantly increase process complexity and are still subject to ongoing research and development. The build time can also be reduced by increasing the number of parts built concurrently on the build plate, as the recoating time is shared across all parts. As such, machine manufacturers have also been increasing build volumes to accommodate larger batches as well as larger sized parts. Typical build L-PBF machine build volumes currently range from 125 × 125 × 125 mm to 600 × 600 × 600 mm (SLM Solutions Group AG, 2021b).

$$T_B = \frac{N_L(T_D N_P + T_R)}{N_P} \tag{4.1}$$

where:

T_B: Approximate build time per part (assuming identical parts).
T_D: Average scanning time per part layer.
T_R: Average powder recoating time.
N_L: Number of layers in a part.
N_P: Number of parts in a build.

Like most AM processes, L-PBF parts featuring geometry prone to distortion may require the use of support structures to ensure a successful build (Figs. 4.2c, 4.3). Support structures comprise of additional structural material which is added into the model during the file preparation process, prior to layer generation (slicing). During the build, support structures ensure the part material layers remain in the correct position, by preventing melt pool sagging of overhanging layers, dissipating heat from hot layers, and reducing geometrical distortions caused by thermomechanical stresses [22]. The structures have intentionally fragile geometry and are fused with processing parameters that result in weaker mechanical strength than the part material, such that they can be separated from the part with relative ease using manual hand tools following the build. Support structures are typically required for overhanging surfaces that are not capable of self-supporting. For L-PBF this is typically for surfaces with an incline of less than approximately 30–40° from the horizontal. With short span distance (typically less than 3–5 mm) it is possible to build intact bridged geometric features at angles between 0 and 30° incline without the need for supports, however the accuracy of the part features and surface roughness may be significantly compromised [23]. Although support structures increase the range of manufacturable geometry, they are generally undesirable due to imposing increased material usage, build times, post-processing effort, as well as residual part surface roughness. Furthermore, for internal part features which are not easily accessible, the removal of support structures may not be feasible. As such, it is desirable to design parts for L-PBF manufacture in order to minimize the need for support structures; by considering modifying overhanging geometry, incorporating self-supporting features or by changing the build orientation of the part.

The part build orientation in particular can have a significant effect on part and process attributes, including:

- The quantity of support structures required and the ease of their removal.
- The build time, due to taller parts with more layers typically requiring longer build times.
- Surface roughness, due to effects such as stair-stepping.
- Internal stress built-up.
- Strength of the part, due to the influence of layer orientation on material microstructure (Sect. 4.3.1).
- The number of parts that can be built concurrently (influenced by the base area of a boundary box enclosing the part).

Generally, the preferred build orientation is one which minimizes the build time and the need for support structures. However, depending on the design requirements, other criteria may take priority, such as the number of parts that can be built concurrently, or the surface finish (for example, Fig. 4.5).

L-PBF parts are subject to large temperature gradients which impose significant thermo-mechanical loads due to differential contraction of hot solidifying layers and colder solid material underneath. This can result in residual stresses within the part, and cause deformation or fracture during or after completion of the build. Residual stress can be mitigated by several methods including preheating, part design, choice of part build orientation, choice of laser scan strategy, as well as post-processing heat treatment. Reducing temperature gradients during the build with a heated build plate can also reduce residual stress, however the preheating temperature is in practice limited to approximately 200 ℃ as higher temperature can increase machine complexity or also result in the onset of solid-state sintering within the powder bed [24]. Powder sintering can increase part surface roughness and reduce powder flowability which compromises the effectiveness of layer recoating and powder removal [25]. Residual stress can also increase in bulk parts with large layer cross-sections, and where possible it is recommended to minimize part thickness during the design phase, or to orientate the part during the build to reduce the layer cross-sectional area. The choice of laser scan strategy can also affect residual stress as outlined in Sect. 4.3. Although the above methods can reduce residual stress, they are often not sufficient to eliminate it, particularly in bulk parts. In these cases, stress reliving heat treatments are recommended.

4.1.2.2 Post-processing Requirements

Parts produced by L-PBF may require a range of post-processing operations before they are suitable for service. The required operations depend on a range of application requirements and can include the following:

Removal of unfused powder

Following the completion of a build, L-PBF parts will be encased in unfused powder which will require removal prior to commissioning the parts into service. The inherently high flowability of L-PBF powders required for effective layer recoating allows the powder to easily flow out from complex internal geometry with narrow channel sizes. However, drain holes (~2–3 mm) may need to be added to the part design to facilitate removal of unfused powder from internal part cavities (for example, Fig. 4.6). Ideally, the powder should be removed without contamination such that it can be reused in subsequent builds.

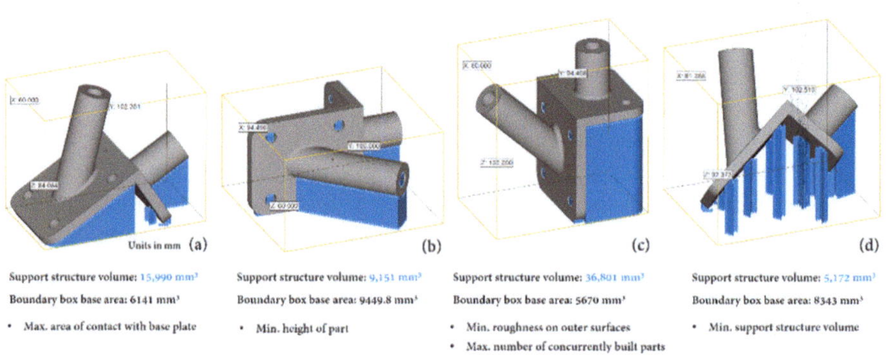

Fig. 4.5 Influence on part orientation on part attributes

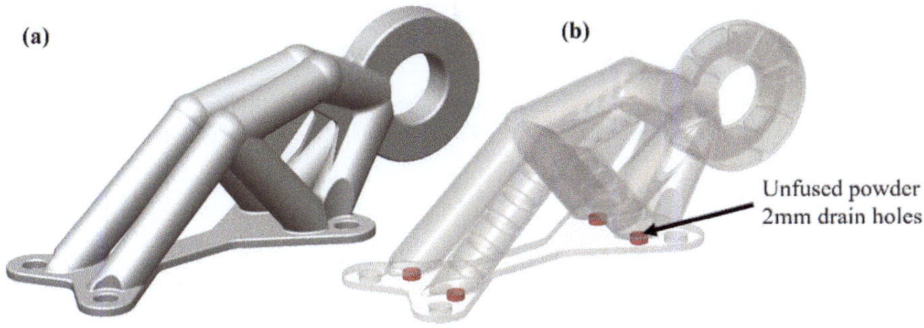

Fig. 4.6 **a** Bracket with hollow features manufacture be L-PBF **b** Drain holes added to the part to removed unfused powder from internal part cavities

Stress-relieving heat treatment

L-PBF can be subject to high residual stresses which can cause part deformation or fracture. Thermal stress reliving heat-treatment may need to be applied to the components prior to or following removal from the build plate in order to eliminate these stresses.

Removal from the build plate

Part removal from the build plate can be accomplished with hand tools if the part is built upon support structures (e.g., Fig. 4.2d) or may require machining if the part is built directly onto the build plate. Typically, Wire Electro Discharge Machining (Wire EDM) is applied for part removal due to the associated fine kerf width, non-contact cutting mechanism and the ability to cut any electrically conductive material.

Removal of support structures

Support structures may need to be removed from components which can be a time-consuming task, requiring manual labour or machining. It is important to consider tool access in the part design and choice of build orientation which will allow for support removal.

Machining of functional surfaces to improve surface finish and dimensional tolerances

Traditional subtractive machining, polishing, and non-traditional abrasive jet machining such as sand or bead blasting can be applied to reduce part surface roughness (Fig. 4.4b). When considering subtractive machining of L-PBF parts, it is necessary to add additional surface material for removal in the part design as well as consider machining constraints such as cutting tool access and work holding requirements.

Metallurgical heat treatment

A range of metallurgical treatments may be required to alter the mechanical properties of as-built components. Metallurgical heat treatments may include annealing,

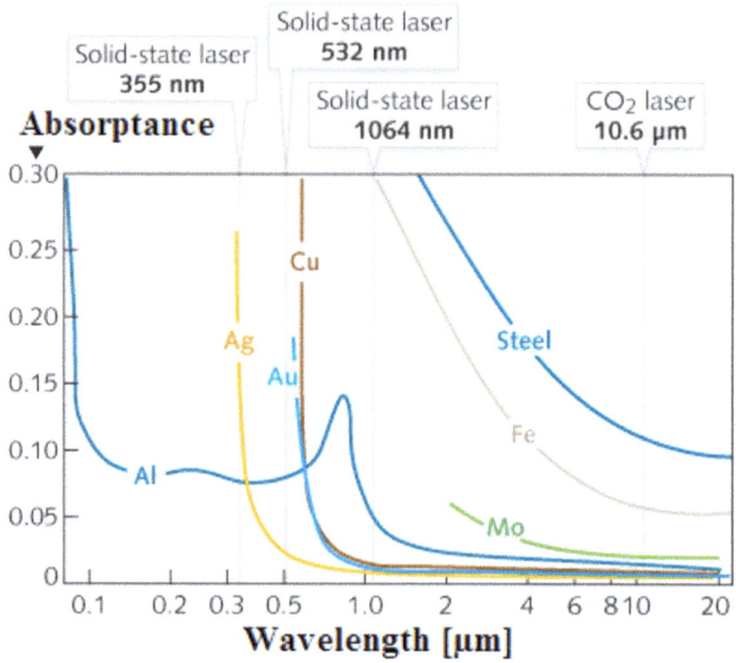

Fig. 4.7 Absorption as a function of wavelength for various solid metals [29]

hardening, and strengthening, depending on the material and service requirements. For example, tool steels built with L-PBF have a high hardness and low ductility state which may not be suitable for service until subject to annealing and controlled hardening [26].

4.1.2.3 Laser Scanning Parameters

The basis of L-PBF is the selective melting of powder particles through the absorption of laser beam heat energy irradiated onto a powder bed surface. The laser beam is focused on the powder bed with a spot size typically ranging from 30 to 100 μm and is scanned by a set of mirror galvanometers with typical scan speeds between 10 and 1000 mm/s (Fig. 4.2) [27]. The laser absorptance (defined as the ratio of energy flux absorbed by the material to the incident energy flux) varies for different wavelengths and materials as shown in Fig. 4.7 (for bulk materials with smooth surfaces). In general, the absorptance of powder materials is higher than bulk solids regardless of the wavelength of irradiation due to higher optical penetration and multiple reflections of the laser beam within the powder bed (Sect. 4.1.3). Most L-PBF machines utilize solid-state fiber lasers with an infrared wavelength of 1064 nm operating in a continuous (most common) or pulsed modes [18]. Lasers in this wavelength are available with relatively high powers (up to ~1000 W) sufficient for melting a broad range of alloys (Table 4.4). However, absorptance in the 1064 nm wavelength is low for certain metals such as copper or silver, which when coupled with the material's high thermal conductivity and heat dissipation in the powder bed, results in difficulties in reliably melting and processing the materials with L-PBF. In an attempt to address these issues, there has been ongoing research into L-PBF with green lasers with 515 nm or 532 nm wavelengths that achieve higher absorbance, however green lasers are not yet broadly available in the L-PBF market [28].

Laser scanning parameters have a critical influence on the material density, surface finish and mechanical properties of L-PBF parts. Of particular importance are the power, scanning speed and adjacent track spacing (also known as hatch spacing, see Fig. 4.9) as well as layer thickness. These parameters are often subject to optimisation with reference to the laser volumetric energy density (E), imparted onto the powder bed (Eq. 4.2) with the objective of identifying a parameter combination which maximizes the density and geometric accuracy of the manufactured parts.

$$E = \frac{P}{h \cdot v \cdot t} \tag{4.2}$$

where:

E : *Volumetric laser energy density [Joules/mm^3].*
P : *Laser power [Watt].*
h: *track/hatch spacing [mm].*
v: *laser scan speed [mm/s].*
t : *powder layer thickness [mm].*

The required heat input is strongly dependent on the material used due to differences in material characteristics such as laser absorptivity (e.g., Fig. 4.7), melting temperature, heat capacity, and thermal conductivity. Optimisation of laser processing parameters for a given material is typically achieved in practice by building a set of test components (typically small prismatic samples) with varying levels of power and scan speed, for a given layer thickness and track spacing, which minimize porosity defects within the part. Figure 4.8a shows example micrographs from several polished prismatic samples (10 × 10 × 5 mm) built from a nickel-based alloy (Inconel 625) with varying laser power and scan speeds (and thereby, energy density). The optimum parameter set which exhibited minimal defects corresponded to E = 60 J/mm^3, P = 200 W, and v = 900 mm/s [30, 31]. As can be observed, the use of suboptimal energy density can lead to porosity defects within parts. Porosity defects are generally in the form of entrapped gas pores (mainly spherical and below approximately 100 μm in diameter), incomplete melting and fusion (characterized by an irregular shape), or micro-cracks (narrows and elongated). Simulated micrographs of these typical defects are shown in Fig. 4.8b. In the example of Fig. 4.8a, gas pores are observed at high energy levels and are caused by vaporization in the melt pool due to excess energy input, due to a combination of high laser power and low scanning speed. Lack of fusion defects are observed in Fig. 4.8a at low energy levels (a combination of low laser power and high scan speed) due to incomplete melting or due to 'balling' phenomena. Balling occurs when the molten material in the weld pool fails to wet the underlying substrate (due to the surface tension), spheroidizing the liquid. This results in a rough and bead-shaped scan track which does not fuse well with adjacent tracks, resulting in porosity [32]. Excess hatch spacing can also result in periodic porosity as adjacent tracks are not fused completely.

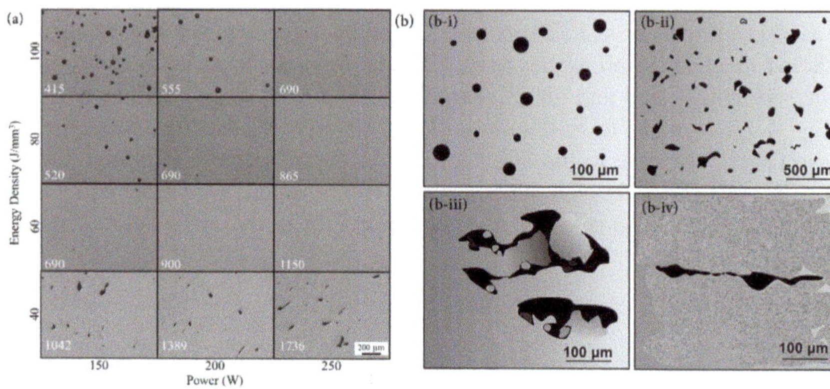

Fig. 4.8 **a** Optical micrographs of L-PBF processed Inconel 625 parts for varying laser powers and volumetric energy density. Laser scan speed (mm/s) is shown on the bottom left on each micrograph. **b** Simulated micrographs depicting typical porosity defects in L-PBF components (b-i) Entrapped gas porosity (b-ii) Incomplete melting and fusion porosity (b-iii) Incomplete melting and fusion detailed view showing unmelted particles inside large irregular pores and (b-iv) Micro-cracks

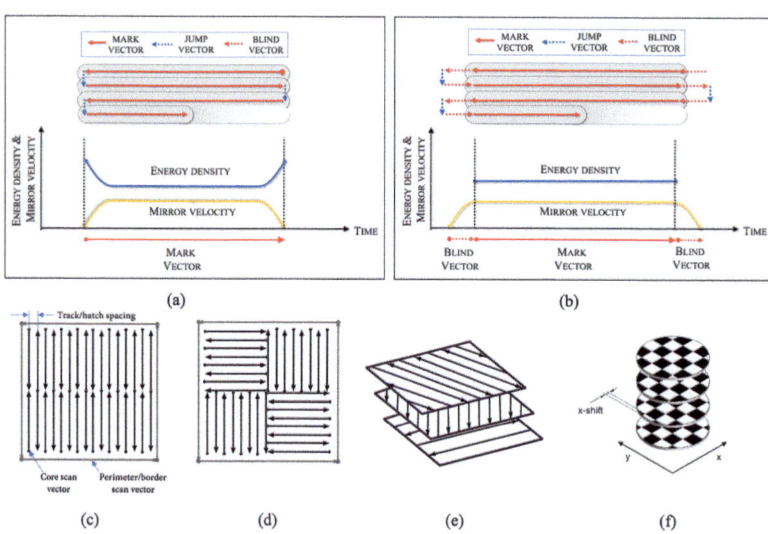

Fig. 4.9 Schematic representations of the energy density, mirror velocity, and scanning vectors when the skywriting function is **a** Off; and, **b** On **c** Stripe scan strategy **d** Checkerboard scan pattern **e** Scan track layer rotation by 66.7° **f** Stripe scan pattern

The formation of porosity defects depends on a broad range of processing and material parameters, which are subject to complex mechanisms related to melting, vaporization, solidification, and thermomechanical stress–detailed studies offer in-depth discussion of these topics [32–34]. In practice, well optimised L-PBF processing parameters can commonly achieve material densities between 99.5 and 99.9%. Achieving 100% density is difficult even with optimised processing parameters as some small-scale gas pores below 5 μm in diameter often remain, however, these have a minor influence on mechanical properties [35]. The material within cross-sectional layers of L-PBF parts is typically fused with a series of laser scan tracks which initially scan the contour permitter of the part (known as the skin), and then fill the interior of the layer (core) with scans comprising of sections of parallel (raster) tracks with some overlap, per Fig. 4.9. The objective being to ensure each scan track is fused completely with the neighboring melt tracks and material in the preceding layers in a time efficient way that minimizes the likelihood of defect formation. A typical L-PBF laser scan strategy consists of 'mark' vectors, during which the laser is active, and 'jump' vectors, during which the laser is inactive, but the mirror system is moving into position (Fig. 4.9a). The scan strategy can have a substantial influence on the key L-PBF part properties such as porosity, residual stress, mechanical properties, and build time.

For example, vaporisation related porosity defects can form when the laser exposure at mark vector ends is increased due to inertial acceleration and deceleration delays of the scanning mirror [36]. To offset such non-uniform laser exposure, a 'skywriting' scan function is typically used in L-PBF systems. When the skywriting function is active, additional 'blind' vectors are added before and after a mark vector to account for abrupt changes of scanning direction when the laser is off, ensuring a constant laser scan speed over the length of the scanned track (Fig. 4.9b). The acceleration and deceleration of the mirrors takes place within the blind vectors, while the scan speed remains constant along the mark vectors [37]. The scan strategy also influences the residual stress experienced in L-PBF parts. Due to differential contraction of solidifying hot tracks and colder solid material underneath, significant residual thermo-mechanical stresses can be imparted to the part. Longer track length increases the amount of residual stress, so scan strategies are typically employed which segment the layer into sections with limited track lengths. Common strategies are comprised of parallel 'stripe' scan vector patterns with alternating directions (Fig. 4.9c), or a combination of parallel and perpendicular 'checkerboard' scan patterns. The checkerboard pattern results in lower residual stress than the stripe pattern but requires a relatively longer layer scan time. Additionally, the relative orientation of adjacent layers is typically rotated between layers (Fig. 4.9e), and the alignment is offset (Fig. 4.9f) to distribute the orientation of any residual stresses more uniformly within the part.

4.1.2.4 Material Parameters—Composition

L-PBF is compatible with a broad range of metal alloys provided that the material is weldable, exhibits sufficient absorption of laser energy (Sect. 4.3), and is available in specific powdered form (Sect. 4.2). A range of materials meet these criteria including high-specific strength Ti alloys [38], Ni alloys with high service temperatures [31, 39, 40], highly thermally conductive alloys of Cu and Al [24, 28, 41], as well as stainless and tool steels [26, 42, 43]. Table 4.4 displays a variety of alloys that are commonly used in L-PBF, their characteristic properties and typical application areas. There is also extensive ongoing research into L-PBF manufacture with other metals and their alloys such as Mn, Au, Ag, Mo, and W, as well as high entropy alloys (HEA). However, these are less commonly employed [44].

In general, most conventional materials compatible with casting or powder metallurgy have been successfully processed using L-PBF following optimization of processing parameters (Sect. 4.2), as these materials typically exhibit good weldability and fluidity as required for L-PBF. The composition of the alloy plays a critical influence on the materials compatibility with L-PBF. For example, in certain alloys, even with careful selection of L-PBF processing parameters, it may not be possible to eliminate defects such as microcracking (Fig. 4.8 (b-iv)) without changing the material composition. Several cracking mechanisms have been identified in L-PBF alloys including solidification, solid phase, ductility-dip, and strain-age cracking. In solidification cracking (also known as hot tearing), the solidifying melt pool may be subject to inadequate compensation of solidification shrinkage by melt flow resulting in voids which act as crack initiation sites, the extent of which is influenced by several factors including material composition, the solidification temperature range [45], the solidification rate [46]and the number of eutectic phases [15]. Solidification cracking can be further exacerbated by compositional segregations at grain boundaries during solidification which can reduce ductility. Liquation cracking is another potential cause of crack formation that takes place away from the melt pool where rapid heating results in localized melting of grain-boundary phases creating potential crack initiation points [47]. In solid phase cracking, the large thermal gradients characteristic of the L-PBF process can induce high residual stresses, which can lead to micro-crack formation where rapid cooling and composition have produced a microstructure with low ductility [48]. Solid-phase cracking may also be related to ductility loss induced by a solid-state grain boundary embrittlement phenomenon, known as ductility-dip cracking, which has been observed in Ni alloys at elevated temperatures between 700 and 900 °C [39, 49]. Finally, a phenomenon known as strain-age cracking can also occur during post-build heat treatment of precipitation-strengthened Ni alloys, where superposition of residual stress, and stress from precipitation strengthening of the alloy can result in strains that exceed the material's ductility [50].

Alloys that are prone to microcracking have been proven difficult to process by L-PBF and can only typically be remedied with compositional modifications. In some Ni alloys such as Inconel 625 and Hastelloy X, even small modification to minor

Table 4.4 Common alloys used in L-PBF and their application areas

	Steel and Stainless steel	Titanium alloys	Aluminium alloys	Nickel alloys	Cobalt alloys
Characteristic properties	High hardness & toughness (tool steels) High corrosion resistance (stainless steels) Good machinability Comparatively low feedstock cost	High specific strength High toughness High bio- compatibility High corrosion resistance Low thermal expansion	Low density Good specific strength High thermal conductivity High electrical conductivity Good corrosion resistance Comparatively low feedstock cost	High strength at elevated temperatures High corrosion resistance High creep resistance Good weldability	High wear resistance High strength Good biocompatibility Good corrosion resistance
Typical L-PBF applications areas	Plastic injection and diecasting tooling Medical devices Maritime components Spindles and screws	Orthopaedic and dental implants Aerospace structural and engine components High performance automotive components Maritime	Aerospace structural components High performance automotive structural components Heat exchangers Hydraulic manifolds	Aerospace engine components Gas turbines Oli and gas processing Rocket engine components Nuclear reactor components Tooling	Dental prosthetics and implants Medical implants High temperature applications
Common alloys used in L-PBF	Maraging: M300 (1.2709) Stainless: 316L (1.4404), 15-5PH (1.4540) 17-4PH (1.4542) Tool: H13 (1.2344)	Pure Titanium Ti6Al7Nb Ti6Al4V	AlSi10Mg AlSi7Mg AlSi12 AlSi9Cu3 AlMg4.5Mn0,4	Inconel 625 Inconel 718 Inconel 939 Inconel 738 Hastelloy X Hastelloy C-276	CoCr28Mo6

alloying elements such a Si and Mn may be sufficient to eliminate micro-crack formation while maintaining the alloy composition within ASTM compositional specification range [31, 51]. The findings highlight that the standard ASTM composition specification range may not be sufficiently narrow to ensure defect-free parts in L-PBF applications. In other materials it may not be possible to eliminate microcrack formation without more extensive alloy modification or effectively the development of new alloys. However, such an undertaking typically requires the manufacture and testing of multiple batches of custom powders with varying compositions, each subject to relatively large minimum volumes of production which can result in high research costs.

4.1.2.5 Material Parameters—Powder

The L-PBF process relies on the ability to spread a thin (20–90 μm) and highly uniform layer of metal powder consistently onto a substrate comprising of either solid or powdered material. Consequently, the powder needs to exhibit desirable flowability characteristics such as low viscosity and low tendency for particle agglomeration. Low powder flowability can result in non-uniformity in the powder bed layer thickness which can cause part defects such as voids, over-melting, or interference damage to the layer recoating components if fused material erroneously protrudes out of the powder layer. Key parameters affecting powder flowability are the powder particle morphology and particle size distribution (PSD). Spherical particle morphology is required in L-PBF to ensure good powder flowability and layer uniformity [52]. Gas or plasma atomization are the most common powder production methods for L-PBF processes as they can produce highly spherical particles in contrast to the irregular morphology produced by a more economical water atomization process [38]. Figure 4.10 demonstrates the typical difference in particle morphology for H13 tool steel powder produced by gas and water atomization processes. L-PBF systems also usually require the powder particle diameters to be within the approximate range of 15–45 μm. This range of particle sizes is desirable as it increases the powder bed packing density by allowing fine particles to fill the voids between the coarser particles [53]. Finer particles are undesirable as they tend to agglomerate together, compromising flowability, whereas coarser particles can exceed the powder bed layer thickness as well as increase part surface roughness. The particle size can also influence the mechanical properties of the parts [54]. The PSD for L-PBF powder is typically characterized by Dv10, Dv50 and Dv90 values which designate the points on a particle diameter distribution curve below which 10, 50 and 90% of the particle's diameters fall, respectively (Fig. 4.11).

L-PBF powders can be made from either pure metal, powder pre-alloyed prior to atomization, or powders made by mixing different powder batches after they are atomized. In general, pure or pre-alloyed powders are preferred due to greater ability to achieve uniform composition and powder properties, which result in parts with more uniform microstructure, higher density and improved mechanical properties [55]. In contrast, mixed metal powder can be a more cost-effective alternative to

Fig. 4.10 Scanning electron microscope (SEM) images of H13 Tool steel powder **a** Water atomized powder characterized by irregular morphology, not suitable for L-PBF **b** Gas atomized powder characterized by nominally spherical morphology, suitable for L-PBF **c** Gas atomized powder including fine particles diameters between approx. 1–45 μm not suitable for L-PBF **d** Gas atomized powder with particles diameter between approx.15–45 μm suitable for L-PBF

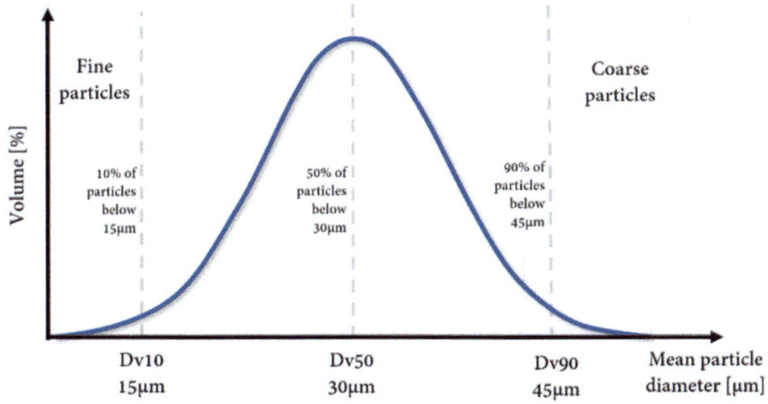

Fig. 4.11 Example Particle Size Distribution (PSD) with characteristic Dv10, Dv50 and Dv90 values for typical L-PBF powder stock

pre-alloyed metal powders. However, as the variable composition of mixed metal powders is required to undergo in-situ alloying during the L-PBF process, it can be challenging to achieve uniform material properties.

Unfused powder in L-PBF systems can typically be reused between builds and changes in morphological, compositional, and rheological properties with use, are minor and do not significantly impact the processability of the powder and the mechanical properties of parts [56, 57]. Safety during powder handling is important as the powders can pose a health risk to the user. The risk can be through inhalation, as the powder contain particle sizes which can enter the respiratory system, as well as abrasion damage to sensitive tissues such as the eyes. Certain alloys can also contain alloying elements which are considered carcinogenic. Additionally, certain powders (particularly aluminium or titanium alloys) are reactive and can pose an explosion risk if not handled properly. As such, appropriate handling of powders is recommended (such as avoiding possible ignition sources), as well as the use of personal protective equipment including dust masks, eye protection, protective clothing, gloves, and safety shoes with protection against electrostatic discharge.

4.1.2.6 Material Parameters—Microstructure and Mechanical Properties

In L-PBF, the local irradiation of the powder bed with concentrated laser energy results in the rapid formation of a melt pool, which is subject to high heat transfer rates, high temperature gradients, complex melt flow dynamics, and rapid solidification. These complex temporally and spatially transient conditions give rise to a range of phenomena which influence the material microstructure and thereby the mechanical properties of L-PBF parts. For example, large thermal gradient within the melt pool results in differences in surface tension which drive fluid currents (Marangoni currents, Fig. 4.2c) from the hot center region to the cooler solid–liquid interface [15]. These can in turn affect the geometry of the melt pool, the laser scan tracks, and ultimately the part porosity and surface finish [58]. Similarly, during solidification the melt pool is subject to high cooling rates (in the order of 10^3–10^8 K/s) and non-equilibrium conditions that have a critical influence on the characteristic of the material's microstructure though effects such as grain refinement, segregation phenomena, or the formation of non-equilibrium phases [15]. As a result, L-PBF processed materials can exhibit characteristics such as fine-grained microstructures, large amounts of dislocations (in the form of elongated cells), small-scale compositional variations and the formation of sub-grain structures with nano-inclusions [59]. Furthermore, heat flux is significantly higher through fused material than the surrounding unfused powder and as such, the directional heat flow is predominantly through the solid part layers to the build plate. The directional heat flux and associated large temperature gradients can result in the formation of heterogeneous columnar grain structures which are typically aligned in the build direction [60]. Additionally, due to the layer-by-layer fusion of material, many cycling heating and cooling stages are experienced by solidified areas of the part throughout the build, which can further alter the evolution of the microstructure. The set of L-PBF process conditions and influential parameters result in complex relationships between the process and the associated material microstructure and mechanical properties, which differ significantly to conventional solidification-based processes such as casting. Developing a comprehensive understating of L-PBF material microstructure formation is the focus of intensive ongoing research [44].

Example mechanical properties for common alloys processed with L-PBF are shown in Table 4.5. During tensile testing L-PBF parts have been observed to exhibit some minor anisotropy in mechanical properties between vertically and horizontally orientated test coupons (Table 4.4). The anisotropy is believed to occur due to differences in the distribution of any residual porosity, heterogenous grain orientation along direction of heat transfer, and differences in states of residual stress [17].

In general, the mechanical properties of L-PBF materials under static loading are superior to conventionally cast materials and similar to wrought materials (which are produced by rolling/forging/extrusion). However, under dynamic loading, the fatigue life of as-built L-PBF parts can be inferior to conventionally manufactured parts. For example, Fig. 4.12 shows cyclic stress against the cycles to failure for L-PBF manufactured Ti-6Al-4 V test coupons in various build orientations (vertical (Z) or horizontal (X–Y)) and subject to different combinations of post-processing, including, as-built, stress relieved, machined, and polished, and treated with Hot Isostatic Pressing (HIP). As fatigue life is highly dependent on the presence of surface or internal stress concentration, the as-built components exhibit the lowest fatigue lives due to comparatively high surface roughness and the presence of residual internal gas pores (since L-PBF parts can only typically achieve 99.95–99.99% density). The fatigue lives for L-PBF parts in the as-built condition are typically similar to that of cast parts. Machining and polishing the surface of the parts improves the fatigue life significantly due to reduced surface roughness. HIP treatment, which closes internal pores, further improves the fatigue lives of the samples and the results are comparable to wrought materials. Samples built in the horizontal orientation can exhibit slightly superior fatigue lives to those built vertically. This is hypothesized to be due to the greater concentration of porosity defects between layers, which result in higher stress concentration when subject to loading applied perpendicular to the layer as is the case of vertically built samples [62]. Figure 4.12 also shows fatigue lives for parts produced by electron beam powder bed fusion (EB-PBF) which is an additive manufacturing process with some similarities to L-PBF [1].

The unique process condition in L-PBF presents challenges but also opportunities for unique control over material microstructure and mechanical properties. For example, with careful selection of energy density and layer thickness, it has been demonstrated that it is possible to significantly affect the microstructure and mechanical properties of L-PBF manufactured Ti-6Al-4 V parts [17]. Figure 4.13 depicts micrographs and tensile stress–strain curves for Ti-6Al-4V L-PBF parts built with different layer thickness and energy densities. For the 60 μm layer thickness sample (Fig. 4.13b and e, designated as sample S3) a near α' martensitic structure forms in each layer which subsequently transforms in-situ into an ultrafine lamellar $(\alpha + \beta)$ structure. The unusual microstructure results in a novel combination of high yield strength and high elongation for the 60 μm sample which is superior to samples produced with 30 μm (sample S3) and 90 μm (sample S7) layer thicknesses.

Table 4.5 Example mechanical, thermal, and surface roughness properties of common alloys processed by L-PBF in the as-built condition with no post-build heat treatment. 3 repeat samples per test [61]

	Units	Build Orientation	Ti6Al4V ELI (Grade 23)	AlSi10Mg	316L Stainless Steel	M300/1.2709 Maraging Tool Steel	Inconel 625	CoCr28Mo6
Material density	[g/cm^3]	–	4.43	2.67	7.99	8.05	8.44	8.47
Component density	[%]		>99.5%	>99.5%	>99.5%	>99.5%	>99.5%	>99.5%
Layer thickness	[μm]	–	30	30	30	30	30	30
Ultimate strength	[MPa]	Horizontal	1281 ± 7	454 ± 5	692 ± 4	1190 ± 20	1072 ± 20	1269 ± 13
		Vertical	1289 ± 17	474 ± 5	618 ± 2	1213 ± 20	945 ± 10	1177 ± 32
Yield strength	[MPa]	Horizontal	1076 ± 30	297 ± 7	591 ± 16	999 ± 8	737 ± 28	824 ± 31
		Vertical	1170 ± 26	271 ± 7	541 ± 2	1076 ± 15	686 ± 11	639 ± 28
Elongation	[%]	Horizontal	8 ± 1	8 ± 1	39 ± 1	5 ± 8	31 ± 1	13 ± 2
		Vertical	9 ± 1	6 ± 1	45 ± 1	2 ± 4	42 ± 5	23 ± 4
Elastic modulus	[GPa]	Horizontal	113 ± 1	73 ± 4	184 ± 20	168 ± 4	178 ± 27	205 ± 49
		Vertical	117 ± 4	74 ± 2	178 ± 9	181 ± 2	153 ± 18	190 ± 16
Thermal conductivity at 20°C	[W/mK]	–	6.8	11.3	15.0	14.2	9.8	13.0
Surface roughness, Ra	[μm]	–	12 ± 1	8 ± 2	8 ± 2	7 ± 1	4 ± 1	16 ± 3
Surface roughness, Rz	[μm]	–	76 ± 6	55 ± 13	68 ± 10	45 ± 5	37 ± 3	99 ± 13

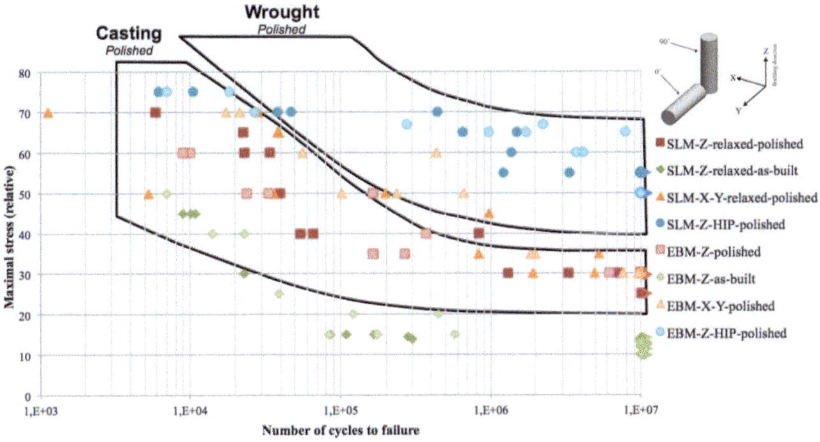

Fig. 4.12 Fatigue of Ti-6Al-4 V manufactured by L-PBF and Electron Beam Powder Bed Fusion (EB-PBF) test samples in various build orientations and subject to differing combinations of post-processing; as-built, stress relieved, machined and polished, and treated with Hot Isostatic Pressing (HIP). For comparison the typical range of values for polished cast or wrought samples are also presented. Stress values are normalized relative to a maximum value. Stress loading ratio of R = −1. From [62]

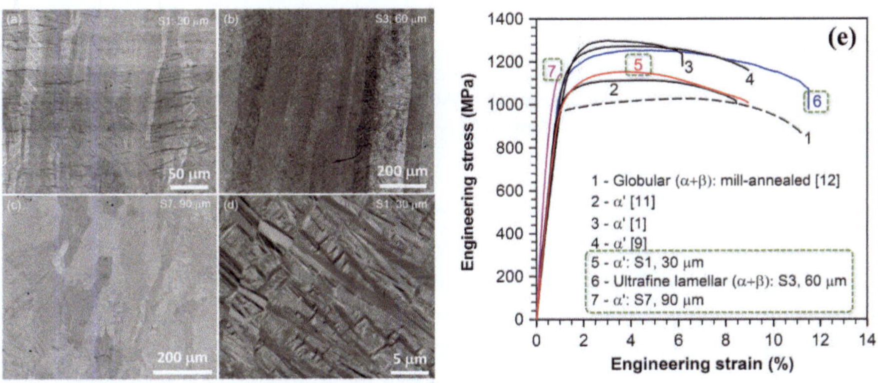

Fig. 4.13 Scanning Electron Microscope (SEM) in backscattered electron (BSE) mode micrographs showing typical microstructures of L-PBF manufactured Ti–6Al–4 V. Columnar prior-β grain structure in parts **a** Built with a layer thickness of 30 μm (designated S1), **b** Layer thickness of 60 μm (designated S3) **c** Layer thickness of 90 μm (designed S7) **d** α′ martensite formed inside the prior-β grains in S1. **e** Engineering tensile stress–strain curves of SLM-fabricated Ti–6Al–4 V comprising ultrafine lamellar (α + β) structure and α′ martensitic structure [17].

This level of control over microstructure presents exciting opportunities for material property optimisation in L-PBF parts and is the subject of ongoing research.

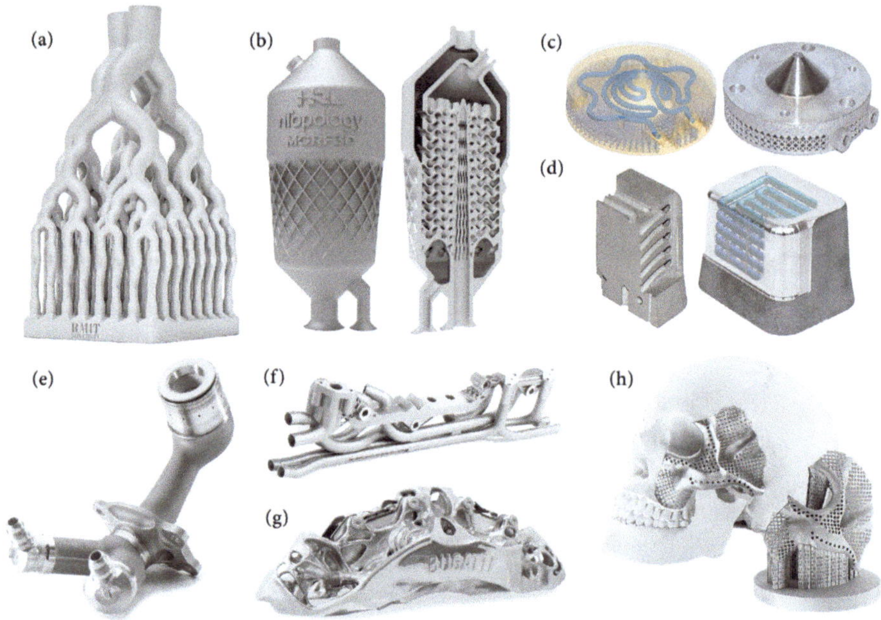

Fig. 4.14 Example parts manufactured by L-PBF **a** High flow uniformity fractal flow distributor and mixer system **b** Heat exchanger incorporating high surface area cellular material structures [63] **c** Heat exchange device for spectroscopy applications incorporating fluid channels and cellular materials **d** Tooling for injection moulding of plastics with conformal cooling channels [26] **e** Jet engine fuel nozzle [64] **f** Fluid manifold [65] **g** Automotive brake caliper (polished) [66] **h** Cranial implant as-built and after post-processing [67]

4.1.3 Laser Powder Bed Fusion—Examples Application

The unique capabilities of L-PBF can enable the manufacture of high-performance components across a variety of application areas. In general, AM techniques such as L-PBF present advantages compared to conventional manufacturing methods when the application can benefit from one or more of the following characteristics:

- Complex geometry
- Customisation of geometry between individual components or small manufacturing quantities
- Part consolidation by integration of features into components rather than assemblies
- Freedom from traditional manufacturing constraints such as tool access in subtractive manufacturing processes or material flow in formative processes

These advantages are particularly beneficial in applications such as gas and fluid flow distributors (Fig. 4.14a, e and f), heat exchangers (Fig. 4.14b, c and d), highly optimised structural components (Fig. 4.14g), and customised medical implants (Fig. 4.14h). Such example parts are generally considered not manufacturable with conventional processes, further highlighting the capabilities for novel products achievable with L-PBF.

The manufacture of cellular material structures further demonstrates the novel opportunities enabled by additive manufacturing with L-PBF. Cellular materials exhibit unique functional characteristics beyond those achievable by solid materials including high specific strength and stiffness, high absorption of mechanical energy, and high surface-to-area ratios for enhanced heat transfer control [68]. Furthermore, as cellular materials have low density, the L-PBF laser scanning time and material use is reduced compared to that of solid sections. Cellular materials are commonly formed from periodic or stochastic arrangements of open or closed cell types, with either two-dimensional cell configurations (such as honeycomb) or, three-dimensional polyhedral layouts (such as lattice structures) [68] (for example Fig. 4.15a–c).

The mechanical performance of cellular structures is dependent on range of parameters such as: cell topology, cell size, wall/strut thickness, material type, and loading conditions [70, 71]. For example, mechanical properties such as compressive strength and stiffness are particularly dependent on whether the cell topology exhibits bending-dominated or stretch-dominated behaviour during loading [20, 23, 72]; with stretch-dominated structures typically exhibiting high strength, and bending-dominated structures exhibiting high compliance. Stochastic open-cell materials such as foams, typically exhibit cell element bending during loading, whereas ordered open-cell lattice structures may exhibit either stretch or bending dominated behavior depending on cell topology [73]. Lattice structures are of particular interest in AM applications, as processes such as L-PBF can reliably produce lattice structures with a high level of control over cell geometry enabling a range of structural properties, including high compliance (for dampening or energy absorption), high surface area (for heat transfer during solid–fluid interaction), and high specific strength (for light weight structural materials). Achieving a range of thermo-fluidic properties with various topologies, densities, surface area-to-volume ratios and permeabilities can result in highly efficient heat exchangers with a desirable combination of high thermal conduction and low flow resistance. An example concept design integrating cellular lattice structures for regenerative cooling and structural load bearing in a rocket nozzle is shown in Fig. 4.15d. Lattice structures additively manufactured with L-PBF also present attractive opportunities for functionalizing surfaces by coating the surface with catalyst following additive manufacture [74] or by building directly with a catalytically active material [75]. The micron scale irregularities associated with the relatively rough surface characteristics of L-PBF parts can aid in bonding with a catalytic material (for example, Fig. 4.15c and d).

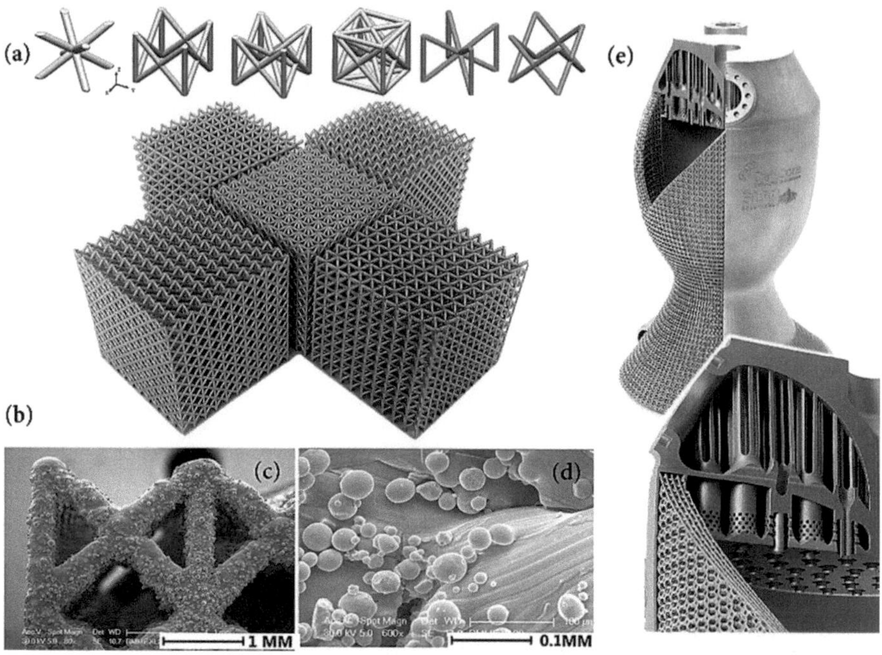

Fig. 4.15 **a** Example lattice structure cells of varying topology **b** L-PBF manufactured lattice structure specimens **c** Magnified view of lattice structure highlighting small feature size **d** Lattice structure surface showing partially fused metal powder and irregular surface **e** Example of direct lattice structure integration into a single piece regenerative cooled rocket nozzle additively manufactured by L-PBF [69]

4.2 Conclusions and Outlook

Laser Powder Bed Fusion (L-PBF) offers exciting new opportunities for the additive manufacture of components with unprecedented levels of design optimisation. As a result, L-PBF manufactured parts are now increasingly being used in range of application areas across aerospace, automotive, chemical, tooling, and medical industries. Continuing capability improvements in AM hardware, as well as broader understanding of the revolutionary benefits of L-PBF technology have driven increasing AM equipment sales over the past several years, suggesting further growth in the application market in the future. Intensive research activities have been addressing knowledge gaps in the physics of laser-powder bed interaction, melt pool dynamics, defects formation and mitigation, microstructure evolution, and alloy development, as well as process-optimised part design. This has further increased the range of L-PBF manufacturable materials and part geometries. However, several remaining challenges are limiting broader adoption of L-PBF technology. The relatively high cost of L-PBF parts currently limits the economic feasibility of L-PBF manufacture to high-value applications and presents a barrier to broader adoption in mass manufactured products. A key driver of high costs is the relatively low productivity of L-PBF systems when compared to conventional subtractive or formative manufacturing processes, with L-PBF parts currently taking up to tens of hours to build and post-process. Increasing the number of multi-laser systems in the market has reduced build times, however, the productivity remains modest and multi-laser systems impose high capital costs. Increasing productivity also requires high manufacturing yield with low defect rates which is difficult to achieve currently with L-PBF, as part defects such as porosity and distortion are relatively common. Existing process monitoring, and in-situ defect resolution technologies currently have limited capability and quality control mainly comprising of time-consuming post-build inspection. Furthermore, L-PBF parts are not able to achieve the low surface roughness of conventionally machined parts which can limit adoption in certain applications. Furthermore, although much has been understood about the complex relationship between L-PBF processing parameters, alloy composition, microstructure evolution, and mechanical properties, there remain significant gaps in understanding which limit the ability to predict and control the material properties of L-PBF manufacture parts, particularly under the spatially and temporally varying conditions present during the build.

Despite these outstanding challenges, significant progress is being made in achieving a better understanding and control of the L-PBF processes, and L-PBF manufactured parts are becoming increasingly common due to their inherent geometric complexity benefits and good material properties. Sustained research and development over the next decade are expected to lead to reduced cost and broader adoption of L-PBF technology, with great potential of high global popularity in the years to come.

4.3 Test Your Knowledge

After reading the above chapter you should be able to answer the following questions:

1. *How do the 'part cost versus part complexity' and 'part cost versus production volume'relationships differ between additive manufacturing and conventional manufacturing processes? What are the reasons for any difference?*
2. *List 3 advantages and 3 challenges associated with the L-PBF additive manufacturing processes.*
3. *State three possible effects of part build orientation on the L-PBF process and on part quality.*
4. *How do the mechanical properties of L-PBF parts typically compare to components manufactured by conventional manufacturing processes?*
5. *Suggest 3 components relevant to chemical engineering application that could be good candidates for redesign for additive manufacture using L-PBF? What would be the advantages and disadvantages of making those parts using additive manufacturing?*

References

1. Gibson I, Rosen D, Stucker B (2013) Additive manufacturing technologies 2nd edn. Rapid Manufacturing Association. Springer. https://doi.org/10.1520/F2792-12A.2.
2. Anderson C (2013) Makers: The New Industrial Revolution. Random House. http://books.google.nl/books?id=_xd69urR7agC
3. Zhai Y, Lados DA, LaGoy JL (2014) Additive manufacturing: making imagination the major limitation. Jom 66(5):808–816. https://doi.org/10.1007/s11837-014-0886-2
4. Cotteleer M, Holdowsky J, Mahto M (2013) The 3D opportunity primer: The basics of additive manufacturing, Deloitte University Press, pp 1–18. http://dupress.com/wp-content/uploads/2014/03/DUP_718-Additive-Manufacturing_Overview_FINAL.pdf
5. Campbell RI, Jee H, Kim YS (2013) Adding product value through additive manufacturing. Paper presented at the proceedings of the international conference on engineering design, ICED, 4 DS75–04(August), pp 259–268
6. Zhang Y, Bernard A, Gupta RK, Harik R (2014) Evaluating the design for additive manufacturing: a process planning perspective. Procedia CIRP 21:144–150. https://doi.org/10.1016/j.procir.2014.03.179
7. Wohlers TaA (2013) Wohlers Report 2013: additive manufacturing and 3D printing. Annual worldwide progress report
8. Tarik Arafat M, Gibson I, Li X (2014) State of the art and future direction of additive manufactured scaffolds-based bone tissue engineering. Rapid Prototyp J 20(1):13–26. https://doi.org/10.1108/rpj-03-2012-0023
9. Gao W, Zhang Y, Ramanujan D, Ramani K, Chen Y, Williams CB, Wang CCL, Shin YC, Zhang S, Zavattieri PD (2015) The status, challenges, and future of additive manufacturing in engineering. Comput Aided Des 69:65–89. https://doi.org/10.1016/j.cad.2015.04.001
10. Tillmann W, Schaak C, Nellesen J, Schaper M, Aydinöz ME, Hoyer KP (2017) Hot isostatic pressing of IN718 components manufactured by selective laser melting. Addit Manuf 13:93–102. https://doi.org/10.1016/j.addma.2016.11.006
11. McMillan M, Leary M, Brandt M (2017) Computationally efficient finite difference method for metal additive manufacturing: a reduced-order DFAM tool applied to SLM. Mater Des 132:226–243. https://doi.org/10.1016/j.matdes.2017.06.058
12. Hussein A, Hao L, Yan C, Everson R (2013) Finite element simulation of the temperature and stress fields in single layers built without-support in selective laser melting. Mater Des 1980–2015(52):638–647. https://doi.org/10.1016/j.matdes.2013.05.070
13. Vilaro T, Colin C, Bartout JD (2011) As-fabricated and heat-treated microstructures of the Ti-6Al-4V alloy processed by selective laser melting. Metall Mater Trans A 42(10):3190–3199. https://doi.org/10.1007/s11661-011-0731-y
14. Leary M, McMillan M, Shidid D, VanToor H, Mazur M, Brandt M (2014) Numerical methods to predict overheating in SLM lattice structures. Paper presented at the International Congress on Applications of Lasers & Electro-Optics
15. Kruth JP, Levy G, Klocke F, Childs THC (2007) Consolidation phenomena in laser and powder-bed based layered manufacturing. CIRP Ann 56(2):730–759. https://doi.org/10.1016/j.cirp.2007.10.004
16. Gusarov AV, Yadroitsev I, Bertrand P, Smurov I (2007) Heat transfer modelling and stability analysis of selective laser melting. Appl Surf Sci 254(4):975–979. https://doi.org/10.1016/j.apsusc.2007.08.074
17. Xu W, Brandt M, Sun S, Elambasseril J, Liu Q, Latham K, Xia K, Qian M (2015) Additive manufacturing of strong and ductile Ti–6Al–4V by selective laser melting via in situ martensite decomposition. Acta Mater 85:74–84. https://doi.org/10.1016/j.actamat.2014.11.028
18. Rehme O, Bachmann FG, Emmelmann C, Hoving W, Lu Y, Washio K (2006) Rapid manufacturing of lattice structures with selective laser melting. Paper presented at the Laser-based Micropackaging

19. Khorasani A, Gibson I, Veetil JK, Ghasemi AH (2020) A review of technological improvements in laser-based powder bed fusion of metal printers. Int J Adv Manuf Technol 108(1–2):191–209. https://doi.org/10.1007/s00170-020-05361-3
20. Mazur M, Leary M, McMillan M, Elambasseril J, Brandt M (2016) SLM additive manufacture of H13 tool steel with conformal cooling and structural lattices. Rapid Prototyp J 22(3):504–518. https://doi.org/10.1108/rpj-06-2014-0075
21. Kasperovich G, Hausmann J (2015) Improvement of fatigue resistance and ductility of TiAl6V4 processed by selective laser melting. J Mater Process Technol 220:202–214. https://doi.org/10.1016/j.jmatprotec.2015.01.025
22. Gan MX, Wong CH (2016) Practical support structures for selective laser melting. J Mater Process Technol 238:474–484. https://doi.org/10.1016/j.jmatprotec.2016.08.006
23. Mazur M, Leary M, Sun S, Vcelka M, Shidid D, Brandt M (2015) Deformation and failure behaviour of Ti-6Al-4V lattice structures manufactured by selective laser melting (SLM). Int J Adv Manuf Technol. https://doi.org/10.1007/s00170-015-7655-4
24. Zhang J, Song B, Wei Q, Bourell D, Shi Y (2019) A review of selective laser melting of aluminum alloys: processing, microstructure, property and developing trends. J Mater Sci Technol 35(2):270–284. https://doi.org/10.1016/j.jmst.2018.09.004
25. Berger U, Merkel M, Liebisch A (2016) The Influence of preheating on laser beam melting. Paper presented at the proceedings of the 2nd international conference on progress in additive manufacturing (Pro-AM 2016), pp 306–311
26. Mazur M, Brincat P, Leary M, Brandt M (2017) Numerical and experimental evaluation of a conformally cooled H13 steel injection mould manufactured with selective laser melting. Int J Adv Manuf Technol 93(1–4):881–900. https://doi.org/10.1007/s00170-017-0426-7
27. Shi W, Liu Y, Shi X, Hou Y, Wang P, Song G (2018) Beam diameter dependence of performance in thick-layer and high-power selective laser melting of Ti-6Al-4V. Materials (Basel) 11(7). https://doi.org/10.3390/ma11071237
28. Kaden L, Matthäus G, Ullsperger T, Engelhardt H, Rettenmayr M, Tünnermann A, Nolte S (2017) Selective laser melting of copper using ultrashort laser pulses. Appl Phys A 123(9). https://doi.org/10.1007/s00339-017-1189-6
29. Wijers JLC (1996) 'Nach CO2 nun Nd:YAG Laser für Blechbearbeitung, Tech. Rundschau, 112
30. Mazur M, Benoit M, Easton M, Brandt M (2020) Selective laser melting of Inconel 625 alloy with reduced defect formation. J Laser Appl 32(2). https://doi.org/10.2351/7.0000093
31. Benoit MJ, Mazur M, Easton MA, Brandt M (2021) Effect of alloy composition and laser powder bed fusion parameters on the defect formation and mechanical properties of Inconel 625. Int J Adv Manuf Technol 114(3–4):915–927. https://doi.org/10.1007/s00170-021-06957-z
32. Li R, Liu J, Shi Y, Wang L, Jiang W (2011) Balling behavior of stainless steel and nickel powder during selective laser melting process. Int J Adv Manuf Technol 59(9–12):1025–1035. https://doi.org/10.1007/s00170-011-3566-1
33. Li Y, Gu D (2014) Parametric analysis of thermal behavior during selective laser melting additive manufacturing of aluminum alloy powder. Mater Des 63:856–867. https://doi.org/10.1016/j.matdes.2014.07.006
34. Zhang B, Li Y, Bai Q (2017) Defect formation mechanisms in selective laser melting: a review. Chin J Mech Eng 30(3):515–527. https://doi.org/10.1007/s10033-017-0121-5
35. Kasperovich G, Haubrich J, Gussone J, Requena G (2016) Correlation between porosity and processing parameters in TiAl6V4 produced by selective laser melting. Mater Des 105:160–170. https://doi.org/10.1016/j.matdes.2016.05.070
36. Tian Y, Tomus D, Rometsch P, Wu X (2017) Influences of processing parameters on surface roughness of Hastelloy X produced by selective laser melting. Addit Manuf 13:103–112. https://doi.org/10.1016/j.addma.2016.10.010
37. Mancisidor AM, Garciandia F, Sebastian MS, Álvarez P, Díaz J, Unanue I (2016) Reduction of the residual porosity in parts manufactured by selective laser melting using skywriting and high focus offset strategies. Phys Procedia 83:864–873. https://doi.org/10.1016/j.phpro.2016.08.090

38. Qian M, Bourell DL (2017) Additive manufacturing of titanium alloys. Jom 69(12):2677–2678. https://doi.org/10.1007/s11837-017-2630-1
39. Lippold JC, Kiser SD, DuPont JN (2011) Welding metallurgy and weldability of nickel-base alloys. Wiley
40. Tomus D, Jarvis T, Wu X, Mei J, Rometsch P, Herny E, Rideau JF, Vaillant S (2013) Controlling the Microstructure of Hastelloy-X components manufactured by selective laser melting. Phys Procedia 41:823–827. https://doi.org/10.1016/j.phpro.2013.03.154
41. Lykov PA, Safonov EV, Akhmedianov AM (2016) Selective laser melting of copper. MSF 843:284–288. https://doi.org/10.4028/www.scientific.net/msf.843.284
42. Rafi HK, Starr TL, Stucker BE (2013) A comparison of the tensile, fatigue, and fracture behavior of Ti–6Al–4V and 15–5 PH stainless steel parts made by selective laser melting. Int J Adv Manuf Technol 69(5–8):1299–1309. https://doi.org/10.1007/s00170-013-5106-7
43. Liverani E, Toschi S, Ceschini L, Fortunato A (2017) Effect of selective laser melting (SLM) process parameters on microstructure and mechanical properties of 316L austenitic stainless steel. J Mater Process Technol 249:255–263. https://doi.org/10.1016/j.jmatprotec.2017.05.042
44. DebRoy T, Wei HL, Zuback JS, Mukherjee T, Elmer JW, Milewski JO, Beese AM, Wilson-Heid A, De A, Zhang W (2018) Additive manufacturing of metallic components–process, structure and properties. Prog Mater Sci 92:112–224. https://doi.org/10.1016/j.pmatsci.2017.10.001
45. Kim WR, Bang GB, Park JH, Lee TW, Lee B-S, Yang S-M, Kim G-H, Lee K, Kim HG (2020) Microstructural study on a Fe-10Cu alloy fabricated by selective laser melting for defect-free process optimization based on the energy density. J Market Res 9(6):12834–12839. https://doi.org/10.1016/j.jmrt.2020.09.051
46. Kellens K, Renaldi R, Dewulf W, Kruth J-p, Duflou JR (2014) Environmental impact modeling of selective laser sintering processes. Rapid Prototyp J 20(6):459–470. https://doi.org/10.1108/rpj-02-2013-0018
47. Zhong M, Sun H, Liu W, Zhu X, He J (2005) Boundary liquation and interface cracking characterization in laser deposition of Inconel 738 on directionally solidified Ni-based superalloy. Scripta Mater 53(2):159–164. https://doi.org/10.1016/j.scriptamat.2005.03.047
48. Mercelis P, Kruth JP (2006) Residual stresses in selective laser sintering and selective laser melting. Rapid Prototyp J 12(5):254–265. https://doi.org/10.1108/13552540610707013
49. Lippold JC, Nissley NE (2008) Ductility-dip cracking in high chromium, Ni-base filler metals. In: Böllinghaus T, Herold H, Cross CE, Lippold JC (eds) Hot cracking phenomena in welds II. Springer, Berlin, Heidelberg, pp 409–425. https://doi.org/10.1007/978-3-540-78628-3_21
50. Donachie MJ, Donachie SJ (2002) Superalloys: a technical guide, 2nd edn. ASM International
51. Tomus D, Rometsch PA, Heilmaier M, Wu X (2017) Effect of minor alloying elements on crack-formation characteristics of Hastelloy-X manufactured by selective laser melting. Addit Manuf 16:65–72. https://doi.org/10.1016/j.addma.2017.05.006
52. Olakanmi EO (2013) Selective laser sintering/melting (SLS/SLM) of pure Al, Al–Mg, and Al–Si powders: effect of processing conditions and powder properties. J Mater Process Technol 213(8):1387–1405. https://doi.org/10.1016/j.jmatprotec.2013.03.009
53. McGeary RK (1961) Mechanical packing of spherical particles. J Am Ceram Soc 44(10):513–522
54. Spierings AB, Schneider M, Eggenberger R (2011) Comparison of density measurement techniques for additive manufactured metallic parts. Rapid Prototyp J 17(5):380–386. https://doi.org/10.1108/13552541111156504
55. Chang I, Zhao Y (2013) Advances in powder metallurgy: properties, processing and applications. Elsevier Science
56. Grainger L (2016) Investigating the effects of multiple re-use of Ti6Al4V powder in additive manufacturing (AM), Renishaw plc
57. Cordova L, Campos M, Tinga T (2019) Revealing the effects of powder reuse for selective laser melting by powder characterization. Jom 71(3):1062–1072. https://doi.org/10.1007/s11837-018-3305-2
58. Yasa E, Deckers J, Kruth JP (2011) The investigation of the influence of laser re-melting on density, surface quality and microstructure of selective laser melting parts. Rapid Prototyp J 17(5):312–327. https://doi.org/10.1108/13552541111156450

59. Godec M, Zaefferer S, Podgornik B, Šinko M, Tchernychova E (2020) Quantitative multiscale correlative microstructure analysis of additive manufacturing of stainless steel 316L processed by selective laser melting. Mater Charact 160. https://doi.org/10.1016/j.matchar.2019.110074
60. Parimi LL, RG A, Clark D, Attallah MM (2014) Microstructural and texture development in direct laser fabricated IN718. Mater Charact 89:102–111. https://doi.org/10.1016/j.matchar.2013.12.012
61. AG SSG (2021) SLM material data 2021a. http://www.slm-solutions.com
62. Chastand V, Quaegebeur P, Maia W, Charkaluk E (2018) Comparative study of fatigue properties of Ti-6Al-4V specimens built by electron beam melting (EBM) and selective laser melting (SLM). Mater Charact 143:76–81. https://doi.org/10.1016/j.matchar.2018.03.028
63. Vlahinos MaOH, R (2020) Unlocking advanced heat exchanger design and simulation with nTop platform and ANSYS CFX. New York. https://ntopology.com/resources/whitepaper-advanced-heat-exchanger-design-and-simulation/
64. Electric G (2018). https://www.ge.com/additive/stories/new-manufacturing-milestone-30000-additive-fuel-nozzles
65. 3D Systems I (2021) 3D Systems, Inc. https://www.3dsystems.com/customer-stories/increasing-performance-productivity-and-reliability-semiconductor-capital
66. S.A.S. BA (2018) Bugatti automobiles S.A.S. https://www.bugatti.com/media/news/2018/world-premiere-brake-caliper-from-3-d-printer/
67. GmbH T (2018). https://www.trumpf.com/en_US/presse/online-magazine/3d-printed-medical-implants-for-half-the-world/
68. Gibson LJ, Ashby MF, Harley BA (2010) Cellular materials in nature and medicine. Cambridge University Press
69. GmbH C (2021). https://cellcore3d.com/en/#1545332928086-ebe2b831-0360
70. Ashby MF (2005) Cellular ceramics: structure, manufacturing, properties and applications. In: Scheffler M, Colombo P (eds). https://doi.org/10.1002/3527606696
71. Wieding J, Wolf A, Bader R (2014) Numerical optimization of open-porous bone scaffold structures to match the elastic properties of human cortical bone. J Mech Behav Biomed Mater 37:56–68. https://doi.org/10.1016/j.jmbbm.2014.05.002
72. Deshpande VS, Ashby MF, Fleck NA (2001) Foam topology bending versus stretching dominaed architectures. Acta mater 49:1035–1040
73. Ashby MF, Evans T, Fleck N, Hutchinson JW, Wadley HNG, Gibson LJ (2000) Metal foams: a design guide. Elsevier Science
74. Zhou X, Liu C-j (2017) Three-dimensional printing for catalytic applications: current status and perspectives. Adv Funct Mater 27(30). https://doi.org/10.1002/adfm.201701134
75. Wei Q, Li H, Liu G, He Y, Wang Y, Tan YE, Wang D, Peng X, Yang G, Tsubaki N (2020) Metal 3D printing technology for functional integration of catalytic system. Nat Commun 11(1):4098. https://doi.org/10.1038/s41467-020-17941-8

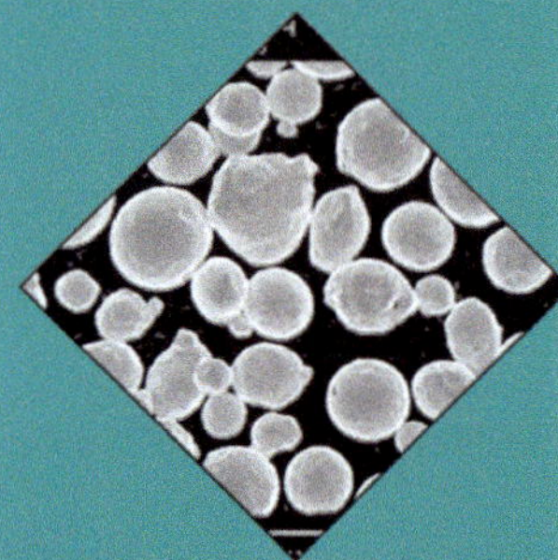

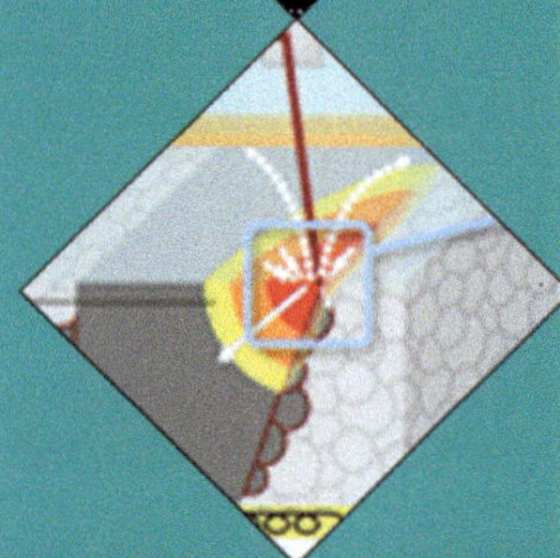

Laser Powder Bed Fusion—Principles, Challenges, and Opportunities

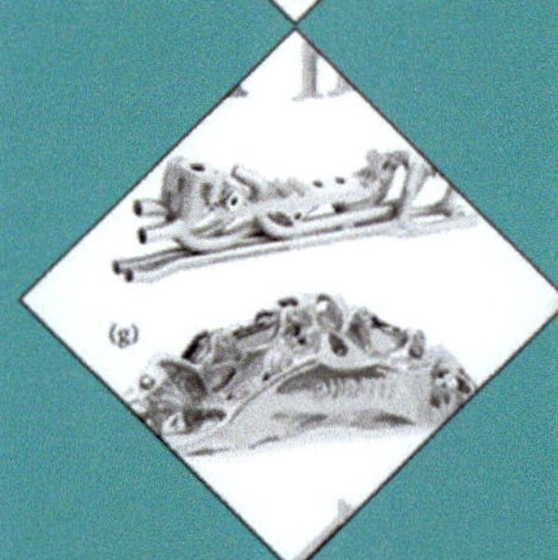

Metal additive manufacturing, laser processing, influential parameters, material properties, common defects, design for additive manufacturing

Maciej Mazur
PR. Selvakannan

4

Chapter 5
Robocasting—Printing Ceramics into Functional Materials

Uzma Malik, PR. Selvakannan, Maciej Mazur, Yongxiang Li, and Suresh K. Bhargava

Contents

Abstract Functional ceramic materials with magnetic, electronic, piezoelectric, superconducting and other dielectric properties constitute an important field of materials research. Indeed, their broad potential enables them to be applied in various applications and are hence considered to be 'smart materials'. However, ceramics are one of the more difficult materials to fabricate into complex morphologies and are challenging to employ for fine surface robocasting. Robocasting is as an emerging AM process, and despite its challenges, is currently the most viable candidate for the printing of complex ceramic structures and morphologies, of which, traditional manufacturing processes are incapable of in their current state. Robocasting differs from

U. Malik · PR. Selvakannan · S. K. Bhargava (✉)
Centre for Advanced Materials and Industrial Chemistry, RMIT University, Melbourne, VIC, Australia
e-mail: suresh.bhargava@rmit.edu.au

M. Mazur · Y. Li
School of Engineering, RMIT University, Melbourne, VIC, Australia

S. K. Bhargava et al. (eds.), *Additive Manufacturing for Chemical Sciences and Engineering*, https://doi.org/10.1007/978-981-19-2293-0_5

other AM techniques as it is capable of printing a precursor ceramic structure, which can be post-processed to produce dense ceramic structures with complex morphologies. As a result, this technique has a unique advantage over other AM technologies, with the freedom to print complex 3D ceramics given that the ceramic formulation (i.e., the ink), is workable. Excitingly, this technique and its applicable scope can be modified to print multi-ceramic functions, and shapable ceramic structures. This chapter explores the current state of robocasting technology and the wide range of potential materials that may be employed with this for fabrication purposes. Expert opinion is provided as to the emerging and future potential of this technology, with an appropriate discussion of the current challenges and drawbacks that may currently limit widespread application.

Keywords Robocasting · Ceramic materials printing · Functional ceramic structures

Abbreviations

3D	Three-dimensional
4D	Four-dimensional
AM	Additive Manufacturing
Al_2O_3	Alumina
BJ	Binder Jetting
B_4C	Boron Carbide
CAD	Computer-Aided Design
FDM	Fused Deposition Modelling
FFE	Freeze-Form Extrusion
HA	Hydroxyapatite
LCM	Lithography-Based Ceramic Manufacturing
PAA	Polyacrylic Acid
PEI	Polyethylenimine
PLGA	Polylactic-Co-Glycolic Acid
SiC	Silicon Carbide
SiO_2	Silica
Si_3N_4	Silicon Nitride
SL	Stereolithography
SLM	Selective Laser Sintering
SLS	Selective Laser Sintering
T_g	Glass Transition Temperature
TiH_2	Titanium Hydride
ZrO_2	Zirconia

5.1 Background—Ceramics Printing

Ceramics have been utilized as construction materials for centuries. Generally, ceramics are produced through well-established processes including casting and extruding, however, due to recent technological advancements, there are now other emerging methods of fabrication [1, 2]. Compared to the significant advances in digital systems for the production and fabrication of polymers and metals, traditional manufacturing practices have limited the potential of ceramics. Whilst most industries have already adapted the latest manufacturing technologies for the manufacture of polymers and metals, newer ceramic production technologies have gained significantly less traction within the industrial setting. Ceramics exhibit unique combinations of thermal, chemical, and mechanical properties and can be engineered to form customized building blocks that may be employed to expand the current range of ceramics production.

The production of complex morphologies of ceramics, microstructures and porosity are still seen as major challenges to conventional techniques. AM, therefore, provides an unlimited design freedom that can overcome the challenges found in traditional manufacturing practices [3]. AM of ceramic materials is a relatively new field in comparison to the established AM processes currently employed for metals and polymer production. It is no surprise that the emergence of ceramic materials printing has gained significant attention in the research sector. With numerous benefits, it would be highly advantageous to rollout this technology on a mass basis for industrial purposes. The increasing demand for producing ceramic materials with geometrically complex structures (e.g., lattice, porous and mesh structures) has shifted the focus to AM technologies was they offer unprecedented rapid fabrication of complex structural components (Fig. 5.1).

A selection of various current AM processes for the manufacture of ceramics is displayed, as per Fig. 5.2. Such processes include fused deposition modeling (FDM) [5], selective laser sintering [6], lithography-based ceramic manufacturing (LCM) [7], stereolithography (SL) [8], freeze-form extrusion (FFE) [9], robocasting [10, 11], and selective laser melting [12]. Powder bed processes such as binder jetting (BJ), selective laser sintering (SLS) and selective laser melting (SLM) lead to the formation of porous structures, that are subsequently sintered to enhance the density of the printed object [12, 13]. AM printing of ceramic materials using stereolithography and fused deposition modelling can be used with high volumes of binder materials (>40 vol%), but exhibit sudden outgassing, crack formation and shrinkage of structures during the removal of binder from the final structures. Slurry-based material deposition techniques such as robocasting are generally preferred over other AM techniques to obtain dense ceramic structures and complex geometries. These net-shaping processes are promising due to their versatility, rapid processing capabilities, and ability to cater to increasing engineering demands. A comparison of these ceramic printing processes in terms of their characteristic features are tabulated in Table 5.1.

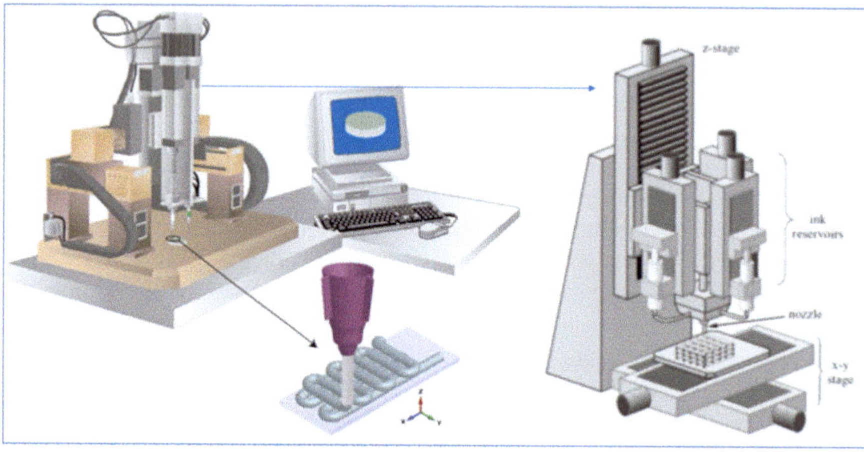

Fig. 5.1 Pictorial illustration of the robocasting process; Extruding the ceramic slurry through a fine nozzle and construct the object layer-by-layer, as directed by the design [4]. Reprinted with permission from Springer Nature 2008

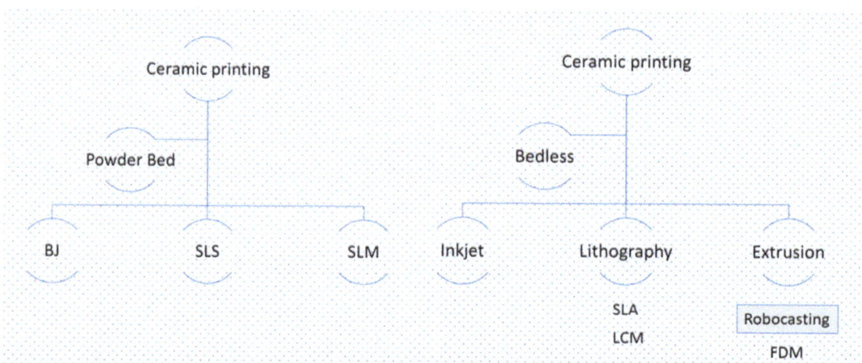

Fig. 5.2 Schematic description of the current ceramic printing technologies

Robocasting is a slurry-based extrusion method, developed in the mid-1990s at Sandia National Laboratories. This process is also known as robot assisted deposition, direct ink writing, or direct write fabrication. Robocasting, when compared to other.

AM processes, has the freedom and suitability to deposit various unique materials, opening a plethora of potential application avenues. This robocasting process commences with the preparation of the pseudoplastic paste filament [14] (i.e., the ink), which consists of ceramic particles and other organic binders. Subsequently, the prepared ink will be extruded through a fine nozzle to construct an object onto a fixed or dynamically controlled platform.

Robocasting also has the potential to produce multi-ceramic materials by including multi-ceramic particles in the ink preparation stage. The nozzle position

Table 5.1 Technical capabilities of the main AM technologies in ceramic printing [19]

Technique	Features of the printed objects				
	Closed cavity	Filling	Floating	Overhanging	Spanning
Robocasting	Possible	Possible	Impossible	Limitations	Limitations
FDM	Possible	Possible	Impossible	Limitations	Limitations
Directed energy deposition	Possible	Possible	Impossible	Limitations	Impossible
Binder jetting	Impossible	Possible	Possible	Possible	Possible
Direct inkjet printing	Possible	Possible	Impossible	Limitations	Impossible
Layered object manufacturing	Impossible	Possible	Possible	Possible	Possible
Selective laser sintering	Impossible	Possible	Possible	Possible	Possible
Stereolithography	Possible	Possible	Impossible	Possible	Impossible

and amount of ink per pass are all controlled by taking the 2D slicing of a model designed by computer-aided design (CAD). Like other AM techniques, the object can be constructed in a layer-by-layer approach, by controlling the addition of ink. FDM and robocasting are extrusion-based AM processes, however, in robocasting, layers are fused through the surface tension of the liquid. Like FDM process, the robocasting process also has technical challenges associated with the spanning, overhanging and all of the other floating features. Another notable difference in robocasting, compared to other established AM technologies, is that the process initially produces a ceramic green structure comprised of plasticizer or binders, which is removed from the final structure by burning-off [15]. The purpose of this post-processing at high temperatures is to sinter and densify the final structure of the product. This chapter presents a detailed discussion on the robocasting process as well as the latest developments on the fabrication of multi-ceramic structures into complex architectures [16] by colloidal ceramic processing [17]. The major challenges associated with ceramic robocasting and the future scope of this technique in printing shape-memory ceramics [18] and self-shaping ceramic objects, are also discussed.

5.2 Principles of Robocasting and Type of Materials

As discussed earlier, robocasting is a slurry-based extrusion process, employed for the manafacture of ceramic objects through the use fine nozzles of which the diameter is generally hundreds of microns or less. Thus, robocasting is also referred as micro-extrusion. The pseudoplastic paste or ink containing ceramic particles is extruded through this fine nozzle and solidifies into a green body upon deposition. Solidification is generally achieved by surface evaporation of water or volatile liquids present

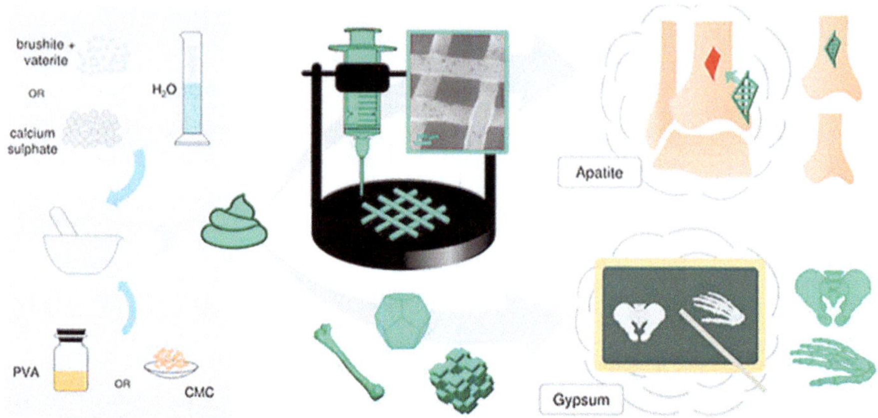

Fig. 5.3 Generic approach to making pseudoplastic ink and the robocasting process to produce ceramic structures [22]. Reprinted with permission from Elsevier 2021

in the slurry. This approach has drawbacks such as brittleness of the printed object during construction, and may also lead to disproportional shrinkage and cracking.

The pseudoplastic paste must have specific rheological properties in order to be printed effectively and smoothly. The composition of the ceramic material, binders and the robocasting method all dictate the rheological properties of interest [20, 21]. To retain its shape during extrusion, the paste needs either the appropriate viscoelastic consistency or is dried in-situ during extrusion. The ink preparation method for assembling bioceramics through the robocasting process is displayed below, as per Fig. 5.3 [22]. Similar to the other AM techniques, ceramic objects may be constructed in a layer-by-layer approach. This is achived by extruding paste for each "slice" of the CAD model, through the nozzle in a specific pattern. Nozzle diameters often vary per the application, ranging from 30 μm for ceramic pastes to 1 μm for polymer pastes. In this printing approach, the final printed shape is determined by the pseudoplasticity of the ink, rather than solidification of ink after the material has been extruded through the nozzle.

5.2.1 Paste Formulation

The volume percentage of the solid content of the ceramic slurry is very important in maintaining the pseudoplasticity [23], as demonstrated in Fig. 5.4 These paste or slurry formulations are loaded into the chamber and extruded through the nozzle using the force exerted by the piston, as shown in Fig. 5.3. The experimental setup displayed, is the generic form of robocasting for layerwise deposition of ceramic slurries through a nozzle. This technique can be further equipped with multiple print-heads to print multi-ceramic materials with less material wastage.

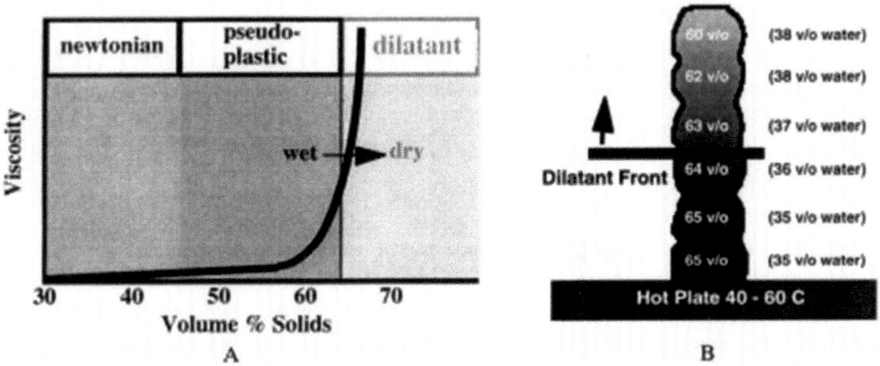

Fig. 5.4 Typical viscosity versus volume percent solids behaviour for optimal robocasting of dispersed alumina slurries [23]

Solid content must be maintained and regulated around 70% or above to obtain better ceramic paste for the appropriate fluidity and drying kinetics. The viscocity of the ceramic paste is a particularly important parameter that must be regulated for smooth nozzle extrusion and shape maintining properties. High solid content needs to be maintained for a dense ceramic structure, better drying, and less porous upon extrusion [24, 25]. Another advantage of using a paste containing high solid content is to achieve high resolution on the micron scale by retaining shape during extrusion [26, 27]. Paste formulations are central to the robocasting process and various synthetic approaches have been followed to obtain a suitable paste with the desired characteristics. Cesarano et al. developed the initial slurry formulations that consist of >60 wt% solids in water with <1 wt% organic dispersants [23].

The inter-particle interactions, surface interactions and binder particle interactions also play a vital role in controlling the required psuedoplastic properties. Required pseudoplastic properties were obtained from the nature of inter-particle interactions and surface interactions in the slurry. Moreover, it was observed that the transition from the pseudoplastic to the dilatant phase provided strength to the underlying layers [28]. As the paste has very high solids loading that enables the densification in post-processing, Lewis et al. have developed a slurry containing polyethylenimine (PEI) functionalized silica particles that provide the slurry "fluid-like" desired rheological properties and the observed gelling behaviour [29].

5.2.2 *Inter-particle Interactions*

The slurry used for robocasting generally consists of complex system comprised of particles within the colloidal size range, hence controlling the inter-particle interaction to obtain a stable slurry is vital to the process. Surface bound ions, electrostatic interactions and steric interactions are the three major inter-particle interactions occuring in colloidal systems, as per Fig. 5.5 [16]. These three kinds of interactions keep the particles dispersed in the solvent medium by the repulsive interaction between the particles. Similar surface charge of the particles tend to repel each other (electrostatic stabilization), which is generally observed when charged particles are dispersed in a polar medium such as water [30]. Polyelectrolytes can be added to further stabilize colloidal sized particles. Some examples include amines [polyethyleneimine (PEI)], carboxylic acid [poly acrylates] and hydroxyl groups [poly alcohols], the polarity of which change as a function of pH [31]. These groups bind strongly with the ceramic particles and have opposite surface charges. Surface bound polymers in the case of polymer-protected ceramic particles stabilize the particles through steric interactions between the polymer chains on the surface [32, 33].

As mentioned earlier, using functionalized ceramic particles for robocasting poses a handful of challenges including stability under a broad pH range, and the addition

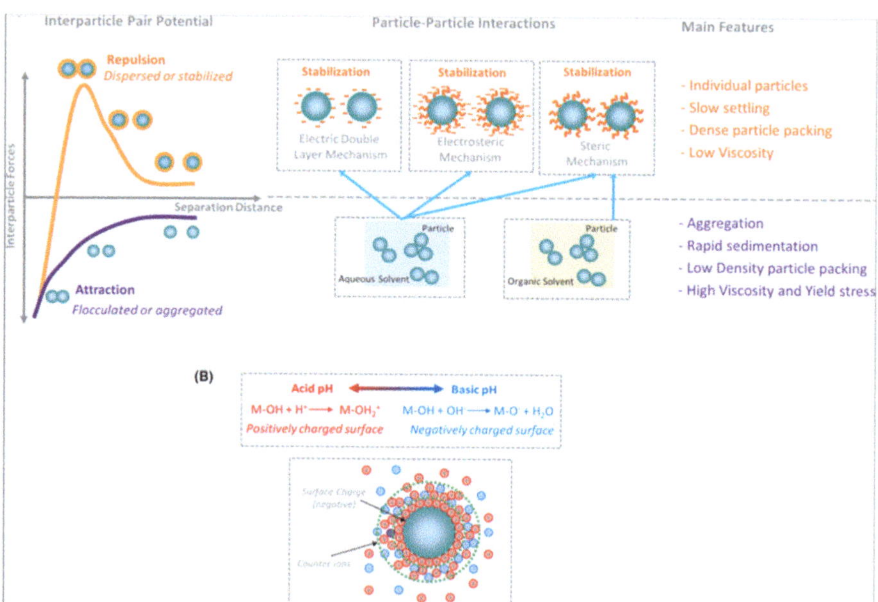

Fig. 5.5 Pictoral representation illustrating the nature of inter-particle forces between ceramic particles and various kinds of inter-particle interactions [16]

of additives that may adversely affect the stability of the system. These functionalized ceramic particles surface chemistry is very sensitive to the pH of the solvent, therefore, any small altercation to the medium greatly influnces the printing process quality, restricting the use of other materials and the paste production quality. Coagulation techniques are followed currently when the slurry contains moderate amounts of solid load ranging from 40–50 wt% cellulose [34, 35] and lead nitrate [36]. Polyacrylic acid (PAA) functionalized ceramic particles were stable in the slurry due to the electrostatic stabilization, which may be reversed by a change in pH or addition of salt which neutralises surface charges, resulting in coagulation [34, 37].

Using hydrogel based carriers alongside ceramic particles eliminates the need for additional surface stabilisation and was demonstrated. This approach enables various kinds of ceramic particles to be utilized for robocasting [38]. As inter-particle interaction become less prominent, slurry formulations can be compatible with virtually any powder regardless of surface functional groups. The stability of these paste formulations was found to be highly stable, lasting for months without any aggregation. Hydrogels were formed by the triblock copolymer F127 (PEO-PPO-PEO) when it was dissolved in a range of 20–30 wt% in water. These polymers form micelle networks [39]. This hydrogel carrier approach ensures the slurry is homogeneous at low temperature, and workable for printing through the robocasting process.

5.2.3 Role of Water (as a Solvent)

The majority of the robocasting slurry formulations use water as the solvent (continuous phase) due to its ability to form stable suspensions with ceramic particles [40–42]. Water is a cheap, readily available, non-toxic and non-flammable option. It's more eco-friendly than a slurry containing organic solvents which are also known to modify the properties of the resultant slurry/suspension/paste [43]. Water has a low viscosity which can be increased with additives but the green pre-sintered ceramics produced from aqueous slurry formulations are not as mechanically strong as those with resin or polymer based green bodied ceramic. The drying of the water from the green-body can also lead to cracks and voids if drying is not sufficiently rapid [45].

5.2.4 Effect of Viscosity

Viscosity of slurry formulations is a key parameter in robocasting. After extrusion, viscosity increases when the liquid in the formulation undergoes solidification. Figure 5.6a. shows that at low shear rates, a yield stress type fluid demonstrates solid-state behaviour, whilst at high shear rates, the viscosity of the yeild stress fluid is close to it's zero shear viscosity [46]. Materials that exhibit this feature are denoted as pastes or gels, and have a greater viscosity at any given shear as compared to any

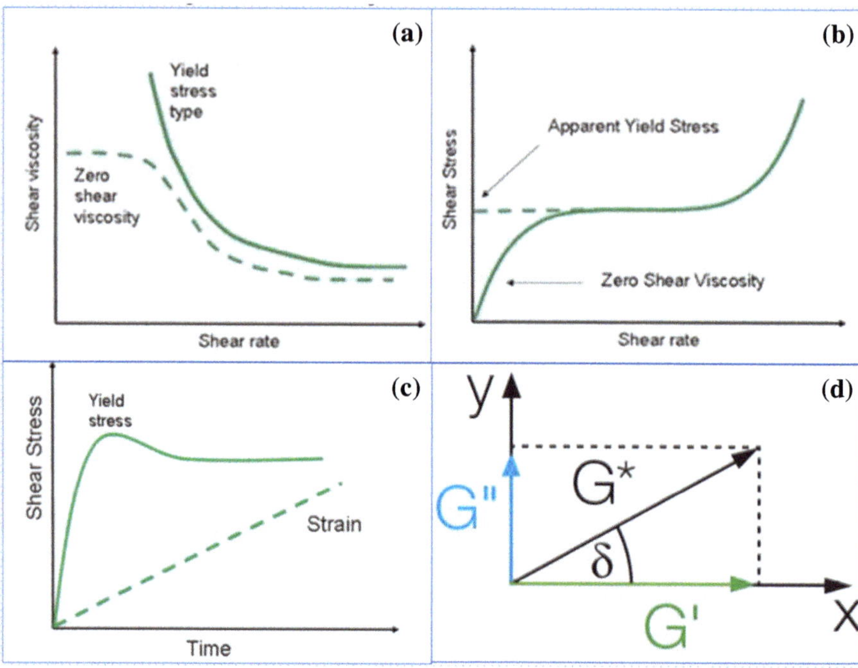

Fig. 5.6 Rheological properties of the ink [44] **a** Characteristic flow curves for materials with a true yield stress and a zero shear viscosity **b** The measurement versus behaviour of a highly viscous material that appears to have a yield stress but shows viscous behvaiour at much lower shear rates rheology for pastes versus fluids on a linear scale; **c** Stress evolution at constant shear rate; **d** Vector diagram of the relationship between storage modulus and loss modulus

Newtonian fluid [42, 47]. Figure 5.6b shows shear stress measured to be zero at zero shear rate despite a yield stress present in the material. This is the primary reason that the true shear stress in the material gives an apparent yield stress as zero at a lower shear rate, although this can be affected by the measuring equipment itself [48].

5.2.5 *Effect of Yield Stress*

The energy required to permit the solid-state viscoelastic paste to flow like a fluid is known as yield stress. Within the elastic limit, stress causes deformation and reforms back to its original limit. Above this limit, the material permanently deforms. Figure 5.6c shows stress measurement under constant strain and an increase in stress in the material until it yields. Thereafter, a constant stress is required to plastically deform the paste. The reduction in stress required for the paste to flow after yielding represents the shear thinning properties of the paste [46]. Materials that have long relaxation times lead to the formation of creep in the material, and in the short term they behave like high viscous liquids without a true yield stress [46]. Figure 5.6d shows the vector diagram where the storage modulus, G' (solid state behaviour) plotted against the loss modulus G" (liquid state behaviour), that produces a complex modulus G* at a phase shift angle (δ). To reain the viscoelastic paste, the phase shift angle needs to be maintained >45° and the storage modulus must be greater than the loss modulus [44].

From detailed analysis, it has been determined that the properties of the slurry are a key component to robocasting. The following additional features need to be maintained for a slurry to be suitable for the robocasting process.

- Slurry formulations must be homogenous and should be free of air bubbles and agglomerates, and;
- The viscocity of the slurry must be optimized to allow for extrusion through a fine nozzle at relatively low pressures, and;
- The slurry formulations yield stress and storage modulus must be optimized, and finally;
- The volume fraction of the solid ceramic component must be high, to reduce shrinkage during drying and maintain high density during sintering.

To improve the dimensional accuracy of printing, the instrument variables that control the robocasting process need to be optimized. Air pressure, printing speed, flow rate of slurry formulation, layer height and nozzle dimensions are the main instrumental variables that need to be varied for each material and method of printing.

5.3 Post-processing

The ceramic green bodies fabricated by the robocasting technique need to be dried and cured in order to remove the binder, and further sintered at high temperature to obtain a high-density ceramic object. Adversely, sintering at high temperatures may lead to cracks and less dense structures, hence the amount of total shrinkage needs to be minimized during the post-processing of the green-body. This is why it is ideal for the slurry to have a high weight percentage of fine ceramic particles and conversely a low weight percentage of organic binders and other additives [35, 50]. Moreover, a slurry used for printing should have the appropriate viscoelastic properties, which are controlled by the amount of organic binder present, and conforms to the Herschel–Bulkley model [29, 35].

Colloidal ceramic materials generally have very high organic binder contents, which are timely to remove and can lead to significant dimensional change in the process. Contrastly, slurries containing lower organic binder and additive contents may be simply removed under ambient conditions, otherwise, the structures tend to crack during non-ambient drying [29]. However, drying and burn-out are the two major post-processing steps that are conducted in order to obtain the final shape of

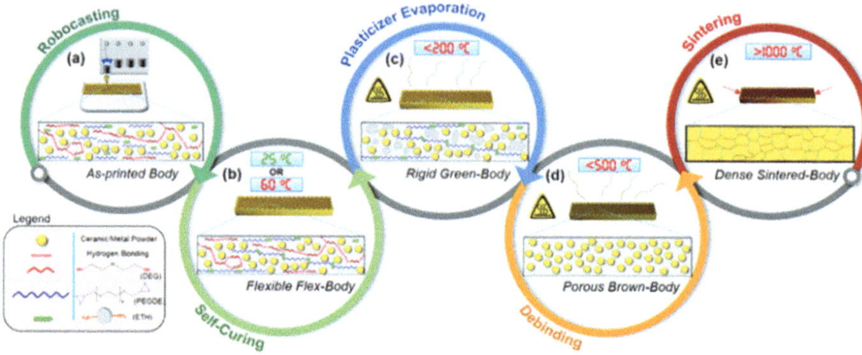

Fig. 5.7 Multistage heating process to achieve crack-free and dense sintered structures [49]. Adapted with permission

the material, as per Fig. 5.7 [49]. These steps are as important as the printing process because inefficient post-processesing may lead to deformation, cracks and shrinkage in the final structure.

5.3.1 *Drying*

Printed products of the robocasting process need to be dried in two defined stages. In the first stage, drying is carried out on the substrate at room temperature for 24 h in a closed system. This is to prevent crack formation that can occur if the samples are dried too rapidly. This step is followed by drying in a convection oven at temperatures above room temperature and below the boiling temperature of the solvent for a further 24 h.

5.3.2 *Burn-out and Sintering*

Evaporation of plasticizer from the dried object introduces pores into the material. Removal of binder from the built parts can be achieved by heating the dried object at 600 °C for an hour. In the case of oxide materials, air is generally used as a carrier gas, whilst argon can be used for non-oxides such as carbides. The resultant material can be sintered at high temperature (>1000 °C) to obtain a highly dense ceramic printed object. Figure 5.8. displays an avionic pod camera printed using SiC powder showing an increase in density after sintering, while the same structure fabrication using conventional manufacturing requires high quantities of material and tedious machining procedures [51]. Figure 5.8b shows that drying of gelatin-bioactive glass scaffolds also reduced their volume and increased their densities [52].

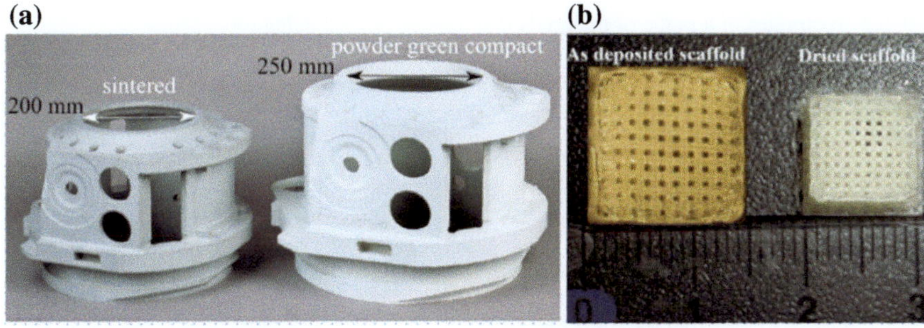

Fig. 5.8 Volume reduction after post-processing. **a** Avionic pod camera made from silicon nitride [51]; **b** Gelatin-Bioglass scaffold as printed and after vacuum drying [52]

5.4 Materials Suitable for Ceramic Printing

Previous sections demonstrated robocasting's suitability for printing ceramic materials in complex geometries as ceramics have high melting points as well as high-temperature post-processing procedures, which restricts other AM processes. Moreover, this technique has suitability with a wide variety of ceramic materials with varying chemical compositions. Any slurry containing ceramic particles of suitable size and viscoelastic properties may be printed using this technique with appropriate conditions and post-processing methodology.

Particle size in the slurry is an important factor that affects viscoelastic properties. It has been shown that smaller particle sizes (nm range) are known to relatively increase viscosity [68], whilst larger particles may block the printing nozzle [35]. To reiterate, robocasting is a wet process and post-processing provides highly dense structure. Thus, the mechanical properties of the highly dense objects produced, surpass those produced by other AM techniques. Furthermore, the mechanical properties of objects printed using robocasting are equivalent to that of the ceramic materials fabricated through well-estaiblished conventional subtractive manufacturing processes [69]. When compared to metals and polymers, ceramic materials are difficult to process into shapes like filaments or sheets. However, robocasting has clearly demonstrated its capability to print ceramics into any shape or morphology. Therefore, robocasting is considered the most versatile AM process with any ceramic material composition [10, 51]. Table 5.2 shows the classification of ceramic materials into oxides, nitrides, carbides, and borides based on chemical composition.

These ceramic materials are further classified into technical ceramic materials and functional ceramic materials. Technical ceramic materials are known for their mechanical properties and chemical inertness. Robocasting can print these materials into various shapes and morphologies, with excellent geometrical accuracy.

Inorganic oxides, carbides and nitrides fall under the category of technical ceramics. Robocasting can print a wide variety of technical ceramic materials such as silica (SiO_2) [68], alumina (Al_2O_3) [53], mullite [55], zirconia (ZrO_2) [61], hydroxyapatite (HA) [70], silicon carbide (SiC) [64], silicon nitride (Si_3N_4) [60], and boron

Table 5.2 Examples of a range of materials that may be printed using robocasting

Technical ceramic materials	Materials	Functional ceramic materials	Materials
Oxides	Al_2O_3 [14, 53]	Ferroelectric materials	$BaTiO_3$ [54] $PbTiO_3$
	Mullite [55]		
	Hydroxyapatite [56–59]	Nitrides	Si_3N_4 [60]
	ZrO_2 [61, 62]		
	Bioglass, SiO_2 [37, 63]		
Carbides	Silicon carbide [64]	Superconductor materials	$YBa_2Cu_3O_7$ [65]
Borides	ZrB_2	Magnetic materials	$NiFe_2O_4$ [66] $BaFe_{12}O_{19}$
	B_4C [67]		

carbide (B_4C) [67]. Figure 5.9 shows some of these materials printed using these techniques. In contrast, functional ceramic materials exhibit characteristic electronic, magnetic and superconducting properties by virtue of their chemical composition, therefore, these structures have associated functionality. Robocasting is capable of printing a wide range of functional ceramic materials including magnetic materials [66], ferroelectric materials [71], superconductors [65] and electro ceramic materials.

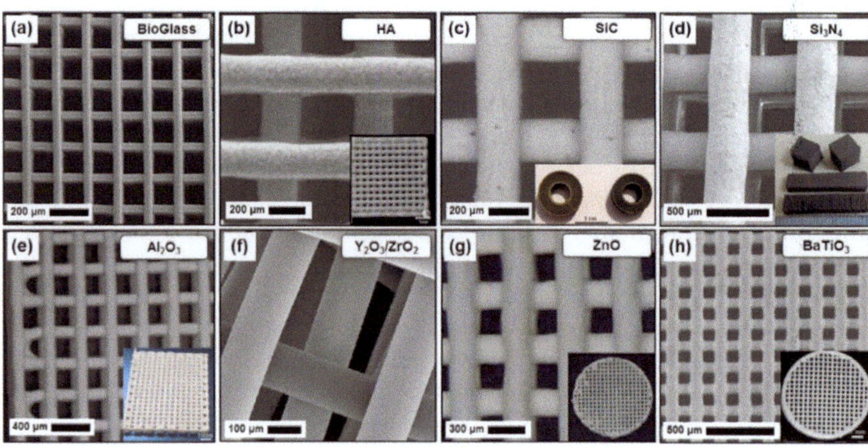

Fig. 5.9 Robocasting printed structures of Bioglass, Hydroxy apatite, Silicon Carbide (SiC), Silicon Nitride (Si_3N_4), Alumina (Al_2O_3), Yittria/Zirconia, Zinc Oxide (ZnO), Barium Titanate ($BaTiO_3$). Reprinted with permission from Refs. [10, 51]

5.4.1 Robocasting Printed Structures as Functional Ceramic Structures

Earlier work on ceramic printing using robocasting was mainly focused on monolithic materials such as alumina, zirconia, and silicon carbide. The scope of this technique in producing woodpile style geometries as shown in Fig. 5.10, has attracted significant attention as this geometry can be easily fabricated with load-bearing 3D architectures and precisely controlled porosity. This woodpile style geometry can be functional as scaffolds, which may have great potential to produce bio-compatible bone grafts. The porous nature of these materials enables the inflow of tissue and ingrowth. These materials exhibit characteristic mechanical properties that were found to have greater compressive strength than the foam scaffolds of the same porosity. In recent years, there has been much progress in printing ceramic composites and multi-ceramic materials using this technique in various fields of science and engineering. In particular, robocasting can be used to print ceramic materials in applications that require complex geometries and fine surfaces, resulting in materials with improved performance. These printed green materials can achieve 90% density upon sintering, which is an advantage of this technique. The following sections highlights the recent developments in robocasting printed ceramic and multi-ceramic materials reported recent literature.

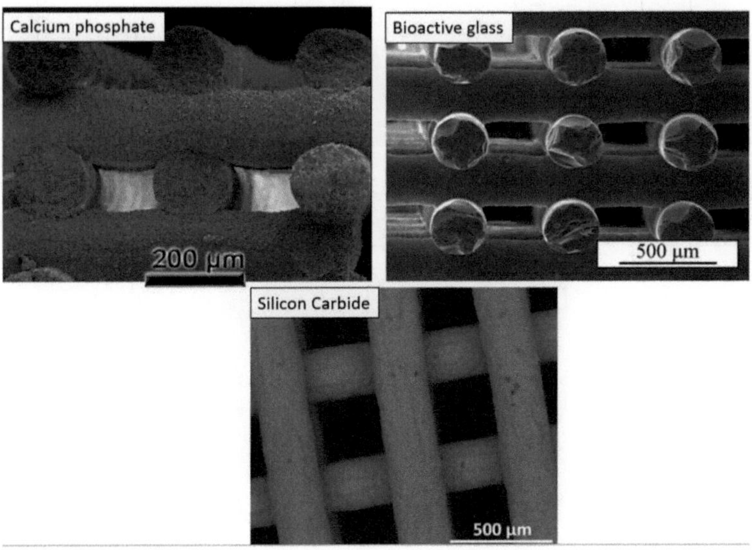

Fig. 5.10 Woodpile style scaffold structures of calcium phosphate [72], bioactive glass [73] and silicon carbide [74]

5.4.1.1 Porous Ceramic Components

For applications that demand high porosity, there will be a compromise on material density, thus a fine balance between the material's porosity and density needs to be maintained for optimal performance [75]. Scaffolds with an anatomical cortical–trabecular structure were fabricated using robocasting and demonstrate its ability to design and print scaffolds with complex architectures, as shown in Fig. 5.11a. The ability to fabricate hydroxyapatite scaffolds with varied porosity provides the same mechanical properties as bone scaffold [75]. Porous mullite structures are considered promising materials, and Stuecker et al. demonstrated the formation of mullite meshes as per Fig. 5.11b, with 100–1000 μm pore sizes and with a sintered density of 96%. The right balance between porosity and density can enable the porous structures to obtain the desired mechanical strength and efficient mass transfer for any scaffold application.

Miranda et al. have used robocasting to produce hydroxyapatite scaffolds of 3D square mesh of interpenetrating rods as shown in Fig. 5.11c, which was shown to be excellent scaffolding material for bio ceramics [77]. The robocasting approach of printing single ceramic materials has been used for other applications in addition to simple scaffold. Figure 5.12 provides a selection of examples of robocasted printed single ceramic structures for various applications.

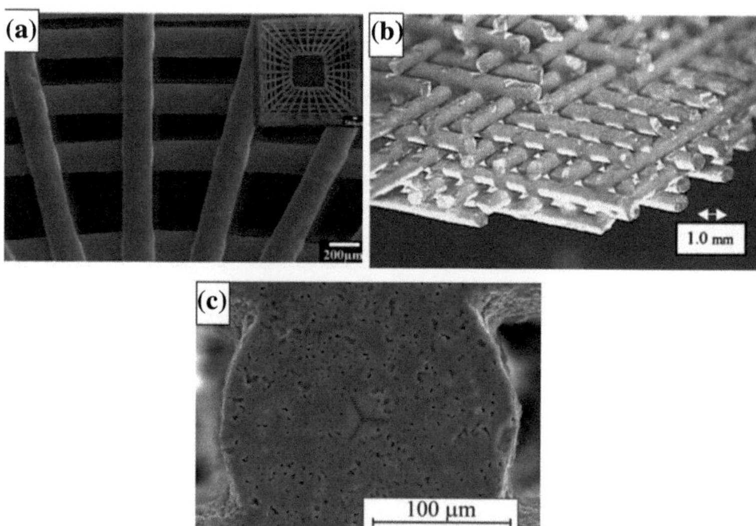

Fig. 5.11 Porous scaffolds printed using robocasting; **a** A glass scaffold with dense outer layer and porous inner layers, with the inset showing the sintered scaffold [75]; **b** Porous mullite [76]; **c** Porous hydroxy apatite scaffold [77]

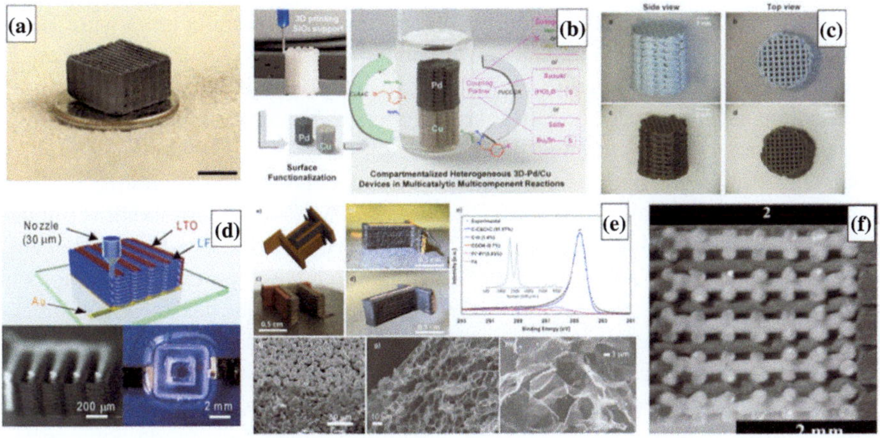

Fig. 5.12 **a** Graphene lattices; **b** Pd and Cu functionalized SiO_2 monoliths; **c** Cu/Al_2O_3 catalysts; **d** Micro Lithium-ion batteries; **e** Graphene electrodes; **f** Ni-$BaTiO_3$ ceramics

5.4.1.2 Catalysts and Catalyst Supports

Robocasting has demonstrated the ability to print graphene-based materials into highly porous catalyst supports as shown in Fig. 5.12a. Fabricating graphene based aerogels is challenging, however, Zhu et al. has been able to develop novel graphene formulations to make porous lattice structures [78], as shown in Fig. 5.12a. Diaz-Marta et al. functionalized silica monoliths with catalysts and demonstrated this composite structure as an excellent heterogeneous catalyst. Figure 5.12b shows Pa and Cu functionalization on the surface of a silica monolith and its application as a bifunctional catalyst [79, 80]. Tubio et al. demonstrated the Cu functionalization on robocast-printed Al_2O_3 monoliths and showed an efficient catalyst for the copper-catalysed Ullman reactions. The strong interaction between the Cu and the alumina matrix prevented leaching of the Cu metal into the reaction medium. The choice of Al_2O_3 was made due to its high mechanical stability following high-temperature treatment [81]. The monolithic catalyst also exhibited high catalytic efficacy and good recyclability, as displayed in Fig. 5.12c.

5.4.1.3 Electro-ceramic Materials

Robocasting has been also shown to produce electro-ceramic materials, showing major potential for the production of materials used in energy based applications. Sun et al. have demonstrated the fabrication of 3D Li-ion microbatteries composed of high-aspect-ratio anode and cathode micro-arrays, as shown in Fig. 5.12d. These microarrays were interdigitated on a micron-scale, which displayed higher energy and power densities [82]. 3D interdigitated microbattery architectures were fabricated using $Li_4Ti_5O_{12}$ (LTO) as the anode and $LiFePO_4$ (LFP) as the cathode materials, which exhibited minimal volumetric expansion, thereby reducing the strain that accompanies the charge and discharge processes. Rocha et al. developed similar ceramic materials with complex geometries, tailored to fit as supercapacitors or anodes for lithium-ion batteries. Using the triblock copolymer Pluronic F127, thermoresponsive ink formulation was made as well as rCMG-based self-standing binder-free electrode was produced. This multi-material printing of electrode components and its electrochemical activity was demonstrated to have important applications in electrochemical energy storage [83]. The robocasting technique has the added advantage of discrete or graded printing capabilities in multi-material printing, which was exploited in developing 3D periodic structures of electro-ceramic and metal electrode compositions. Ternary mixtures consisting of $BaTiO_3$, $BaZrO_3$, and $SrTiO_3$ were mixed in an active mixing nozzle, wherein the nickel electrode components specifically interact with the $BaTiO_3$ phase. Therefore, the printed multi-layer structures consist of alternating layers of metal and ceramic materials, and exhibit better capacitance [71].

5.4.2 Current Progress in Multi-ceramic Material Printing

Mullti-ceramic materials printed using robocasting were briefly mentioned in the previous sections, however, this section presents the processes in greater detail alongside its current developments. Multi-ceramic material printing is an important extension of ceramic robocasting technology, that was made possible because a single robocasting printing platform can accommodate multiple ceramic feeding systems. Therefore, extruding more than one material can be possible in a single printing procedure. Printing such multi-ceramic materials in one single printing process combine the advantages of many materials' unique properties or graded compositions within a single printed structure, clearly demonstrating the versatile approach of multi-material ceramic robocasting. Figure 5.13 illustrates the typical printing setup with multi-material dispensing function. This setup can be operated in two different ways. In the first approach, a single mixing nozzle can be integrated with multiple feeding systems [23], while the second approach involves the use of discrete multi-nozzle systems [84].

In the first approach, the single mixing nozzle blends the materials received from the multiple feeding systems at specific feed-stock proportions to form a final mixture, which can be extruded as single filament. As mentioned earlier, this approach has the capability of printing functionally graded materials. Moreover, by operating one feeder material at one time, structures can be printed with two discrete individual materials in the same time it takes to set up a dual robocasting setup. In the case of discrete multi-nozzle systems, they have the advantage of allowing automatic interchangeable printing heads during printing [41], that allow multi-material printing.

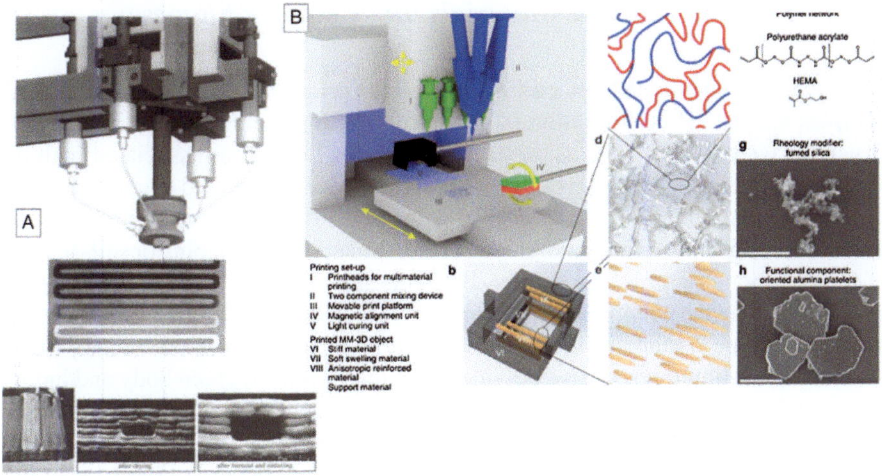

Fig. 5.13 **A** Mixing head of four slurries [85], graded transition between two slurries and kaolin slurry with a fugitive material; **B** Robocasting with multiple dispensers setup

With single and multiple nozzle systems alike, it is possible to print alternating materials of different ceramic composition as well as printing each component in a different spatial geometry. The concept of graded printing using multi-materials through a single mixing nozzle was developed by Cesarano et al. As portrayed in Fig. 5.13A, this was the first demonstration of printing different kinds of ceramic materials into a single material, wherein grading of single materials was obtained with a gradual transition from one composition to another. Using a similar approach, an alternating structure of $Pb(Zr_{0.95}Ti_{0.05})O_3$ with different densities was fabricated [86] by Tuttle et al. The printed ceramic structures were obtained without any cracks at the interface between the layers. Similar to the printing of multimaterial of electroceramic materials presented in the last section, Smay et al. used an approach to print 3D periodic structures of the metal electrode and the electroceramic material in the form of multi-layer capacitors [87]. In this approach, separate nozzles for each ceramic and ceramic/metal suspensions were assembled that enabled the deposition of the ternary mixture of ceramic suspensions into 3D multi-layer lattice structures. The resultant green body structures were calcined in an inert atmosphere for sintering and inert gas conditions were provided to prevent oxidation of the metal components used for the electrodes. The major advantage of this techniques as compared to the other slurry based ceramic techniques is its ability to achieve the layered structured structure with high resolutions in combination with the shaping capabilities. Significant developments in printing soft materials and other cellular structures using the multi-nozzle printing systems, Robocasting can be used to print multi-ceramic material printing and bioprinting platforms [84].

5.4.3 Printing Ceramic Materials into Complex Shapes

All of the afore-mentioned robocasting process can print sintered ceramic bodies with complex shapes, while formation of ceramic thin films, ceramic foils and origami structures were limited. This is probably because of limitations associated with the secondary shaping of the green body as the final shape of the sintered structure entirely depends upon the shape of the as-printed green body. A recent development denoted as "4D printing" in the case of polymers, where the secondary shaping or self-shaping of the polymer structures can be possible. This self-shaping can be achieved using stimulus responsive polymers, where the stimuli such as temperature, pH, solvent can cause the change in shape, properties, and function. This concept is extended by incorporating such stimuli responsive polymers into the as-printed green body and induce the secondary shaping or self-shaping to obtain ceramic thin films and foils. Ahn Bok et al. demonstrated this concept first time and showed the formation of 3D printed origami structures based on titanium hydride (TiH_2), as shown in Fig. 5.1a. In this work, titanium hydride (TiH_2) materials were printed using a triblock copolymer of poly(methyl methacrylate)–poly(n-butyl acrylate)–poly(methyl methacrylate) [88] and showed the self-shaping process. As it shown in Fig. 5.14, as-printed TiH_2 green body was directly shaped into the cylindrical structure and folded into the form of origami and eventually into crane structure. Similar finding was observed even in the case of Robocasting printed metal powders and an elastomeric binder. Adam et al. also reported the formation of metallic cellular structures using the similar approach and induce the secondary shaping of the stimuli responsive polymer [89]. Using hydrogen as reductant, these metal oxides were reduced that lead to the formation of metallic architectures. This work uses dichloromethane solvent and dibutyl phthalate as a plasticizer to enhance the drying process as well as improve the flexibility of the green-body for shaping process. The printed cellular architectures were found to have high ductility, leading to excellent elastic and plastic energy absorption, when subjected to uniaxial compression.

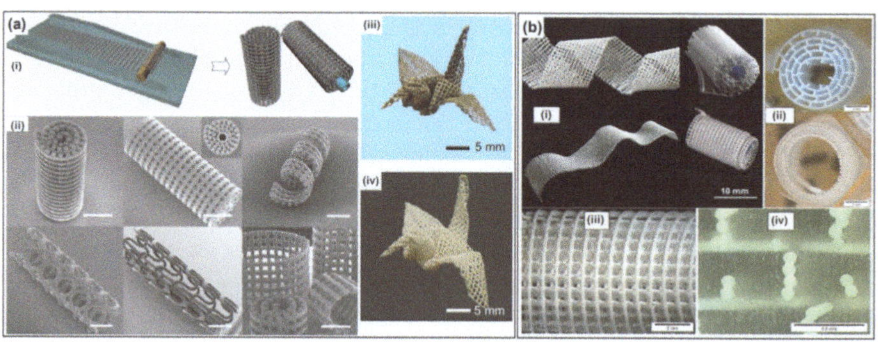

Fig. 5.14 Geometrically complex structures of **a** Titanium hydride (TiH_2) **b** Two layered periodic structures of Hydroxy apatite and Alumina Reprinted with permission [10]. Copyright 2018 John Wiley & Sons

Hazan et al. demonstrated this concept of secondary shaping by using flexible backbone using PLGA and photocurable resins to produce flexible hydroxy apatite and alumina structures, as shown in Fig. 5.14b. PLGA backbone provides the flexibility and photocurable monomers such as 4-hydroxybutyl acrylate and ethoxylated bisphenol A diacrylate that provides low glass transition temperature. The secondary shaping process of these as-printed green bodies such as folding, rolling and bending can be used to obtain the flexible shapes of HA and alumina [32, 90]. The conceptual development of secondary shaping or the shapeable ceramic green-bodies in the near future will open up more opportunities to find a suitable combination for self-shapeable ceramic green bodies to fabricate geometrically complex structures with fine textural features while retaining high density even after sintering.

5.5 Challenges and Outlook

The current trends and progress in the ceramic robocasting technologies for printing ceramics into functional structures demonstrate the potential of this technology in transforming the century old ceramics into modern world needs. Among the slurry based AM techniques, ceramic robocasting is the most capable technology to print wide variety of functional ceramic materials into dense shapes with fine microstructures and complex morphologies. The maximum achievable resolution using the advanced robocasting processes are ranging from 60 to 100 μm, whereas the density of the printed objects can be as high as 90%. Recent developments in multi-material printing in combination with 3D spatial material deposition, this technique can provide significant opportunities in producing shapeable ceramic structures that are shown various applications in energy research. Similar to the other AM techniques, stair-step finishing is the main disadvantage of the surface finish in the in robocasting process as well. However, the use of fine nozzle for the printing process, equipment calibration, height setting and optimizing the slurry formulation, the surface finish of these ceramic structures can be made into high resolution.

To obtain the high resolution printing characteristic, it is important to develop a slurry formulation that consist of agglomerate free particles with narrow size distribution. The main challenge in any robocasting process is to to prepare the ceramic slurry with high solid content of ceramic particles as well as retaining the the pseudoplastic consistency of the ceramic slurry. The other major components of the pseudoplastic ink are binders, organic additives, and plasticizers, which are mixed with ceramic particles in a solvent or dispersion medium. The rheological behavior of the pseudoplastic ink is mainly controlled by the nature of interaction among the various components present in the slurry, therefore the ceramic slurry used for the robocasting process are generally made with the major considerations involving the suitable dispersant, gelling agent, binders, and plasticizers.

Identification of suitable dispersant is the most important parameter to print a dense ceramic structure because the medium need to disperse the high content ceramic particles. Solvents that can undergo faster evaporation rate under standard conditions may not be suitable because use of these solvents lead to aggregation during printing when fine nozzles are used in a low humidity setting. Addition of non-volatile hydrophilic solvents to the aqueous formulations can address this drying kinetics mediated agglomeration. Addition of triblock copolymers such as Pluronic F127 to the ceramic slurry form a hydrogel that was shown to retain the pseudo-plastic behavior. Another potential advantage of hydrogel formulation is that it can retain specific percentage of water and avoids its complete evaporation and drying behavior of the ceramic slurry. The main role of suitable binder and plasticizer is to bind the particles together to reduce the void formation, which other lead to the formation of cracks and delamination during the drying process. As it was mentioned earlier, the capability of printing multi-ceramic materials using robocasting broadens the scope these materials application for the fabrication of battery, capacitor, and sensor, which generally consist of multiceramic components. However, there are significant technical challenges associated with printing of multi-ceramic materials and affect the final finish of the materials and its properties. In addition to the challenges associated to the monomaterial printing, there are other challenges related to multiceramic printing that include the relative drying kinetics of the multi ceramic paste, cracking or delamination due to mismatch between the layers, the resultant cracking/delamination, and any undesired phase formation between two or more components during the sintering process. Shrinkage of both materials may be different during the drying process that may lead to delamination, cracks, and failures. It would be important to choose materials that have similar sintering behavior to avoid this shrinkage. At high temperature sintering, there is a possibility of chemical reaction between the ceramic components and formation of any undesired chemical phase, which may lead to structural and functional failure. These are the current challenges reported in the literature about the multiceramic printing using the robocasting process. The future of this ceramic robocasting process is focused on developing new ceramic slurry formulations, which will be compatible with all kinds of ceramic powder, shapeable ceramic green structures, or the potential of self-shaping ceramic structures during sintering. This demands advanced level understanding of interparticle interactions and inter-particle forces, which can help in developing self-shaping technology for designing complex ceramic pieces.

5.6 Test Your Knowledge

After reading the above chapter you should be able to answer the following questions:

1. *What would be the major difference between Robocasting printed structure to the other 3D printed structures?*
2. *What is the role of solvent in controlling the viscosity of the pseudoplastic ink? Is water a good solvent for making the pseudoplastic ink?*
3. *What will be the general composition of pseudoplastic ink? Comment about the volume fraction of solids and its role in printing the ceramic green body?*

4. *How post-processing of ceramic green body is significant in getting a dense ceramic structure?*
5. *How multi-ceramic printing process can be operated?*

References

1. Slowik T (2019) Additive manufacturing of ceramics
2. Faes M, Valkenaers H, Vogeler F, Vleugels J, Ferraris E (2015) Extrusion-based 3D printing of ceramic components. Procedia CIRP 28:76–81. https://doi.org/10.1016/j.procir.2015.04.028
3. Feilden E (2017) Additive manufacturing of ceramics and ceramic composites via robocasting. Ph.D. Thesis, Imperial College London
4. Munch E, Franco J, Deville S, Hunger P, Saiz E (1989) Tomsia AP (2008) Porous ceramic scaffolds with complex architectures. JOM 60(6):54–58. https://doi.org/10.1007/s11837-008-0072-5
5. Jafari MA, Han W, Mohammadi F, Safari A, Danforth SC, Langrana N (2000) A novel system for fused deposition of advanced multiple ceramics. Rapid Prot J 6(3):161–175. https://doi.org/10.1108/13552540010337047
6. Bertrand P, Bayle F, Combe C, Goeuriot P, Smurov I (2007) Ceramic components manufacturing by selective laser sintering. Appl Surf Sci 254(4):989–992. https://doi.org/10.1016/j.apsusc.2007.08.085
7. Felzmann R, Gruber S, Mitteramskogler G, Tesavibul P, Boccaccini AR, Liska R, Stampfl J (2012) Lithography-based additive manufacturing of cellular ceramic structures. Adv Eng Mater 14(12):1052–1058. https://doi.org/10.1002/adem.201200010
8. Gmeiner R, Mitteramskogler G, Stampfl J, Boccaccini AR (2015) Stereolithographic ceramic manufacturing of high strength bioactive glass. Int J Appl Ceram Technol 12(1):38–45. https://doi.org/10.1111/ijac.12325
9. Huang T, Mason MS, Hilmas GE, Leu MC (2006) Freeze-form extrusion fabrication of ceramic parts. Virtual Phys Prot 1(2):93–100. https://doi.org/10.1080/17452750600649609
10. Peng E, Zhang D, Ding J (2018) Ceramic robocasting: recent achievements, potential, and future developments. Adv Mater 30(47):1802404. https://doi.org/10.1002/adma.201802404
11. William R (2019) Additive manufacturing of advanced ceramics for demanding applications. https://doi.org/10.26174/thesis.lboro.8319917.v1
12. Sing Swee L, Yeong Wai Y, Wiria Florencia E, Tay Bee Y, Zhao Z, Zhao L, Tian Z, Yang S (2017) Direct selective laser sintering and melting of ceramics: a review. Rapid Prot J 23(3):611–623. https://doi.org/10.1108/RPJ-11-2015-0178
13. Dowling L, Kennedy J, O'Shaughnessy S, Trimble D (2020) A review of critical repeatability and reproducibility issues in powder bed fusion. Mater Des 186:108346. https://doi.org/10.1016/j.matdes.2019.108346
14. Feilden E, Blanca EG-T, Giuliani F, Saiz E, Vandeperre L (2016) Robocasting of structural ceramic parts with hydrogel inks. J Eur Ceram Soc 36(10):2525–2533. https://doi.org/10.1016/j.jeurceramsoc.2016.03.001
15. Yu J, Yang J, Huang Y (2011) The transformation mechanism from suspension to green body and the development of colloidal forming. Ceram Int 37(5):1435–1451. https://doi.org/10.1016/j.ceramint.2011.01.019
16. Franks GV, Tallon C, Studart AR, Sesso ML, Leo S (2017) Colloidal processing: enabling complex shaped ceramics with unique multiscale structures. J Am Ceram Soc 100(2):458–490. https://doi.org/10.1111/jace.14705
17. Lewis JA (2002) Direct-write assembly of ceramics from colloidal inks. Curr Opin Solid State Mater Sci 6(3):245–250. https://doi.org/10.1016/S1359-0286(02)00031-1

18. Sabahi N, Chen W, Wang C-H, Kruzic JJ, Li X (2020) A review on additive manufacturing of shape-memory materials for biomedical applications. JOM 72(3):1229–1253. https://doi.org/10.1007/s11837-020-04013-x
19. Feilden E (2017) Additive manufacturing of ceramics and ceramic composites via robocasting materials. Imperial College London
20. Wätjen AM, Gingter P, Kramer M, Telle R (2014) Novel prospects and possibilities in additive manufacturing of ceramics by means of direct inkjet printing. Adv Mech Eng 6:141346. https://doi.org/10.1155/2014/141346
21. Travitzky N, Bonet A, Dermeik B, Fey T, Filbert-Demut I, Schlier L, Schlordt T, Greil P (2014) Additive manufacturing of ceramic-based materials. Adv Eng Mater 16(6):729–754. https://doi.org/10.1002/adem.201400097
22. Paterlini A, Le Grill S, Brouillet F, Combes C, Grossin D, Bertrand G (2021) Robocasting of self-setting bioceramics: from paste formulation to 3D part characteristics. Open Ceramics 5:100070. https://doi.org/10.1016/j.oceram.2021.100070
23. Cesarano J (1998) A review of robocasting technology. MRS Proc 542:133. https://doi.org/10.1557/PROC-542-133
24. Liu H, Liu J, Leu MC, Landers R, Huang T (2013) Factors influencing paste extrusion pressure and liquid content of extrudate in freeze-form extrusion fabrication. Int J Adv Manuf Technol 67(1):899–906. https://doi.org/10.1007/s00170-012-4534-0
25. Scheithauer U, Schwarzer E, Richter H-J, Moritz T (2015) Thermoplastic 3D printing—an additive manufacturing method for producing dense ceramics. Int J Appl Ceram Technol 12(1):26–31. https://doi.org/10.1111/ijac.12306
26. Greulich M, Greul M, Pintat T (1995) Fast, functional prototypes via multiphase jet solidification. Rapid Prot J 1(1):20–25
27. Hoy CV, Barda A, Griffith M, Halloran JW (1998) Microfabrication of ceramics by co-extrusion. J Am Ceram Soc 81(1):152–158. https://doi.org/10.1111/j.1151-2916.1998.tb02307.x
28. Cesarano J, King B, Denham H (1998) Recent developments in robocasting of ceramics and multimaterial deposition
29. Lewis JA, Smay JE, Stuecker J, Cesarano J (2006) Direct ink writing of three-dimensional ceramic structures. J Am Ceram Soc 89(12):3599–3609. https://doi.org/10.1111/j.1551-2916.2006.01382.x
30. Jakus AE, Taylor SL, Geisendorfer NR, Dunand DC, Shah RN (2015) Metallic architectures from 3D-printed powder-based liquid inks. Adv Func Mater 25(45):6985–6995
31. Sun Y, Peng C, Wang X, Wang R, Yang J, Zhang D (2017) Rheological behavior of Al_2O_3 suspensions containing polyelectrolyte complexes for direct ink writing. Powder Technol 320:223–229
32. de Hazan Y, Thänert M, Trunec M, Misak J (2012) Robotic deposition of 3d nanocomposite and ceramic fiber architectures via UV curable colloidal inks. J Eur Ceram Soc 32(6):1187–1198. https://doi.org/10.1016/j.jeurceramsoc.2011.12.007
33. Tohver V, Morissette SL, Lewis JA, Tuttle BA, Voigt JA, Dimos DB (2002) Direct-write fabrication of zinc oxide varistors. J Am Ceram Soc 85(1):123–128
34. Li Q, Lewis JA (2003) Nanoparticle inks for directed assembly of three-dimensional periodic structures. Adv Mater 15(19):1639–1643. https://doi.org/10.1002/adma.200305413
35. Smay JE, Cesarano J, Lewis JA (2002) Colloidal inks for directed assembly of 3-D periodic structures. Langmuir 18(14):5429–5437
36. Tuttle BA, Smay JE, Cesarano J, Voigt JA, Scofield TW, Olson WR, Lewis JA (2001) Robocast $Pb(Zr0.95Ti0.05)O_3$ ceramic monoliths and composites. J Am Ceram Soc 84(4):872–874. https://doi.org/10.1111/j.1151-2916.2001.tb00756.x
37. Smay JE, Gratson GM, Shepherd RF, Cesarano J III, Lewis JA (2002) Directed colloidal assembly of 3D periodic structures. Adv Mater 14(18):1279–1283. https://doi.org/10.1002/1521-4095(20020916)14:18
38. Munch E, Franco J, Deville S, Hunger P, Saiz E, Tomsia AP (2008) Porous ceramic scaffolds with complex architectures. JOM 60(6):54–58. https://doi.org/10.1007/s11837-008-0072-5

39. Vadnere M, Amidon G, Lindenbaum S, Haslam J (1984) Thermodynamic studies on the gel-sol transition of some pluronic polyols. Int J Pharm 22:207–218
40. Jean J-H, Wang H-R (1998) Dispersion of aqueous barium titanate suspensions with ammonium salt of poly(methacrylic acid). J Am Ceram Soc 81(6):1589–1599. https://doi.org/10.1111/j.1151-2916.1998.tb02521.x
41. Prabhakaran K, Ananthakumar S, Pavithran C (2002) Preparation of extrudable alumina paste by coagulation of electrosterically stabilized aqueous slurries. J Eur Ceram Soc 22(2):153–158. https://doi.org/10.1016/S0955-2219(01)00249-7
42. Liu H, Li Y, Li D (2016) Research on rheological properties and extrusion behavior of aqueous alumina paste in paste-extrusion-based SFF processes. Int J Adv Manuf Technol 83(9):2039–2047. https://doi.org/10.1007/s00170-015-7720-z
43. Bourret J, El Younsi I, Bienia M, Smith A, Geffroy P-M, Marie J, Ono Y, Chartier T, Pateloup V (2018) Micro extrusion of innovative alumina pastes based on aqueous solvent and eco-friendly binder. J Eur Ceram Soc 38(7):2802–2807. https://doi.org/10.1016/j.jeurceramsoc.2018.02.018
44. Rowlands W (2019) Additive manufacturing of advanced ceramics for demanding applications. Loughborough University
45. Zhou W, Li D, Wang H (2010) A novel aqueous ceramic suspension for ceramic stereolithography. Rapid Prot J 16:29–35
46. AZoM MP (2019) Determining and understanding the yield stress of complex fluids. https://www.azom.com/article.aspx?ArticleID=12448
47. http://soft-matter.seas.harvard.edu/index.php/Rheological_behavior
48. Hellström LHO, Samaha MA, Wang KM, Smits AJ, Hultmark M (2014) Errors in parallel-plate and cone-plate rheometer measurements due to sample underfill. Meas Sci Technol 26(1):015301. https://doi.org/10.1088/0957-0233/26/1/015301
49. Zhang D, Jonhson W, Herng TS, Ang YQ, Yang L, Tan SC, Peng E, He H, Ding J (2020) A 3D-printing method of fabrication for metals, ceramics, and multi-materials using a universal self-curable technique for robocasting. Mater Horiz 7(4):1083–1090. https://doi.org/10.1039/C9MH01690B
50. Cesarano J (1998) A review of robocasting technology. MRS Online Proc Library Arch 542
51. Zocca A, Colombo P, Gomes CM, Günster J (2015) Additive manufacturing of ceramics: issues, potentialities, and opportunities. J Am Ceram Soc 98(7):1983–2001. https://doi.org/10.1111/jace.13700
52. Gao C, Rahaman MN, Gao Q, Teramoto A, Abe K (2013) Robotic deposition and in vitro characterization of 3D gelatin−bioactive glass hybrid scaffolds for biomedical applications. J Biomed Mater Res, Part A 101A(7):2027–2037. https://doi.org/10.1002/jbm.a.34496
53. Wang X, Sun Y, Peng C, Luo H, Wang R, Zhang D (2015) Transitional suspensions containing thermosensitive dispersant for three-dimensional printing. ACS Appl Mater Interfaces 7(47):26131–26136. https://doi.org/10.1021/acsami.5b07913
54. Nadkarni SS, Smay JE (2006) Concentrated barium titanate colloidal gels prepared by bridging flocculation for use in solid freeform fabrication. J Am Ceram Soc 89(1):96–103. https://doi.org/10.1111/j.1551-2916.2005.00646.x
55. Stuecker JN, Cesarano J, Hirschfeld DA (2003) Control of the viscous behavior of highly concentrated mullite suspensions for robocasting. J Mater Process Technol 142(2):318–325. https://doi.org/10.1016/S0924-0136(03)00586-7
56. Michna S, Wu W, Lewis JA (2005) Concentrated hydroxyapatite inks for direct-write assembly of 3-D periodic scaffolds. Biomaterials 26(28):5632–5639. https://doi.org/10.1016/j.biomaterials.2005.02.040
57. Lin K-F, He S, Song Y, Wang C-M, Gao Y, Li J-Q, Tang P, Wang Z, Bi L, Pei G-X (2016) Low-temperature additive manufacturing of biomimic three-dimensional hydroxyapatite/collagen scaffolds for bone regeneration. ACS Appl Mater Interfaces 8(11):6905–6916. https://doi.org/10.1021/acsami.6b00815
58. Maazouz Y, Montufar EB, Guillem-Marti J, Fleps I, Ohman C, Persson C, Ginebra MP (2014) Robocasting of biomimetic hydroxyapatite scaffolds using self-setting inks. J Mater Chem B 2(33):5378–5386. https://doi.org/10.1039/C4TB00438H

59. Saiz E, Gremillard L, Menendez G, Miranda P, Gryn K, Tomsia AP (2007) Preparation of porous hydroxyapatite scaffolds. Mater Sci Eng, C 27(3):546–550. https://doi.org/10.1016/j.msec.2006.05.038
60. Zhao S, Xiao W, Rahaman MN, O'Brien D, Seitz-Sampson JW, Sonny Bal B (2017) Robocasting of silicon nitride with controllable shape and architecture for biomedical applications. Int J Appl Ceram Technol 14(2):117–127. https://doi.org/10.1111/ijac.12633
61. Liao J, Chen H, Luo H, Wang X, Zhou K, Zhang D (2017) Direct ink writing of zirconia three-dimensional structures. J Mater Chem C 5(24):5867–5871. https://doi.org/10.1039/C7TC01545C
62. Grida I, Evans JRG (2003) Extrusion freeforming of ceramics through fine nozzles. J Eur Ceram Soc 23(5):629–635. https://doi.org/10.1016/S0955-2219(02)00163-2
63. Nguyen DT, Meyers C, Yee TD, Dudukovic NA, Destino JF, Zhu C, Duoss EB, Baumann TF, Suratwala T, Smay JE, Dylla-Spears R (2017) 3D-printed transparent glass. Adv Mater 29(26):1701181. https://doi.org/10.1002/adma.201701181
64. Cai K, Román-Manso B, Smay JE, Zhou J, Osendi MI, Belmonte M, Miranzo P (2012) Geometrically complex silicon carbide structures fabricated by robocasting. J Am Ceram Soc 95(8):2660–2666. https://doi.org/10.1111/j.1551-2916.2012.05276.x
65. Wei X, Nagarajan RS, Peng E, Xue J, Wang J, Ding J (2016) Fabrication of YBa_2Cu_3OY x (YBCO) superconductor bulk structures by extrusion freeforming. Ceram Int 42:15836–15842
66. Peng E, Wei X, Herng TS, Garbe U, Yu D, Ding J (2017) Ferrite-based soft and hard magnetic structures by extrusion free-forming. RSC Adv 7(43):27128–27138. https://doi.org/10.1039/C7RA03251J
67. Costakis WJ Jr, Rueschhoff LM, Diaz-Cano AI, Youngblood JP, Trice RW (2016) Additive manufacturing of boron carbide via continuous filament direct ink writing of aqueous ceramic suspensions. J Eur Ceram Soc 36(14):3249–3256
68. Fisher ML, Colic M, Rao MP, Lange FF (2001) Effect of silica nanoparticle size on the stability of alumina/silica suspensions. J Am Ceram Soc 84(4):713–718
69. Ghazanfari A, Li W, Leu M, Watts J, Hilmas G (2017) Mechanical characterization of parts produced by ceramic on-demand extrusion process. Int J Appl Ceram Technol 14(3):486–494
70. Maazouz Y, Montufar EB, Guillem-Marti J, Fleps I, Öhman C, Persson C, Ginebra MP (2014) Robocasting of biomimetic hydroxyapatite scaffolds using self-setting inks. J Mater Chem B 2(33):5378–5386. https://doi.org/10.1039/C4TB00438H
71. Smay JE, Nadkarni SS, Xu J (2007) Direct writing of dielectric ceramics and base metal electrodes. Int J Appl Ceram Technol 4(1):47–52. https://doi.org/10.1111/j.1744-7402.2007.02118.x
72. Franco J, Hunger P, Launey ME, Tomsia AP, Saiz E (2010) Direct write assembly of calcium phosphate scaffolds using a water-based hydrogel. Acta Biomater 6(1):218–228
73. Eqtesadi S, Motealleh A, Pajares A, Miranda P (2015) Effect of milling media on processing and performance of 13–93 bioactive glass scaffolds fabricated by robocasting. Ceram Int 41(1):1379–1389
74. Román-Manso B, De Pablos A, Belmonte M, Osendi M, Miranzo P (2014) Microstructural designs of spark-plasma sintered silicon carbide ceramic scaffolds. Bol Soc Esp Ceram Vidr 53(2):93–100
75. Fu Q, Saiz E, Tomsia AP (2011) Direct ink writing of highly porous and strong glass scaffolds for load-bearing bone defects repair and regeneration. Acta Biomater 7(10):3547–3554
76. Stuecker JN, Cesarano Iii J, Hirschfeld DA (2003) Control of the viscous behavior of highly concentrated mullite suspensions for robocasting. J Mater Process Technol 142(2):318–325
77. Miranda P, Pajares A, Saiz E, Tomsia AP, Guiberteau F (2007) Fracture modes under uniaxial compression in hydroxyapatite scaffolds fabricated by robocasting. J Biomed Mater Res Part A 83A(3):646–655. https://doi.org/10.1002/jbm.a.31272
78. Zhu C, Han TY-J, Duoss EB, Golobic AM, Kuntz JD, Spadaccini CM, Worsley MA (2015) Highly compressible 3D periodic graphene aerogel microlattices. Nat Commun 6(1):6962. https://doi.org/10.1038/ncomms7962

79. Díaz-Marta AS, Tubío CR, Carbajales C, Fernández C, Escalante L, Sotelo E, Guitián F, Barrio VL, Gil A, Coelho A (2018) Three-dimensional printing in catalysis: combining 3D heterogeneous copper and palladium catalysts for multicatalytic multicomponent reactions. ACS Catal 8(1):392–404. https://doi.org/10.1021/acscatal.7b02592
80. Mehla S, Das J, Jampaiah D, Periasamy S, Nafady A, Bhargava SK (2019) Recent advances in preparation methods for catalytic thin films and coatings. Catal Sci Technol 9:3582–3602. https://doi.org/10.1039/C9CY00518H
81. Tubío CR, Azuaje J, Escalante L, Coelho A, Guitián F, Sotelo E, Gil A (2016) 3D printing of a heterogeneous copper-based catalyst. J Catal 334:110–115. https://doi.org/10.1016/j.jcat.2015.11.019
82. Sun K, Wei T-S, Ahn BY, Seo JY, Dillon SJ, Lewis JA (2013) 3D printing of interdigitated li-ion microbattery architectures. Adv Mater 25(33):4539–4543. https://doi.org/10.1002/adma.201301036
83. Rocha VG, García-Tuñón E, Botas C, Markoulidis F, Feilden E, D'Elia E, Ni N, Shaffer M, Saiz E (2017) Multimaterial 3D printing of graphene-based electrodes for electrochemical energy storage using thermoresponsive inks. ACS Appl Mater Interfaces 9(42):37136–37145. https://doi.org/10.1021/acsami.7b10285
84. Kokkinis D, Schaffner M, Studart AR (2015) Multimaterial magnetically assisted 3D printing of composite materials. Nat Commun 6(1):8643. https://doi.org/10.1038/ncomms9643
85. Cesarano J (1998) a review of robocasting technology. MRS Online Proc Libr 542(1):133–139. https://doi.org/10.1557/PROC-542-133
86. Tuttle BA, Smay JE, Cesarano J, Voigt JA, Scofield TW, Olson WR, Lewis JA (2001) Robocast Pb (Zr0. 95Ti0. 05) O3 ceramic monoliths and composites. J Am Ceramic Soc 84(4):872–874
87. Smay JE, Nadkarni SS, Xu JJ (2007) Direct writing of dielectric ceramics and base metal electrodes. Appl Ceramic Technol 4 (1):47–52
88. Ahn BY, Shoji D, Hansen CJ, Hong E, Dunand DC, Lewis JA (2010) Printed origami structures. Adv Mater 22(20):2251–2254. https://doi.org/10.1002/adma.200904232
89. Taylor SL, Jakus AE, Shah RN, Dunand DC (2017) Iron and nickel cellular structures by sintering of 3D-printed oxide or metallic particle inks. Adv Eng Mater 19(11):n/a. https://doi.org/10.1002/adem.201600365
90. de Hazan Y, Thänert M, Trunec M, Misak JJ (2012) Robotic deposition of 3d nanocomposite and ceramic fiber architectures via UV curable colloidal inks. J Eur Ceramic Soc 32(6):1187–1198

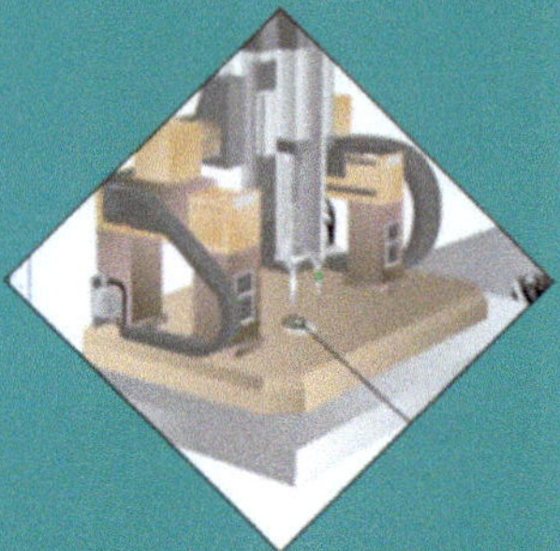

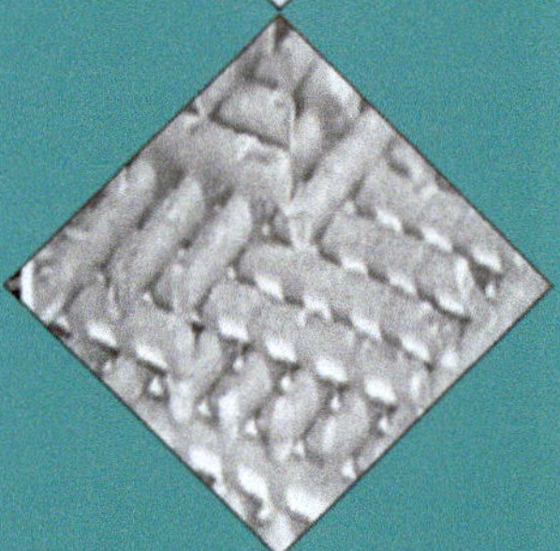

Robocasting—Printing Ceramics into Functional Materials

Extrusion of ceramic paste, robocasting, post-processing of green ceramic, multi-ceramic materials printing

Uzma Malik, PR. Selvakannan, Maciej Mazur, Yongxiang Li, Suresh K. Bhargava

5

Chapter 6
Surface Modification of Additively Manufactured Materials: Adding Functionality as Fourth Dimension

Roxanne Hubesch, Uzma Malik, PR. Selvakannan, Lakshmi Kantam Mannepalli, and Suresh K. Bhargava

Contents

Abstract The emergence of additive manufacturing (AM) as an industry standard has brought unprecedented opportunity to many fields. Some examples of areas that particularly benefit from this technology include biological implantation, biomedicine, aerospace, advanced materials, automotive, tooling, and many chemical industries. This vast and growing sphere of application has prompted the need for specialised technology that enables tailor-made products. Nonetheless, the majority of printable materials currently available on the market are restricted to proprietary polymers, metal powders and ceramics, which are quickly proving to be inadequate for many applications. Typical AM processes build monofunctional structures with single materials. Hence, post-processing and surface chemical modification of AM printed materials are critical in providing a range of new functional and diverse chemical functionalities. Surface functionalization of 3D printed materials is quickly gaining significant traction as a sub-discipline, as this approach can address the technical challenges associated with optimizing the performance of AM end-user

R. Hubesch · U. Malik · PR. Selvakannan · S. K. Bhargava (✉)
Centre for Advanced Materials and Industrial Chemistry, RMIT University, Melbourne, VIC, Australia
e-mail: suresh.bhargava@rmit.edu.au

L. K. Mannepalli
Department of Chemical Engineering, Institute of Chemical Technology, Matunga, Mumbai, India

S. K. Bhargava et al. (eds.), *Additive Manufacturing for Chemical Sciences and Engineering*, https://doi.org/10.1007/978-981-19-2293-0_6

products. This chapter presents an overview of the novel post-physical treatments and surface chemical functionalization required to customize AM printed materials for a wide variety of applications. This chapter further gives indication as to the challenges associated with this field of study, including the status of research and the future avenues that may be able to exploit the true potential of this technology.

Keywords Additive manufacturing · Surface functionalization · Post-processing of AM printed materials

Abbreviations

3D	Three-dimensional
AM	Additive Manufacturing
ATRP	Atom Transfer Radical Polymerization
BMP	Bone Morphogenic Protein
DGEA	Asparagine-glycineglutamate-alanine
DMLS	Direct Metal Laser Sintering
DSC	Differential Scanning Calorimetry
EBM	Electron Beam Melting
EDM	Electrical Discharge Machining
ESD	Electrospray Deposition
FDM	Fused Deposition Modelling
GFOGER	glycine-phenylalanine-hydroxyprolineglycine-glutamate-arginine
HDI	Hexamethylenediisocyanate
HIP	Hot Isostatic Pressing
NP	Nanoparticles
PEM	Polymer Electrolyte Membrane
PEO	Plasma Electrolytic Oxidation
PLA	Polylactic acid
RAFT	Reversible Addition—Fragmentation chain-transfer
RC	Robocasting
SEM	Scanning Electron Microscopy
SI-ATRP	Surface-initiated Atom Transfer Radical Polymerization
SLA	Stereolithography
SLED	Self-limiting Electrospray Deposition
SLM	Selective Laser Melting
UNSM	Ultrasonic Nanocrystalline Surface Modification
UV	Ultraviolet
XPS	X-Ray Photoelectron Spectroscopy

6.1 Background

This unique capability of transforming polymers, metals, and ceramics into functional devices of having complex geometry using additive manufacturing (AM) technologies has stimulated significant interest in various fields including biology [1–3], engineering [4, 5] and materials science. The class of AM techniques including SLM, FDM, RC and SLA have demonstrated their capability in prototyping functionalized chemical catalytic reactors [6], flow devices [7], sensor devices [8], thermoelectric materials [9, 10], mechanochromic materials [11], energy materials [12–14], and biological implants with complex geometries [15–19]. The design challenges in fabricating materials into prototypes using conventional manufacturing practices can be addressed using the appropriate AM technologies and materials. Despite the advantages in making complex designs with unique characteristics, AM practices need to overcome numerous challenges including the availability of limited choice of printable materials, structural anisotropy, spatial inhomogeneity, surface roughness, low resolution, and limited production speed. There are also several improvements that can be made to this technology that will broaden its scope of use, inducing adding new functional dimensions to improve device function and tailoring the microstructures of products.

Specifically, there are number of technical challenges in integrating AM technologies into conventional chemical/chemical engineering applications. The limited choice of available materials currently used for AM processes limits the number of potential applications due to lack of robustness to temperature, pressure, pH, and reactive atmospheres [6]. The first challenge is the choice of material of construction, which is dictated by the operating conditions of the specific chemical processes, as per Fig. 6.1.

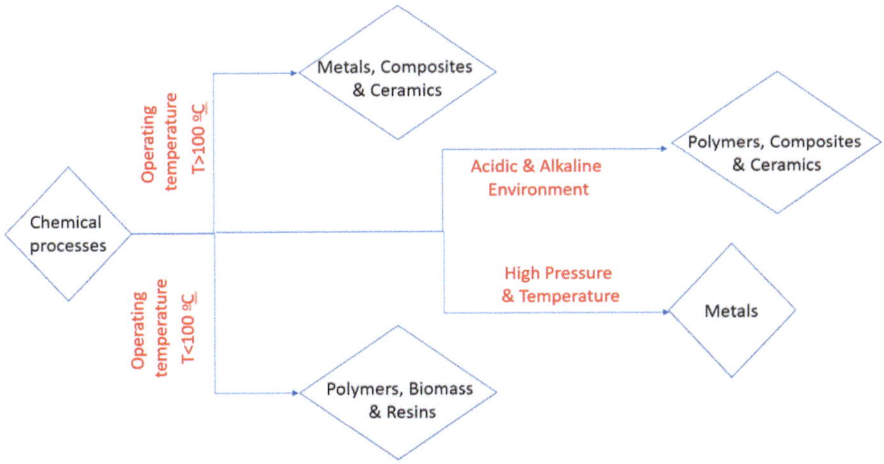

Fig. 6.1 Choice of materials used for AM based on the operating conditions

The limited variety of currently available materials that are workable with AM technology are displayed above, see Table 6.1. The printed materials that are produced are chemically inert hence do not have any chemical function unless their surfaces are modified to achieve specific functionality. Surface chemical functionalization of the AM printed materials is the promising approach in developing 3D printed structures for tailor-made applications. However, surface functionalization of these materials is highly challenging due to their chemical inertness, surface roughness and design dictated non-uniform functionalization. Therefore, the development of post-processing physical methods such as surface functionalization using physical and chemical methods is critical to render the specific functionalization on the AM printed materials surface. Various kinds of post-processing techniques and surface chemical modifications are required to alter printed materials to withstand specific chemical applications that may involve harsh, corrosive, and non-ambient process conditions. As per Fig. 6.1, a range of various conditions alongside the general material(s) required to withstand them are outlined. This chapter presents the challenges involved in existing surface modification and post-processing techniques as well as exploring the various chemical modifications currently used for functionalization, as described in literature.

6.2 Challenges in Functionalization of AM Printed Surfaces

Rendering any chemical functionality to the surface of a printed structure depends upon the nature of the printed material, surface composition, surface roughness and its micro-architecture. As presented in Table 6.1, each AM technique has its own unique printing resolution and surface roughness. These two parameters also depend on the printable material and its numerous characteristics. Therefore, various kinds

Table 6.1 Various AM technologies, materials used for printing, ability to modify their surfaces

Material type	AM Technique	Surface Chemical Modification	
		Pre-printing	Post-printing
Metal and polymer powder	Selective laser melting and selective laser sintering	No	Yes
Photopolymer Resin	Stereolithography	Yes	Yes
Polymer Resin	Two-photon	Yes	Yes
Polymer and metal powder blended with binder ink	3D powder printing	Yes	Yes
Hydrogel ink	Direct Ink writing/Robocasting	Yes	Yes
Thermoplastic	Fused Deposition modelling	Yes	Yes
Ink	Inkjet printing	Yes	Yes

of functionalization strategies must be developed to obtain uniform functionalization on the entire surface of the material, which eventually provides the desired chemical functionality for end-user purposes. These materials have a wide range of chemical and chemical engineering applications. Surface functionalization of 3D printed materials works to improve the porosity, mass transport properties, mechanical properties, and biocompatibility for use in biological applications. However, the following attributes that these AM based materials exhibit, provide challenges for chemical functionalization.

Surface roughness

SLM technique produce metal surfaces that are considerably rougher than those generated by subtractive manufacturing techniques. Figure 6.2 shows SEM images of SLM printed Inconel materials, indicating the presence of micron scale metal structures on its surface. It has been demonstrated that rough surfaces are too difficult to be functionalized uniformly. Therefore, any kind of post-processing polishing methods need to be done to reduce the surface roughness. Else, the chemical functionalization on these surfaces will lead to formation of dis-continuous and non-uniform surface functionalization.

Many of the post-processing strategies involving polishing and etching generally lead to changes in the microstructures of surfaces without affecting chemical composition.

Chemical anisotropy

These materials are anisotropic in nature due to their layer-by-layer deposition; they undergo major structural changes at higher temperatures in different gas atmospheres. For example, SLM printed titanium substrates were heated in an air atmosphere at 500 °C, leading to a colour change as shown in Fig. 6.3. The titanium metal

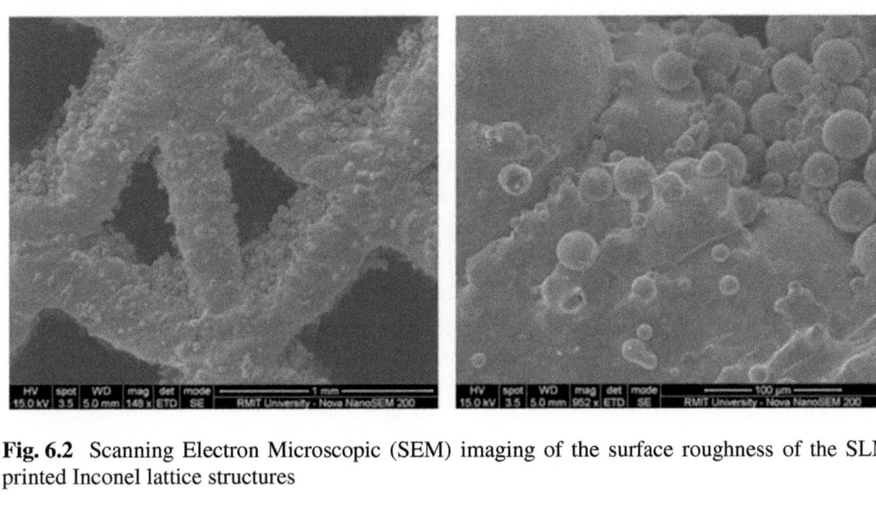

Fig. 6.2 Scanning Electron Microscopic (SEM) imaging of the surface roughness of the SLM printed Inconel lattice structures

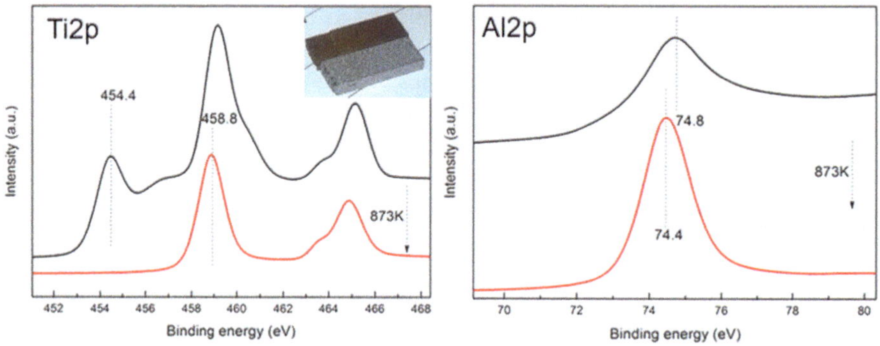

Fig. 6.3 XPS analysis of Ti 2p and Al2p core level collected from TiAlV printed using SLM and after its calcination at 500 °C

powder used for the printing purposes, consisted of titanium, aluminium, and vanadium (TiAlV). X-Ray photoelectron spectroscopy (XPS) investigation of the printed structures before and after heating, revealed that aluminium atoms migrated to the surface to form alumina, which exhibit metallic lustre due to the incorporation of titanium ions in the alumina framework.

Structural anisotropy

Commercially available 3D printers utilize thermoplastic polylactic acid (PLA) to print structures and this technique is classified under either FDM or fused filament fabrication. PLA exists in semi-crystalline and amorphous forms; therefore, the functionality of printed PLA-based structures depends on which form is present. A recent study on PLA printed materials explored how the electrical properties were dependent upon its structure. This study used differential scanning calorimetry (DSC) and dielectric measurements to show that the dielectric properties were found to be from the amorphous domains of as-printed PLA. Figure 6.4 clearly shows the DSC results of the as-printed PLA structure was found to be like the amorphous phase of PLA.

Chemical Inertness

Most printing techniques involve the formation of structures at higher temperatures by melt-processing, UV-crosslinking, and fusing of metal powders. Therefore, these structures are chemically inert and need to be activated prior to any functionalization. Physical or chemical vapor deposition, self-assembly, surface grafting and self-assembly processes are the most common surface functionalization strategies, hence, a chemical activation, adlayer formation or high temperature post-processing needs to be achieved prior, in order to modify the chemical composition of the surface.

Hydrophilicity, Hydrophobicity and Wettability

3D printed metal and polymer structures are widely used in bio-medical sciences, therefore it is important to control the wettability of the surface by tuning the balance between hydrophilicity and hydrophobicity [20]. Most printed objects have rough

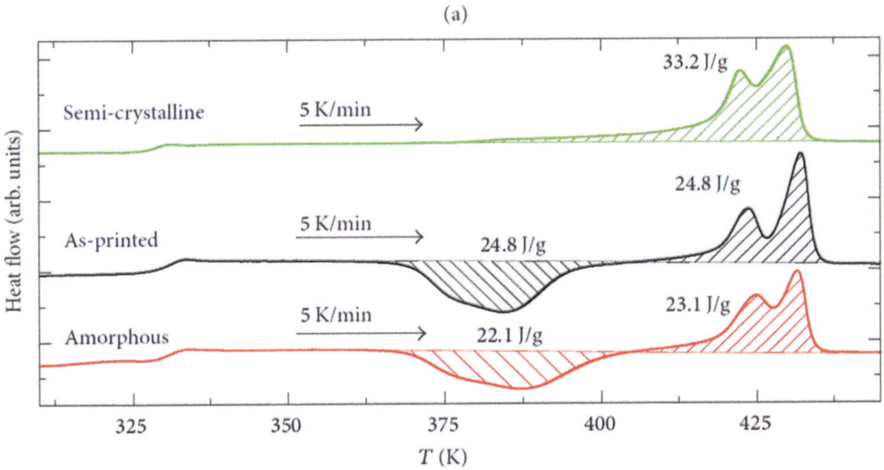

Fig. 6.4 DSC patterns of the Amorphous, semi-crystalline and as printed PLA structures (Reference from Advances in material science engineering 2017, Article ID 6,913,835)

surfaces and require appropriate functionalization strategies either in converting them into hydrophilic surfaces for biomedical applications, cellular growth or converting them into hydrophobic surfaces for self-cleaning application.

6.3 Post-processing Physical Techniques

Many of the AM printed structures are anchored on the base build plate which needs to be removed for post-processing, prior to its application. As discussed in the previous section, surface roughness, chemical anisotropy, structural anisotropy, and inertness render the surface of the printed materials resistant against further chemical functionalization. Therefore, the following post-processing physical techniques are currently used, to enable surface chemical functionalization to be successfully achieved.

Stress relief

The printing process involves the layer-by-layer approach, which may cause thermal stress due to many cycles of heating and cooling. This thermal stress needs to be relieved prior to the removal of the printed component from the build plate. Failure to do this may cause the printed object to become warped or cracked. Stress-relieving of the printed part can be done by heating the printed part in an oven or furnace under inert atmosphere to minimize the oxidation of metals present on the surface.

Heat treatment

The microstructure of the printed materials and mechanical properties of the whole object of the materials can be improved using high temperature treatment under a controlled atmosphere. The rate of heating and cooling must be controlled to optimize the mechanical properties and microstructure.

Powder removal

In powder-bed fusion system, the printed parts are generally covered in metal powder, as the printing approach involves formation of new layers added to the top. After completion of the printing cycle, the powder covering the printed structure needs to be removed and recycled for later use.

Part removal

Electrical discharge machining (EDM) is another post-processing step, which is used to remove the printed parts from the build part.

Hot isostatic pressing

To improve the fatigue life of the cast parts printed using AM techniques, hot isostatic pressing (HIP) is employed, as widely recommended by casting industries.

Surface treatments

This treatment provides the required surface finishing to the printed part to reduce the surface roughness, clean up the inner channels, and ultimately improve surface finish quality.

Machining

The dimensional accuracy of the printed component can be optimized by machining the surfaces, interfaces, support structures, and channels.

Ultrasonic treatment

As shown in Fig. 6.2, AM printed metal structures generally have poor surface finish, which lowers resistance against wear, fatigue and corrosion, restricting their application. Impressively, Chang et al. have developed an ultrasonic nanocrystal surface modification that uses an ultrasonic device attached to a tungsten carbide ball that scans the printed metal structure's rough surface with the high frequency [21]. This high frequency causes the rough metal surfaces to undergo plastic deformation, thereby resulting in smoothing. Figure 6.5 shows the experimental setup used for the surface modification processes by an ultrasonic nanocrystal and demonstrates the change in surface morphology before and after treatment.

Laser based post-processing techniques

To ensure the printed parts have the required physical attributes and quality, many of the laser based post-processing techniques are currently practiced at different stages and Fig. 6.6 provides the schematic of various processes.

The SLM technique discussed in Chap. 4, generally prints significantly rough metal structures due to the layer-by-layer approach. Strong adherence of metal powder grains onto the printed metal surface is known to create a staircase effect along the direction of the printing axis. Therefore, the post-physical processing techniques including mechanical polishing, sandblasting and traditional machining of

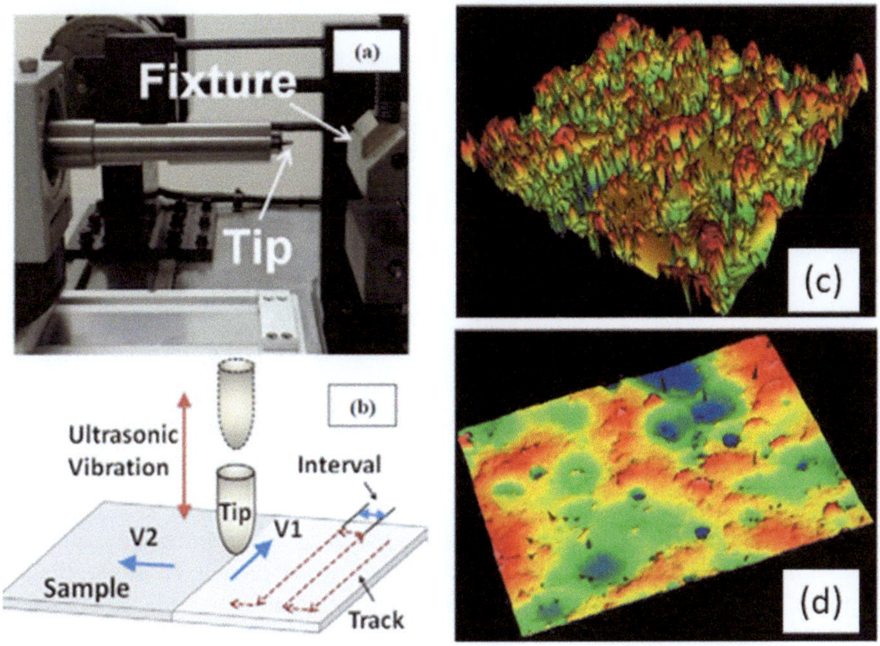

Fig. 6.5 **a** Image of the ultrasonic nanocrystal surface modification (UNSM) unit; **b** UNSM process overview **c** 3D imaging of the untreated sample's surface morphology **d** UNSM treated sample (Reprinted with permission from Ref. [21] Elsevier)

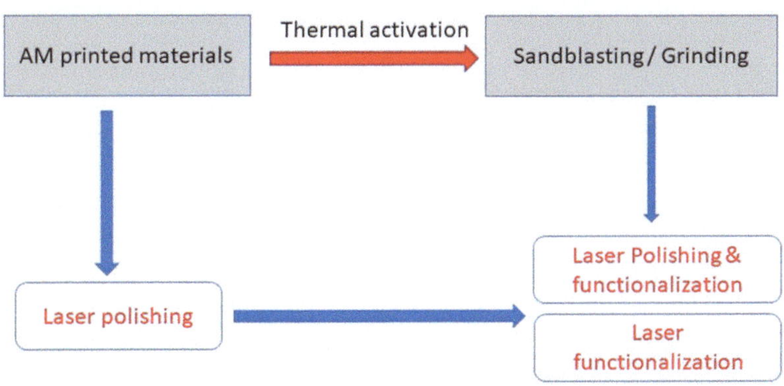

Fig. 6.6 Schematic description of laser based post-processing techniques

printed materials in combination with heat treatment are frequently employed used to smoothen the rough surface as an approach for microstructural modifications. Laser polishing of SLM printed materials have also been shown as an effective physical process for the reduction of surface roughness. This approach involves the laser induced melting of a thin layer of those metal grains resulting in metal uniformity on the surface due to liquified level settling, made possible by its surface tension. Thus, laser polishing is considered as a potential alternative to polishing and mechanical grinding, as there are no residual mechanical stresses in addition to its short processing time. Especially, the advent of ultrafast lasers for the polishing process significantly reduces the laser-material interaction interval, which eventually leads to much lower thermal modification of the printed materials. Moreover, these laser-mediated surface topologies are known to modify the functional properties of printed parts such as surface wettability and compatibility. These two-step laser post-processing methods were demonstrated to modify the printed parts of metal powder identically to that of printed parts made using different metal powder compositions.

6.4 Surface Chemical Functionalization of AM Printed Surfaces

Applications of AM in chemical sciences and chemical engineering are expanding rapidly, mainly driven by the application directed AM approach. However, limited emphasis has been provided on the basic understanding related to surface functionalization of printed materials, as it is important to interface the printed materials for end-user applications. Modelling and simulation of specific processes involving printed materials are required to realize the full potential of AM in chemical processes and technologies. Without further focus and advanced understanding in these areas, this exciting AM technology would print the components of dimensional complexity but with inferior chemical and functional properties. Therefore, the following sections provide detailed description of the various functionalization strategies, their effects on the surface composition and microstructure and the resultant functionality for the desired chemical function.

Surface functionalization of tailor made 3D printed objects contribute significantly to the process of technological development and have the potential to be used in various industrial applications. There are four major strategies of surface functionalization of the AM printed structures, as shown in Fig. 6.7. These include chemical impregnation, electrochemical/electroless plating procedures, physical/chemical vapor deposition and surface initiated polymerization. All of these currently employed functionalization strategies can be used on their own, in in conjunction with one another. Generally, these functionalization strategies refer to the processes that are compatible with the surface of the printed material of construction, without compromising its characteristic mechanical and physical properties. Apart from the post-functionalization strategies, there are also advanced printing process that can use composite materials, which provide a direct approach to produce printed materials. However, this direct functionalization approach does not currently have a broad scope, most likely due to compatibility issues of these composite materials with all of the AM technologies, limited availability of composite materials and the heterogeneity of the printed surfaces.

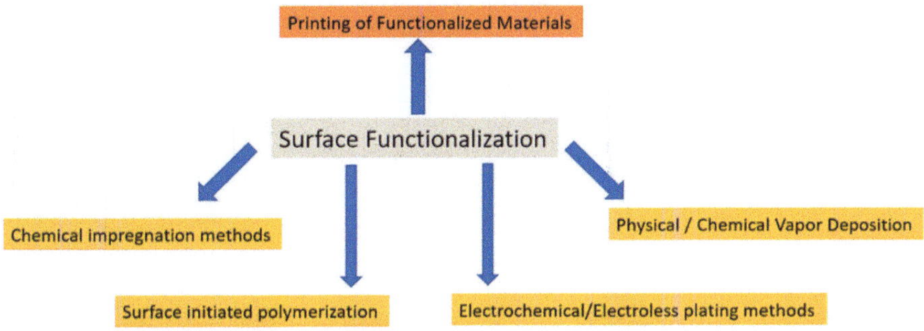

Fig. 6.7 Schematic description of laser based post-processing techniques

6.4.1 Direct Surface Functionalization

Printing of functionalized materials using AM techniques involves the suitable preparation of feed polymer or resins consisting of pure polymer, or resin and the additive in the case of polymer printing processes. In robocasting, a suitable ink formation needs to be prepared, to prepare the functionalized ceramic materials, while direct printing of metals along with additives are not very common. The fused deposition modelling (FDM) technique has been demonstrated as a simple, low-cost extrusion technique to print directly functionalized materials. Li et al. have developed an efficient process of blending multi-walled carbon nanotubes and thermo electric material BST ($Bi_{0.5}Sb_{1.5}Te_3$) with PLA [22], as shown in Fig. 6.8. This composite material was printed as wires, which have excellent mechanical and thermoelectric properties. Robocasting is a highly promising AM technology for the direct printing of ceramic structures that exhibit catalytic activity. Albeit the mechanical strength of these materials was found to be less than materials synthesized through other existing methods. Prompted by this fact, Yu et al. further developed a novel ink formulation that consists of halloysite nanotubes and zeolites. The printed zeolite monolithic strictures as shown in Fig. 6.8 was found to have excellent mechanical strength and catalytic properties [23]. Robocasting of these functionalized ceramic materials results in a product that has the general morphology of stacked ceramic fibres oriented vertically to each other and every subsequent layer [24–27]. The common configurations followed in ceramic printing are the 1–1 [24–26] and 1–3 [24, 26, 27] woodpile configurations. In contrast, few other works reported the '1.3.5' configuration [25], a less common technique. These kinds of catalytically functionalized structures can be prepared using robocasting and the printed structures were known to retain the textural properties of the catalyst materials including surface area and porosity over long periods of time [24–27].

Several ceramic compositions are difficult to print directly using robocasting, one of which is manganese hexaaluminate ($BaMn_2Al_{10}O_{19-\alpha}$). Therefore, Stuecker et al. [27] used a multi-step approach that involves the co-precipitation of Al, Ba and Mn

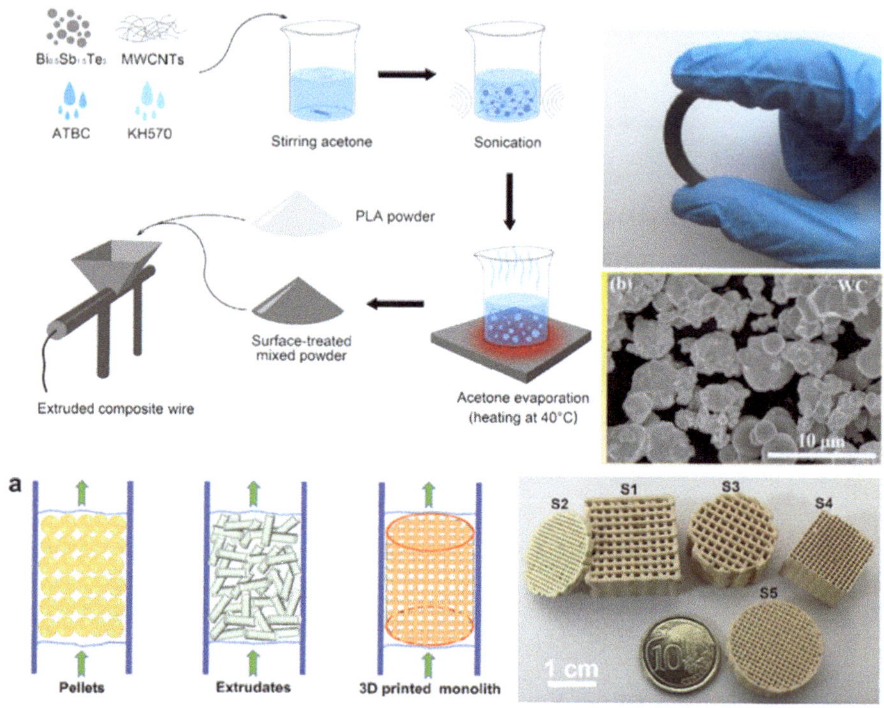

Fig. 6.8 **a** Printing of BST and CNT modified PLA composite materials reprinted with permission @Elsevier 2018 [22]; **b** Direct printing of binder free zeolites, reprinted with permission @Wiley 2019 [23]

ions, subsequent dispersion of ammonium polyacrylate and calcination of the resultant materials at 1300 °C. The same group made barium manganese hexaaluminate wash coated alumina. The former was found to be a very active catalyst compared to the wash coated catalyst, hence the direct functionalization of these kinds of ceramic materials can be effectively synthesised using the described approach. Catalyst activity was measured as a function of temperature and it clearly demonstrated that the direct printing approach retains the catalyst function, even after the material was made at very high temperature. Kotz et al. has developed a strategy to print a fused silica functionalized structure. They utilized stereo lithographically to print a mixture containing UV curable monomer and silica particles. This printing process generated the structured polymer composite consisting of Silica particles dispersed on a photopolymer, as presented in Fig. 6.9 [28]. As this composited sintered at 1300 °C, functionalized fused silica glass was formed.

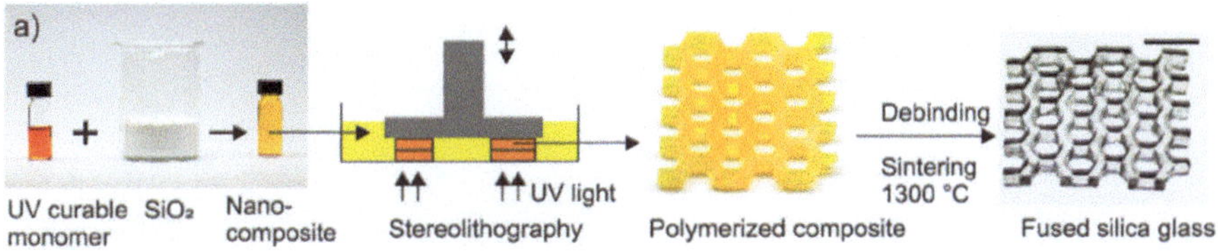

Fig. 6.9 Printing of composite materials consist of PLA, reprinted with permission from Nature Publishing Group @ 2017 [28]

6.4.2 Electrochemical Functionalization

Electrochemical deposition procedures are mainly used for depositing metal structures on the surface of 3D printed plastics, polymers, and ceramics. Selective laser-melting-based-printing of metals has several limitations, including the availability of few metal powder compositions with specific textural properties that are suitable to be printed using SLM. Moreover, printing precious metals including Pt, Au, Pd, and Rh are expensive and require large quantities. All of these metals exhibit catalytic, optical, and sensing properties. Functionalization of 3D printed objects including polymers and ceramics have received significant attention because the deposited metal layers can enhance their mechanical strength, electrical conductivity, thermal resistance, and biocompatibility. Various approaches have been reported for the functionalization 3D metal printed objects including the direct deposition of metal nanostructures, chemical vapor deposition of organometallic precursor and physical vapor deposition [29]. However, these methods are limited by non-uniform coating, deposition in small regions and difficult to control deposition. Metals can be further electrodeposited or electroless plated on the surface of a printed object to obtain metal functionalized ceramics and polymers [33]. Combination of 3D printed polymers or ceramics with the functional properties of a metal surface is a potential approach for the fabrication of functional microelectronic devices. Therefore, this electrochemical or electroless deposition of metals on a printed polymer and ceramics is a unique functionalization strategy, which leads to the formation of custom made electronic circuitry on the surface of various substrates as shown in Fig. 6.10. Electrochemical deposition, electroless plating and electrospray deposition are the major electrochemical methods currently used for the surface functionalization of 3D printed objects. The electroless deposition method has the added advantage of depositing metals even on a non-conducting substrate, therefore this method can be used very well for the non-conducting "printed" materials including polymers and ceramics. In this approach, the printed object needs to be immersed in a solution containing the metal ions, a reductant and complexing agent. A redox reaction between the metal ions and the reductant leads to the reduction of metal ions and causes the formation of metallic film on the surface of a 3D printed object. Pip et al. used this electroless plating approach to deposit bimetallic Ni–Fe films on to the 3D microstructures printed

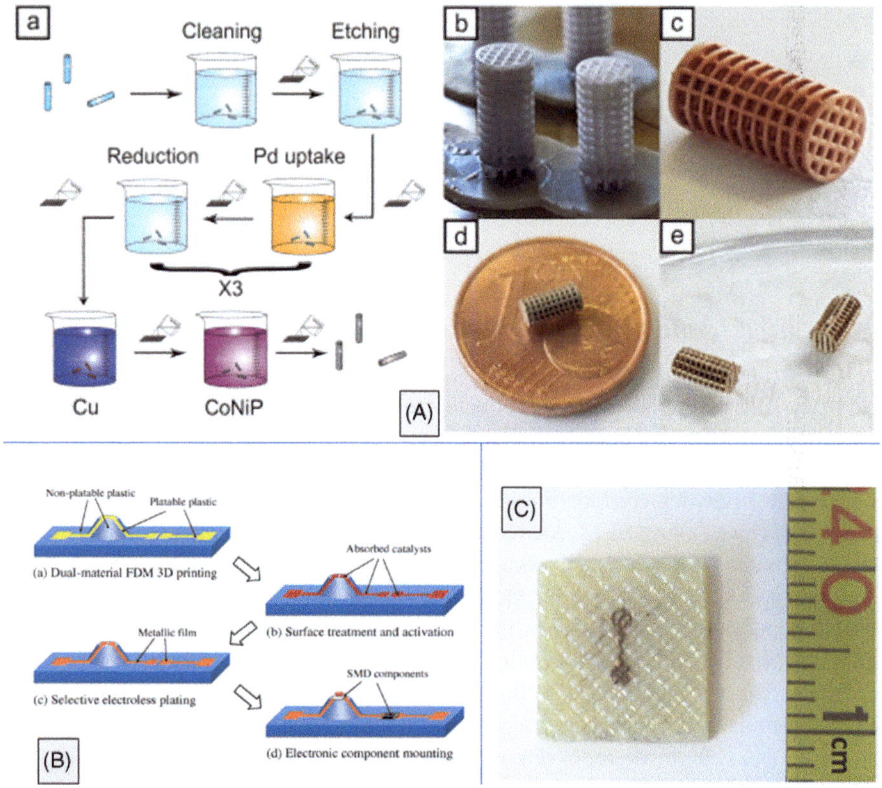

Fig. 6.10 (**A**) Electroless deposition of transition metals (**a**), images of the printed objects (**b**) Surface fucntionalization using metals such as Cu (**c**) CoNi (**d**); Au NP [30] (**B**) Selective electroless deposition of metals on the surface of FDM printed materials (Reprinted with permission from Ref. [31] (**C**) Selective deposition of Copper metal on 3D printed sample [32]

using lithographic techniques. These bimetallic surfaces render the printed object to exhibit nanomagnetism [34, 35]. Kim et al. developed a strategy to functionalize the FDM printed PLA structures using copper.

Electroless plating of copper on the printed polymer structures was used to fabricate toroidal shaped inductors [36]. Bernasconi et al. developed a modified approach to electroless metallization to produce functional 3D printed micro devices [37, 38]. Figure 6.10A shows the functional devices prepared using the aforementioned approach. Stereolithographic printed structures were modified by sequentially deposited metal layers using electroless printing and demonstrated the uniform deposition of copper, cobalt, and gold nanoparticles, as illustrated in Fig. 6.10A. Co nanoparticle functionalization renders 3D printed devices to be precisely magnetically actuated, which enables for efficient control over its motion. Au nanoparticle functionalization renders the 3D printed structures biocompatible, which allows for the growth of cells. Ji Li et al. developed free-form circuitry on 3D printed plastic structures using electroless plating method. This approach integrates the advantage of dual material based fused deposition modelling. Jose Marques-Hueso et al. used FDM printed PEI filaments as substrate on which silver nanostructures were photo printed [32]. This approach involves the initial deposition of a photosensitive AgCl layer, followed by the photo-reduction of AgCl into silver nanoparticles to form conductive tracks. These silver nanoparticles act as a seed layer for the electroless deposition of copper, resulting in the selective formation of conductive tracks on the 3D printed structures, as shown in Fig. 6.10C. This kind of selective patterning of metal structures has the potential to be applied in the microelectronics industry. Complex architectures printed using 3D printing can also be functionalized with metal nanostructures using this approach. Wang et al. demonstrated that 3D printed Eiffel tower structures can be fully functionalized using Ni nanoparticles [40].

Electrochemical deposition of metals is another approach to functionalize the 3D printed objects surface using metals. In contrast to the electroless plating, this electrochemical deposition process requires conductive substrates. A wide variety of materials can be deposited on any 3D printed electrochemical object through this method [41–43]. Ambrosi et al. used the electrochemical deposition process to transform a cell printed using metal and polymer in a self-contained electrochemical system [44], as shown in Fig. 6.11. This consists of electrodes with enhanced electrocatalytic properties. A stainless-steel built cell was printed using the selective laser melting process and this material had poor electrocatalytic properties. Using electrochemical deposition methods, Ni, Pt and iridium oxide (IrO_2) materials were deposited as thin layer. Surface functionalization of the SLM printed electrode was carried out using cyclic voltammetry to deposit these electrocatalysts to improve their electrocatalytic behaviour [45]. In this approach, aqueous solution of IrO_2 precursor of ($IrCl_4.H_2O$) was mixed aqueous solution of oxalic acid. To this mixture, hydrogen peroxide solution was added dropwise, as an oxidant and the pH of this solution was adjusted to alkaline (10.5) by adding aqueous solution of weak base of $Na_2(CO_3)$.

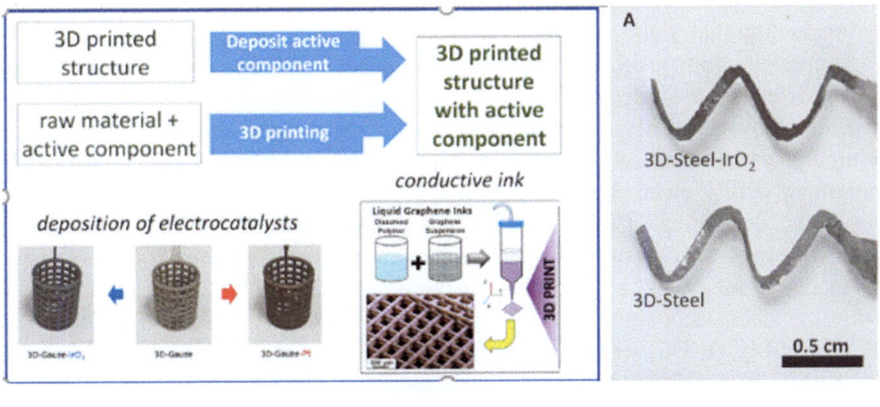

Fig. 6.11 Electrochemical deposition of metals and IrO_2 deposited metal structures [33]

The SLM printed SS electrode was immersed in this deposition bath and electrochemical potential was scanned over a range for the efficient deposition of IrO_2 on the surface of the SS electrode, as shown in Fig. 6.11.

Integrating 3D printing techniques with electrodeposition provides a unique approach in designing electrodes and electrochemical cells, which were shown as efficient catalysts towards splitting of water to produce hydrogen. Shah et al. blended polymer structures with conductive graphene materials, with the resultant materials composites having the potential to be electrodeposited.

Electrospray deposition (ESD) or electrostatic spray deposition is another electrochemical approach of functionalizing 3D printed object. This process is similar to the spray coating process of microdroplets, which are formed by atomizing the solution into charged droplets by applying high electric potential, as illustrated in Fig. 6.12a. As these are similar charged microdroplets, they don't agglomerate due to their electrochemical repulsion. These charged droplets will be passed on the target 3D printed object, which results in the deposition of solids dispersed in the microdroplets. Significant efforts were developed to prevent charge dissipation as well as optimizing the thickness of the deposited film. The major advantage of this process is its capability to conformally deposit the solids onto the surface of printed objects, without making much changes in the experimental setup including the orientation of the object and relocation of the spray needle [46]. Using this technique, films of thickness ranging from 1 nm to 50 μm can be deposited on any kind of printed objects that are having conductivity. Chun Huang et al. used the electrospray deposition of Al particles on PVDF film [39], with very high metal loadings, as shown in Fig. 6.12a. Combination of high metal loadings with controlled thickness demonstrated by this technique, has the potential to 3D print rocket motors. Kovacevich et al. investigated the influence of target geometry in controlling the thickness of the deposited layer using the electrospray deposition method. A printed hydrogel was functionalized using this technique, as shown in Fig. 6.12b. The hydrophobic and antifouling properties of these materials were studied, and it was concluded that electrospray deposition can be a potential way of functionalizing 3D printed objects by conducting materials.

Simple electrodeposition of metals on 3D printed polymer substrates were also shown as effective strategies for the fabrication of compact electrochemical reactors. Lee Chronin et al. used this approach for the metal functionalization of 3D printed flow plates and demonstrated its application to manufacture proton exchange membrane (PEM) electrolyser, as shown in Fig. 6.13a [47]. Fabrication of such flow plates and deposition of expensive Pt catalysts to fabricate the PEM electrolyser for the electrolysis of water using the conventional manufacturing is an expensive approach. In this work, a 3D printed polypropylene flow plate was functionalized using the electrodeposition of silver and this setup was shown as flow plates of a PEM electrolyser. This setup was considered an economic and adaptable strategy for the manufacture of PEM devices. A recent work extended this approach to fabricate the membrane free electrochemical reactor [48], printed using stereolithography, as

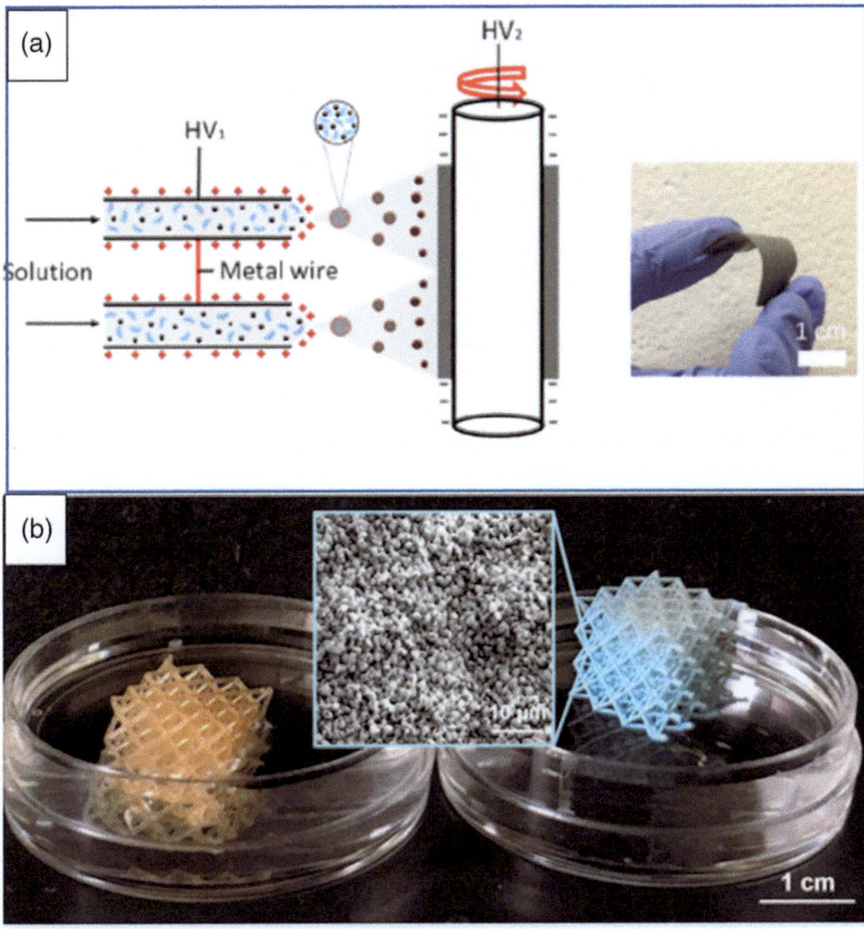

Fig. 6.12 **a** Pictorial representation of the Electrospray deposition process and deposition of Al on printed PVDF film [39] **b** Printed lattice structures before and after the electrospray deposition [39]

shown in Fig. 6.13b. In this work, the electrodes were functionalized using oxides of Ir, Ru and Ti and assembled inside the 3D printed electrochemical reactor. This setup used a membrane free electrochemical reactor for the electrolysis of water.

6.4.3 Chemical Modification Procedures

Most of the functionalized 3D printed materials can be produced using chemical modification methods. These chemical methods include deposition processing, surface grafting and etching.

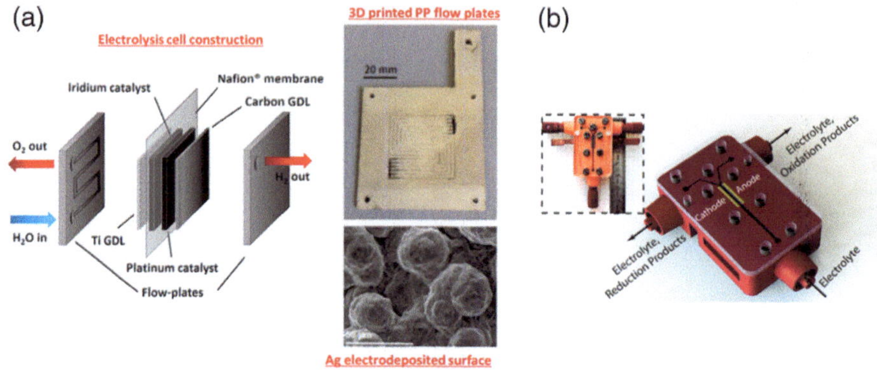

Fig. 6.13 **a** Characteristic components of a PEM electrolyzer, 3D printed PP flow plates functionalized using silver nanoparticles (Adapted with permission from Ref. [47]. Copyright 2014 Royal Society of Chemistry). **b** Stereolithography technique based printing of membrane-free electrolysis cell, functionalized using electrodeposition of silver (Adapted with permission from Ref. [48]. Copyright 2014 Royal Society of Chemistry)

Deposition processing

This is a generic chemical method of functionalizing 3D printed objects, which involves deposition of the precursor material of interest onto the surface, followed by a chemical reaction or subsequent temperature treatment, which renders the desired functionality to the surface. Zeolites are widely used catalysts in petrochemical industries and many other chemical industries. Thus, the surface functionalization of the SLM printed metal objects using zeolites is considered as a potential approach for fabricating and implementing AM printed reactors for chemical industries. Functionalizing the 3D printed metal surfaces using zeolite was developed by Hubesch et al. recently, using the deposition processing method [49]. In this method, Y-Zeolite was blended with organic and inorganic binders as well as peptizing agent, in water, to form a uniform dispersion. This dispersion was deposited over the surface of the SLM printed Inconel (Alloy of Ni, Cr and Mo) open framework structures, dried and calcined at 550 °C to deposit the layer of zeolite catalyst on the surface of SLM printed lattice structure. Thermal activation of framework step was conducted prior to the deposition of this zeolite dispersion, because calcination of the framework structure led to the formation of their metal oxides and enhance the adsorption of zeolite, as shown in Fig. 6.14a–d. Similarly, deposition of electrode metals on 3D printed polymer electrodes enable the fabrication of micro-structured composite electrodes, as shown in Fig. 6.14e. Streb et al. developed a three-stage approach of chemically functionalizing the 3D printed ABS polymer based meshes with the electrode metals such as Ni and further modifying the Ni–Fe layered double hydroxide [50]. The three-stage approach involves the chemical oxidation of the surface, followed by the electroless deposition of nickel and subsequent deposition of Ni–Fe layered double hydroxide to fabricate a micro-structured composite electrode. These electrodes were shown as efficient catalysts towards oxygen evolution reaction (OER)

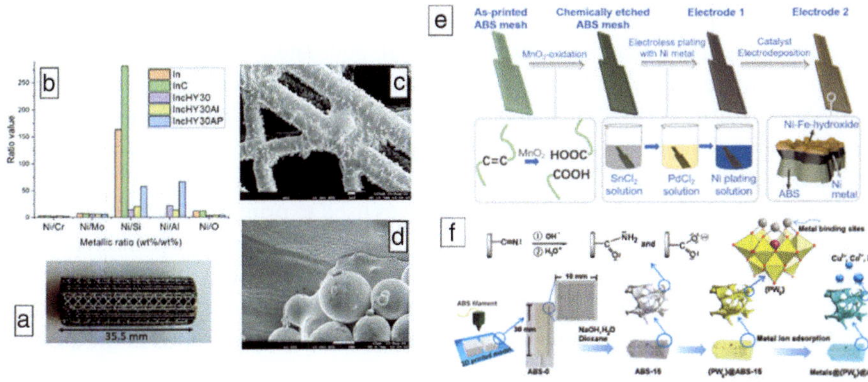

Fig. 6.14 **a–d** SLM printed inconel lattice structure, elemental composition, SEM images of the lattice and surface morphology [49]; **e** Three stage approach of modifying the ABS mesh electrodes using the chemical etching, electroless plating with Ni metal and deposition of a Ni–Fe hydroxide catalyst [50]; **f** Surface modification of ABS, POM deposition and the POM functionalized ABS for heavy-metal removal [51]

over lower potentials, exhibiting exceptional stability. Metal oxide functionalization on 3D printed polymers was demonstrated by the same group of researchers, that displayed heavy metal ion binding sites on the surface, and highlighted its potential to decontaminate water containing heavy metal ions [51]. Figure 6.14f shows the functionalization approach that involves the chemical activation of the surface using the alkali treatment, followed by the adsorption of polyoxometalates (POM) using non-covalent interactions. These surface bound polyoxometalates were shown to bind with common transition metal ions, which causes the metal ion pollution in drinking water.

Surface Chemical Etching Methods

These methods involve the use of chemical treatments on that can selectively remove or etch certain components from the surface of 3D printed objects to activate the surface for subsequent functionalization or enabling it for specific application. This approach tends to activate the surface by creating defective sides and it is generally carried out by immersing the printed structures either in acidic or alkaline solution [52]. Alkaline solutions such as aqueous solution of NaOH can hydrolyse specific chemical functional groups, therefore these new functional groups can enhance the application. This approach was used in activating the printed structures of poly caprolactam scaffolds by NaOH solution, which showed enhanced cellular attachment and proliferation. This was probably attributed to alkaline treatment was found to enhance surface porosity and roughness. Solvent treatment [53] and the utilization of ultrasound were shown to modify the surface without compromising the structural integrity of printed objects. This kind of surface modification of polymer scaffolds strongly influence the macroscopic and microscopic properties that eventually resulted in high surface area, enhanced roughness, and increased hydrophilicity.

Most of the characteristic functionalization was done on one substrate and was shown for a single application. Kitson, P.J. et al. generated an efficient approach of developing multiple catalyst functionalization a 3D printed reactionware, as shown in Fig. 6.15. Two different kinds of catalysts were introduced into the two reaction chambers and starting materials could pass through these catalyst functionalized chambers. In this approach, a multi-step reaction was conducted in a single reactionware containing multiple chambers. FDM process was used to print the reaction chambers using polypropylene because of its inertness towards organic solvents and reactants. Catalytic functionalization of these materials was done using robocasting. The catalyst ink formulation consists of acetoxy-silicone polymer and the catalysts Pd/C and K10 Montmorillonite clay were separately mixed and deposited over the printed

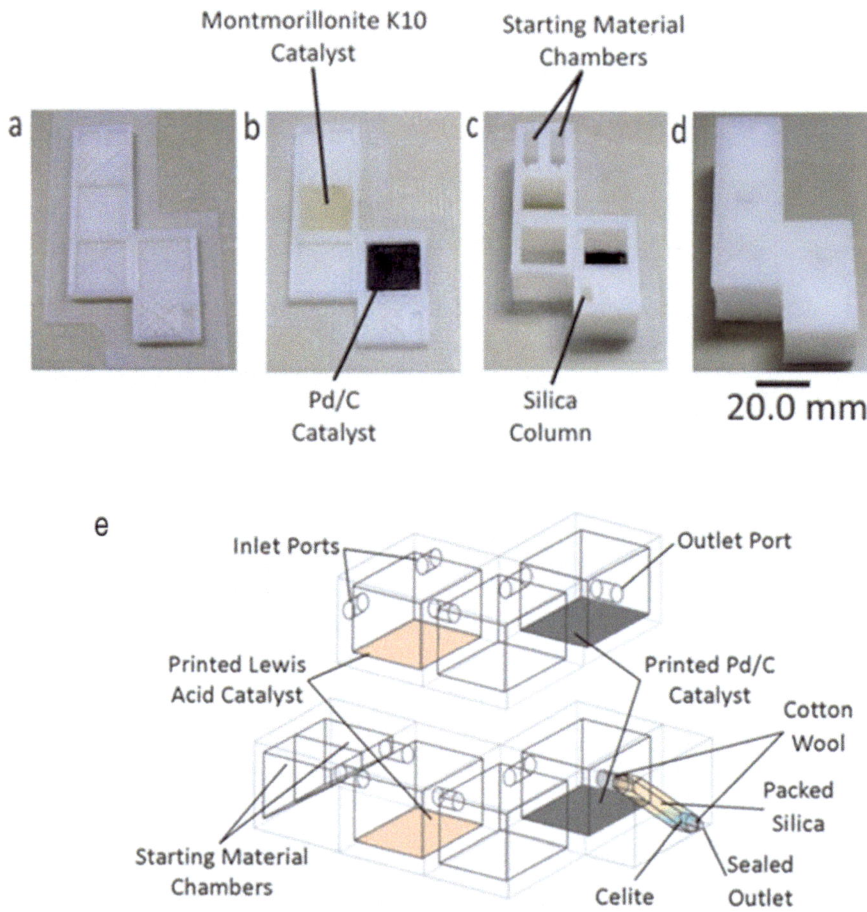

Fig. 6.15 Illustration of catalyst functionalized 3D printed sequential reactors [48, 49]; **a** Printed reactors before the catalyst deposition, catalyst functionalization, chambers for the reactants the sealed reactor and overall diagram of the reactionware

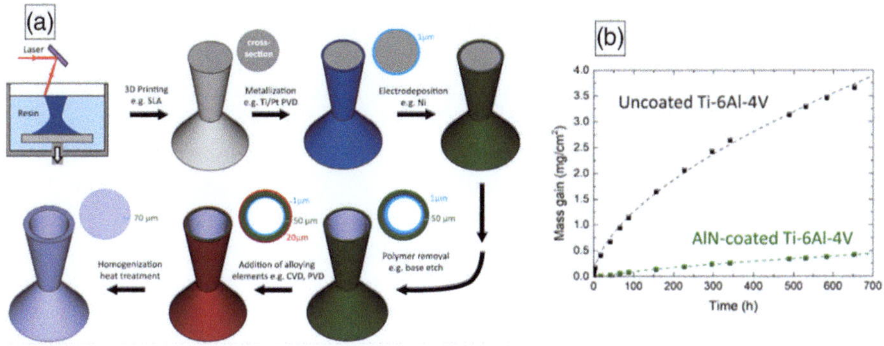

Fig. 6.16 (**a**) FDM/SLA printed polymer templated metallization and the polymer removal to obtain a thin-walled rocket engine thruster [54]; (**b**) Aluminium nitride (AlN) functionalization on the surface of Ti-6Al-4 V to improve its oxidation resistance at much higher temperatures [55]

reactor using robocasting. Using FDM and robocasting, functionalized reactionware were integrated for a chemical process.

Templated Approach

To obtain thin-walled metal structures using AM such as SLM technique is limited by the resolution of the instrument and such fine resolution using powder bed process is hard to achieve. Schaedler et al. developed an interesting approach of templating to obtain such thin-walled metal structure, used in rocket engines [54]. As presented in Fig. 6.16a, the first step involves the printing of polymer using SLA or FDM technique to fabricate thin-walled polymer structure. The second step involves the metal deposition using electrodepostion, electron beam deposition and CVD onto the polymer structures. This step was followed by the removal of the polymer template and this high temperature curing led to the alloy formation and harden the thin-walled structures.

Chemical vapor deposition (CVD) method

CVD is another standard strategy to obtain functionalized thin film structures. standard method of functionalizing thin films. Product material thickness can be varied from hundreds of nm to a few microns. 3D printed plastics and polymers are difficult to functionalize because these polymers are thermoplastic materials. Gupta et al. used a modified CVD approach for the functionalization of hydrophobic and hydrophilic polymers [56]. Acrylonitrile butadiene styrene (ABS) and poly- (lactic acid) were used to print the structures using FDM and the printed objects were placed on the heating stage of a closed chamber. Hydrophilic/Hydrophobic monomer was blended with a polymerization initiator and introduced in the chamber. The heated 3D printed component tend to decompose the polymerization initiator and started the polymerization of the monomer on the surface of 3D printed object. This functionalization was shown to modify the 3D printed objects to either be hydrophilic or hydrophobic. CVD of hydrocarbons on the metal surface generally leads to the formation of carbon filaments on the metal surface. This approach was used to form graphene on a metal surface. Yang et. al. showed the graphene formation on the surface of 3D printed Ni scaffold [57]. CVD of styrene at high temperature on the surface of Ni led to the formation of graphene on its surface. Ti-6Al-4 V is the commonly used material used for printing structures using SLM technique, however the printed structures using Ti alloy are known to be oxidized above 400 °C. To extend its working limit and increase the resistance towards oxidation, the printed materials are generally functionalized using other materials to improve its thermal stability. Therefore, Moll et al. developed an approach that combines Atomic Layer Deposition (ALD) and CVD to deposit Aluminium Nitride (AlN) coating on the surface of SLM printed Ti-6Al-4 V materials [55]. As a result of this functionalization, AlN functionalized Ti alloy showed significant resistance towards oxidation, as compared to the as-printed Ti-6Al-4 V structure, as shown in Fig. 6.16b.

Surface grafting approach

In this approach, the surface of a 3D printed object was functionalized by grafting specific functional molecules on the surface, which enhances the stability of the surface or modifies the reactivity of the surface. Stereo lithographically printed polymer materials are known to react with any organic solvent, that leads to swelling of the printed objects and eventually degrade the material. Thus, these materials are difficult to be used as 3D printed reactors, especially housing reactions using organic solvents. Recently, Griesbeck et al. developed an approach of using a modified resin containing isocyanate groups as a monomer in vat photopolymerization [58]. The as-printed object was reacted with 4-aminobenzophenone, which was found to graft a catalytically active surface as well as showing resistance against organic solvents, as shown in Fig. 6.17. This kind of surface grafting approach is known to transform the stereo lithographically printed objects into photochemical reactors.

Bio-functionalization

Latest developments in the AM technologies are currently focused on the fabrication of biomedical structures, bio-compatible devices, medical implants, organ replacements, bone tissue engineering, antibiofouling materials, bioanalytical devices, bio recognition devices and biocatalysts. Most of these biomedical devices and implants obtain their functionalities mainly from the way it was printed or post-functionalization strategies that involve biochemical functionalization or surface chemical treatments of the printed devices. Therefore, a significant interest was developed for the bio-compatible 3D printed objects for biomedical applications, as well suitable surface functionalization strategies to enable biocompatibility for biological applications. The materials currently used for such applications are printed using Ti, Mg metal alloys, PEEK, and PLA, which are all biocompatible. Unfortunately, all of these printed objects possess the risk of infection and may cause an undesired immune response. Therefore, surface functionalization of AM printed objects constitute an important frontier of research, that is collectively called biofunctionalization. This field is known to improve the expected biological performance and biocompatibility of objects through physical and biochemical mechanisms. As this

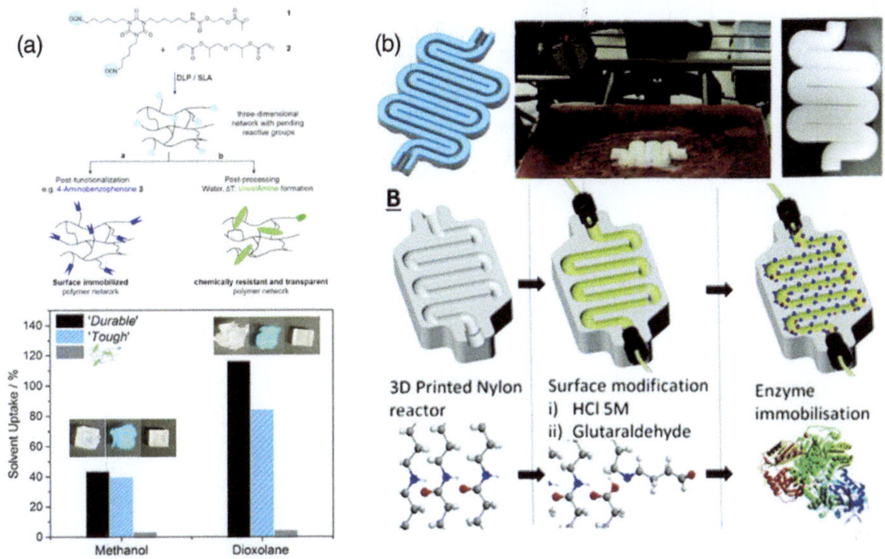

Fig. 6.17 Modification of resin and the amine modification of the printed object & solvent uptake of the printed objects (Adapted with permission from Ref. [58] Copyright 2014 Royal Society of Chemistry); **B** Enzyme functionalized 3D printed reactor (Adapted with permission from Ref. [59] Copyright 2014 Royal Society of Chemistry)

field is expanding, it is important to develop suitable biofunctionalization strategies that can provide improved tissue regeneration performance, suppression of immune response, and infection prevention and biocompatibility.

3D printed implants may get infected within a biological system, posing the implication of health hazards that mat require repeated treatment. Amin Yavari et al. addressed this challenge by functionalizing the printed Ti biomaterials with nanotubular titanium oxide (TiO_2) and Ag nanoparticles to provide resistance against implant-associated infection [60]. Ag nanoparticles are known antibacterial and antibiofouling agents, hence, functionalizing of these nanoparticles provides an avenue for resistance against infection. Introducing small peptides, enzymes, and extracellular components into the microarchitectures of AM printed objects is one of the biofunctionalization approaches in inducing the growth of specific tissues and cells. These biomolecules provide a functional site that cab enable the growth of specific cells and tissues. This concept was demonstrated using nitinol functionalization of porous printed substrate to enhance the interaction between the scaffolds and the tissue as shown in Fig. 6.18b.

The complex micro-architectures, porosity and larger surface area present in the AM printed objects provide a handle to control their mechanical properties, diffusion, and mass transport, which eventually contribute to tissue regeneration performance. Fox et al. developed a functionalization strategy to deposit polycrystalline diamond coatings on SLM printed titanium surfaces. This surface modification was found to

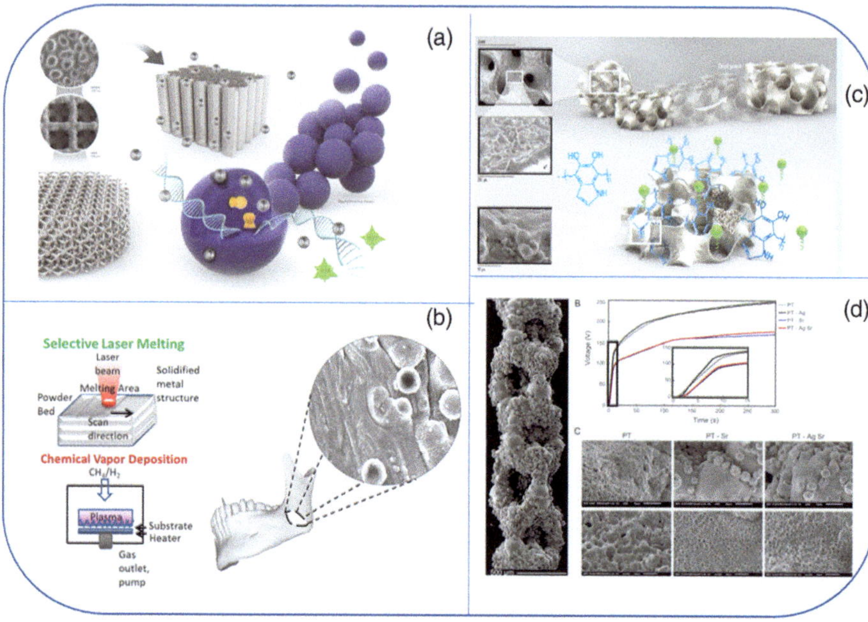

Fig. 6.18 **a** The porous Titanium and its release of silver ions: Antibacterial behavior of the SLM printed porous titanium [60]; **b** Nitinol biofunctionalization of AM printed reactor; **c** Deposition of the polycrystalline diamond on the surface of SLM printed titanium to improve its applicability in biomedical applications [61]; **d** SEM images of the SLM printed biological implants [62]

improve the applicability as medical implant by enhancing the biological scaffold interactions [61]. The polycrystalline diamond coatings have excellent bonding to the Ti substrate, as shown in Fig. 6.18c. The functionalized SLM printed materials were shown to enhance mammalian cell growth, deposition of hydroxy apatite, and suppressing microbial contamination.

Oxide layers on the printed materials surface and their porosity was considered as an effective functionalization strategy, which was found to increase the biocompatibility and provide sites for the growth of tissues. The technique, Plasma Electrolytic Oxidation (PEO) was known to create a functional oxide layer with interconnected porosity. This interconnected porous system can serve as preferential sites for cellular growth that will result in efficient integration of bone and the scaffold. This surface layer is beneficial for increasing cell affinity and growth, which in turn results in better bone-scaffold integration. This kind of surface modification to form a porous structure was known to strongly influence the mechanical properties and virtue of their high surface area. In a recent study, this PEO technique used to form a functional oxide layer that was shown to release both osteogenic (such as strontium ions) and antibacterial active silver ions [62]. Similar biofunctionalization processes also used hydroxyapatite functionalization into the surface of the implants and these functionalized materials exhibited an enhanced osteogenic behavior coupled with strong antibacterial behavior, due to the higher levels of alkaline phosphatase (ALP) activity compared with non-biofunctionalized implants, as shown in Fig. 6.18d. These biofunctionalization approaches provide new opportunities for the AM printed objects in general and emerge as potential candidates for improved biomedical implants.

6.4.4 Surface Initiated Polymerization

Polymer functionalization, by grafting polymers on the surface of a 3D printed objects opens up a new class 3D printed functionalized materials, as the characteristic properties of the polymer started appearing on the surface. Photopolymerization techniques including stereolithography can be modified to produce polymer functionalized printed materials that were demonstrated to have strong polymer networks, resistance to mechanical degradation and swelling against organic solvents. Griesbeck et al. developed a hybrid resin consisting of hexamethylenediisocyanate based isocyanurate (HDI-trimer) with hydroxyethylmethacrylate and printed a transparent flow reactor using stereolithography [58]. This functionalized reactor device was demonstrated to be chemically resistant to organic solvents and did not show any kind of swelling behaviour. This surface functionalized reactor was demonstrated as photocatalytic reaction, used for the photooxygenation reaction. Yang et al. used the atom transfer radical polymerization (ATRP) approach to functionalized 3D printed polymer brushes to render these materials to have antimicrobial properties for biomedical applications [63], as shown in Fig. 6.19a. This interesting approach involves the addition of a radical initiator to the ink used for the 3D printing and this hybrid ink was used to print to obtain functionalized polymer brushes. 3-sulfopropyl methacrylate potassium salt (SPMA) was grafted onto the polymer brushes surface

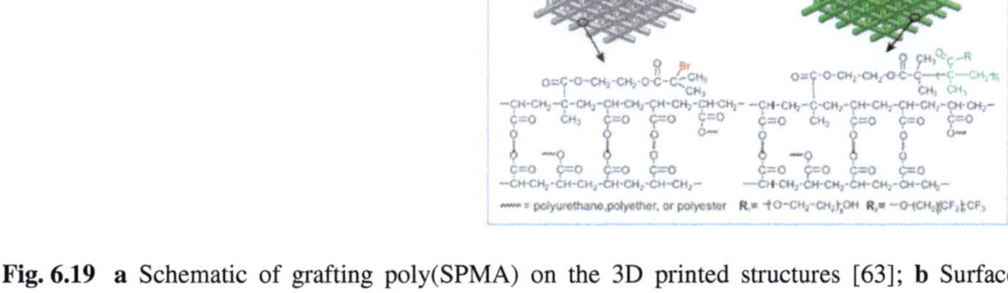

Fig. 6.19 **a** Schematic of grafting poly(SPMA) on the 3D printed structures [63]; **b** Surface modified lattice structure using ATRP and RAFT assisted photopolymerization

and the bacterial cells adhesion of bacteria were significantly inhibited. Wang et al. has developed a versatile approach to modify the surface of the 3D printable structural materials. This photo-initiated polymerization approach was demonstrated to print complex 3D architectures with desired surface properties. Development of surface-initiated atom transfer radical polymerization (SI-ATRP) approach was considered to be a robust methodology to print surface functionalized 3D printed objects and is illustrated in Fig. 6.19b. Another approach involves the addition of Vinyl-terminated initiator to the photo-curable (Ultra-violet light UV) resin to print functional polymer objects. This kind of 3D initiator integrated objects (i3DP) can be further transformed into functional materials by allowing the polymerization to print polymer brushes using surface-initiated polymerization.

Majority of the afore-mentioned application, light is used to initiate the radical polymerization or photo-cross linking to obtain polymer functionalized 3D printed materials. Use of light can initiate free radical or cationic photopolymerization systems, that create subsequent functionalization. Using this photopolymerization approach, a broad range of monomer families (epoxide, acrylate, vinyl ethers, etc.) can be used, that eventually lead to fabrication of thermosetting, elastomeric and temperature sensitive materials.

6.5 Summary

AM is a new manufacturing practice that enables us to create complex geometries and material properties. Integrating these technologies with the recent developments and the current knowledge in fabricating reactors, membranes, sensors, electrodes, catalysts, and flow devices can potentially translate these AM technologies into a single manufacturing process for afore-mentioned chemical applications. Surface chemical functionalization of AM printed components is the major strategy to achieve the development of functional 3D printed objects, which can be modified further to extend their suitability for wide variety of practical applications. AM technologies can print materials of specific composition, those are limited and proprietary materials, therefore surface chemical functionalization of the printed objects can only extend the scope and capabilities of this AM process and becomes crucial for customization and process-intensification of any chemical processes. Despite the recent developments, surface functionalization of 3D printed objects remains challenging and need insights. Thus, developing novel functionalization methods is the key to enable chemical functionality as fourth dimension to the 3D printing. Surface roughness, chemical inertness, thermal and mechanical stability of AM printed materials at non-ambient physical conditions are the major factors that need to be addressed for efficient functionalization. Moreover, the fundamental interactions that govern the interfacial adhesion, stability and durability of surface functionalization need to be studied to overcome the challenges associated with the surface functionalization of AM printed objects. This chapter summarized the recent developments and present the methodologies currently used for surface functionalization of AM printed components. Direct functionalization during the printing process has major limitations, therefore post-physical and chemical surface modifications such as chemical etching, deposition, self-assembled polymerization, surface grafting of polymers, electroless plating and electrodeposition have been used to create functionality on the surface of a AM printed object or component. These methodologies can potentially lead to the development of a functional AM devices that can be equipped with chemical features including catalysis, sensor, adsorption, conductivity, heat transfer, wettability, electronic and electrochemical properties. Depending upon the nature of the AM printed component, its surface roughness, desired surface chemical functionalization, any of the afore-mentioned methods combined with suitable surface activation are used to create strong interfacial bonding for efficient functionalization. To establish such strong interfacial bonding between the both materials as well as achieving uniform functionalization, further studies are needed to establish underlying mechanism that govern the surface/interfacial engineering, when applied to surface functionalized 3D printing. Designing and development of Imaginative exploration of designs of 3D printed materials, reactionwares and reactors and advances in printing technologies, along with a greater fundamental understanding of the chemical, physical, and mechanical properties of printed materials, and their correlation to functional applications are essential for developing next-gen 3D-printed components for chemical applications.

6.6 Test Your Knowledge

1. *What would be the methodology used to functionalize the lattice structure printed using SLM?*
2. *Stereo lithographically printed structures swell in organic solvents. What would be the best way to use these printed objects in organic media?*
3. *FDM printed objects are made up of thermoplastics. How would you deposit metals to make electronic circuits on the surface of FDM printed objects?*
4. *What are the post-processing physical methods available for reducing the surface roughness of the SLM printed metal objects?*
5. *Why is chemical or thermal activation of 3D printed objects required for functionalization?*

References

1. Boydston AJ, Cui J, Lee C-U, Lynde BE, Schilling CA (2020) 100th anniversary of macromolecular science viewpoint: integrating chemistry and engineering to enable additive manufacturing with high-performance polymers. ACS Macro Lett 9(8):1119–1129. https://doi.org/10.1021/acsmacrolett.0c00390

2. Sanchez-Rexach E, Johnston TG, Jehanno C, Sardon H, Nelson A (2020) Sustainable materials and chemical processes for additive manufacturing. Chem Mater 32(17):7105–7119. https://doi.org/10.1021/acs.chemmater.0c02008
3. Ghanem MA, Basu A, Behrou R, Boechler N, Boydston AJ, Craig SL, Lin Y, Lynde BE, Nelson A, Shen H, Storti DW (2020) The role of polymer mechanochemistry in responsive materials and additive manufacturing. Nat Rev Mater Ahead Print. https://doi.org/10.1038/s41578-020-00249-w
4. Choi J-W, Lu Y, Wicker RB (2021) Projection microstereolithography as a micro-additive manufacturing technology: processes, materials, and applications. In, 2016. vol Copyright (C) 2021 American Chemical Society (ACS). All Rights Reserved. CRC Press, pp 101–129. doi:https://doi.org/10.1201/b19360-5
5. Guerra MG, Volpone C, Galantucci LM, Percoco G (2018) Photogrammetric measurements of 3D printed microfluidic devices. Addit Manuf 21:53–62. https://doi.org/10.1016/j.addma.2018.02.013
6. Hurt C, Brandt M, Priya SS, Bhatelia T, Patel J, Selvakannan PR, Bhargava S (2017) Combining additive manufacturing and catalysis: a review. Catal Sci Technol 7(16):3421–3439. https://doi.org/10.1039/C7CY00615B
7. Femmer T, Flack I, Wessling M (2016) Additive manufacturing in fluid process engineering. Chem Ing Tech 88(5):535–552. https://doi.org/10.1002/cite.201500086
8. Ruan X, Wang Y, Cheng N, Niu X, Chang Y-C, Li L, Du D, Lin Y (2020) Emerging applications of additive manufacturing in biosensors and bioanalytical devices. Adv Mater Technol (Weinheim, Ger) 5(7):2000171. https://doi.org/10.1002/admt.202000171
9. Du Y, Chen J, Meng Q, Dou Y, Xu J, Shen SZ (2020) Thermoelectric materials and devices fabricated by additive manufacturing. Vacuum 178:109384. https://doi.org/10.1016/j.vacuum.2020.109384
10. Du Y, Chen J, Meng Q, Xu J, Lu J, Paul B, Eklund P (2020) Flexible thermoelectric double-layer inorganic/organic composites synthesized by additive manufacturing. Adv Electron Mater 6(8):2000214. https://doi.org/10.1002/aelm.202000214
11. Peterson GI, Larsen MB, Ganter MA, Storti DW, Boydston AJ (2015) 3D-printed mechanochromic materials. ACS Appl Mater Interfaces 7(1):577–583. https://doi.org/10.1021/am506745m
12. Lin KJ, Gu DD, Du L, Liu J (2018) Additive manufacturing (AM) technologies for clean energy applications: a review with a perspective on the role played by lasers. Lasers Eng 41(4–6):235–262
13. Zhakeyev A, Wang P, Zhang L, Shu W, Wang H, Xuan J (2017) Additive manufacturing: unlocking the evolution of energy materials. Adv Sci (Weinheim, Ger) 4 (10):n/a. https://doi.org/10.1002/advs.201700187
14. Mahmood N, De Castro IA, Pramoda K, Khoshmanesh K, Bhargava SK, Kalantar-Zadeh K (2019) Atomically thin two-dimensional metal oxide nanosheets and their heterostructures for energy storage. Energy Storage Mater 16:455–480. https://doi.org/10.1016/j.ensm.2018.10.013
15. Fan D, Li Y, Wang X, Zhu T, Wang Q, Cai H, Li W, Tian Y, Liu Z (2020) Progressive 3D printing technology and its application in medical materials. Front Pharmacol 11:122. https://doi.org/10.3389/fphar.2020.00122
16. Galante R, Figueiredo-Pina CG, Serro AP (2019) Additive manufacturing of ceramics for dental applications: a review. Dent Mater 35(6):825–846. https://doi.org/10.1016/j.dental.2019.02.026
17. Gao C, Wang C, Jin H, Wang Z, Li Z, Shi C, Leng Y, Yang F, Liu H, Wang J (2018) Additive manufacturing technique-designed metallic porous implants for clinical application in orthopedics. RSC Adv 8(44):25210–25227. https://doi.org/10.1039/C8RA04815K
18. Pekkanen AM, Mondschein RJ, Williams CB, Long TE (2017) 3D printing polymers with supramolecular functionality for biological applications. Biomacromol 18(9):2669–2687. https://doi.org/10.1021/acs.biomac.7b00671

19. Sabahi N, Chen W, Wang C-H, Kruzic JJ, Li X (2020) A review on additive manufacturing of shape-memory materials for biomedical applications. JOM 72(3):1229–1253. https://doi.org/10.1007/s11837-020-04013-x
20. Liu X, Yuk H, Lin S, Parada GA, Tang T-C, Tham E, de la Fuente-Nunez C, Lu TK, Zhao X (2018) 3D Printing of living responsive materials and devices. Adv Mater 30(4):1704821. https://doi.org/10.1002/adma.201704821
21. Ma C, Dong Y, Ye C (2016) Improving surface finish of 3D-printed metals by ultrasonic nanocrystal surface modification. Procedia CIRP 45:319–322. https://doi.org/10.1016/j.procir.2016.02.339
22. Wang J, Li H, Liu R, Li L, Lin Y-H, Nan C-W (2018) Thermoelectric and mechanical properties of PLA/Bi0·5Sb1·5Te3 composite wires used for 3D printing. Compos Sci Technol 157:1–9. https://doi.org/10.1016/j.compscitech.2018.01.013
23. Wang S, Bai P, Sun M, Liu W, Li D, Wu W, Yan W, Shang J, Yu J (2019) Fabricating mechanically robust binder-free structured zeolites by 3D printing coupled with zeolite soldering: a superior configuration for CO2 capture. Adv Sci 6(17):1901317. https://doi.org/10.1002/advs.201901317
24. Danaci S, Protasova L, Lefevere J, Bedel L, Guilet R, Marty P (2016) Efficient CO2 methanation over Ni/Al2O3 coated structured catalysts. Catal Today 273:234–243. https://doi.org/10.1016/j.cattod.2016.04.019
25. Van Noyen J, De Wilde A, Schroeven M, Mullens S, Luyten J (2012) Ceramic processing techniques for catalyst design: formation, properties, and catalytic example of ZSM-5 on 3-dimensional fiber deposition support structures. Int J Appl Ceram Technol 9(5):902–910. https://doi.org/10.1111/j.1744-7402.2012.02781.x
26. Lefevere J, Gysen M, Mullens S, Meynen V, Van Noyen J (2013) The benefit of design of support architectures for zeolite coated structured catalysts for methanol-to-olefin conversion. Catal Today 216:18–23. https://doi.org/10.1016/j.cattod.2013.05.020
27. Stuecker JN, Miller JE, Ferrizz RE, Mudd JE, Cesarano J (2004) Advanced support structures for enhanced catalytic activity. Ind Eng Chem Res 43(1):51–55. https://doi.org/10.1021/ie030291v
28. Kotz F, Arnold K, Bauer W, Schild D, Keller N, Sachsenheimer K, Nargang TM, Richter C, Helmer D, Rapp BE (2017) Three-dimensional printing of transparent fused silica glass. Nature 544(7650):337–339. https://doi.org/10.1038/nature22061
29. Mehla S, Das J, Jampaiah D, Periasamy S, Nafady A, Bhargava SK (2019) Recent advances in preparation methods for catalytic thin films and coatings. Catal Sci Technol 9:3582–3602. https://doi.org/10.1039/C9CY00518H
30. Bernasconi R, Cuneo F, Carrara E, Chatzipirpiridis G, Hoop M, Chen X, Nelson BJ, Pané S, Credi C, Levi M, Magagnin L (2018) Hard-magnetic cell microscaffolds from electroless coated 3D printed architectures. Mater Horiz 5(4):699–707. https://doi.org/10.1039/C8MH00206A
31. Li J, Wang Y, Xiang G, Liu H, He J (2019) Hybrid additive manufacturing method for selective plating of freeform circuitry on 3D printed plastic structure. Adv Mater Technol (Weinheim, Ger) 4 (2):n/a. https://doi.org/10.1002/admt.201800529
32. Marques-Hueso J, Jones TDA, Watson DE, Ryspayeva A, Esfahani MN, Shuttleworth MP, Harris RA, Kay RW, Desmulliez MPY (2018) A Rapid photopatterning method for selective plating of 2D and 3D microcircuitry on polyetherimide. Adv Func Mater 28(6):1704451. https://doi.org/10.1002/adfm.201704451
33. Lee C-Y, Taylor AC, Nattestad A, Beirne S, Wallace GG (2019) 3D printing for electrocatalytic applications. Joule 3(8):1835–1849. https://doi.org/10.1016/j.joule.2019.06.010
34. Pip P, Donnelly C, Döbeli M, Gunderson C, Heyderman LJ, Philippe L (2020) Electroless deposition of Ni–Fe alloys on scaffolds for 3D Nanomagnetism 16(44):2004099. https://doi.org/10.1002/smll.202004099
35. Mukherjee S, Gupta R, Garg A, Bansal V, Bhargava S (2010) Influence of Zr doping on the structure and ferroelectric properties of BiFeO3 thin films. J Appl Phys 107(12). https://doi.org/10.1063/1.3436593

36. Jian JR, Kim T, Park JS, Wang J, Kim WS (2017) High performance 3D printed electronics using electroless plated copper. 7(3):035314. https://doi.org/10.1063/1.4979173
37. Bernasconi R, Cuneo F, Carrara E, Chatzipirpiridis G, Hoop M, Chen X, Nelson BJ, Pane S, Credi C, Levi M, Magagnin L (2018) Hard-magnetic cell microscaffolds from electroless coated 3D printed architectures. Mater Horiz 5(4):699–707. https://doi.org/10.1039/C8MH00206A
38. Rai A, Chaudhary M, Ahmad A, Bhargava S, Sastry M (2007) Synthesis of triangular Au core–Ag shell nanoparticles. Mater Res Bull 42(7):1212–1220. https://doi.org/10.1016/j.materresbull.2006.10.019
39. Huang C, Jian G, DeLisio JB, Wang H, Zachariah MR (2015) Electrospray deposition of energetic polymer nanocomposites with high mass particle loadings: a prelude to 3D printing of rocket motors. Adv Eng Mater 17(1):95–101. https://doi.org/10.1002/adem.201400151
40. Wang X, Guo Q, Cai X, Zhou S, Kobe B, Yang J (2014) Initiator-integrated 3D printing enables the formation of complex metallic architectures. ACS Appl Mater Interfaces 6(4):2583–2587. https://doi.org/10.1021/am4050822
41. Mehla S, Kandjani A, Coyle V, Harrison CJ, Low MX, Kaner RB, Sabri Y, Bhargava SK (2022) Gold sunflower microelectrode arrays with dendritic nanostructures on the lateral surfaces for antireflection and surface-enhanced raman scattering. ACS Appl Nano Mater 5(2):1873–1890. https://doi.org/10.1021/acsanm.1c03501
42. Plowman BJ, Jones LA, Bhargava SK (2015) Building with bubbles: the formation of high surface area honeycomb-like films via hydrogen bubble templated electrodeposition. Chem Commun (Camb) 51(21):4331–4346. https://doi.org/10.1039/c4cc06638c
43. Plowman BJ, Bhargava SK, O'Mullane AP (2011) Electrochemical fabrication of metallic nanostructured electrodes for electroanalytical applications. Analyst 136(24):5107–5119. https://doi.org/10.1039/c1an15657h
44. Ambrosi A, Pumera M (2018) Self-contained polymer/metal 3D printed electrochemical platform for tailored water splitting. Adv Funct Mater 28(27):1700655-n/a. https://doi.org/10.1002/adfm.201700655
45. Ambrosi A, Moo JGS, Pumera M (2016) Helical 3D-printed metal electrodes as custom-shaped 3D platform for electrochemical devices. Adv Func Mater 26(5):698–703. https://doi.org/10.1002/adfm.201503902
46. Kovacevich DA, Lei L, Han D, Kuznetsova C, Kooi SE, Lee H, Singer JP (2020) Self-limiting electrospray deposition for the surface modification of additively manufactured parts. ACS Appl Mater Interfaces 12(18):20901–20911. https://doi.org/10.1021/acsami.9b23544
47. Chisholm G, Kitson PJ, Kirkaldy ND, Bloor LG, Cronin L (2014) 3D printed flow plates for the electrolysis of water: an economic and adaptable approach to device manufacture. Energy Environ Sci 7(9):3026–3032. https://doi.org/10.1039/C4EE01426J
48. Hashemi HSM, Karnakov P, Hadikhani P, Chinello E, Litvinov S, Moser C, Koumoutsakos P, Psaltis D (2019) A versatile and membrane-less electrochemical reactor for the electrolysis of water and brine. Energy Environ Sci 12(5):1592–1604.https://doi.org/10.1039/C9EE00219G
49. Hubesch R, Mazur M, Föger K, Selvakannan PR, Bhargava SK (2021) Zeolites on 3D-printed open metal framework structure: metal migration into zeolite promoted catalytic cracking of endothermic fuels for flight vehicles. Chem Commun 57(75):9586–9589. https://doi.org/10.1039/D1CC04246G
50. Liu S, Liu R, Gao D, Trentin I, Streb C (2020) A 3d-printed composite electrode for sustained electrocatalytic oxygen evolution. Chem Commun 56(60):8476–8479. https://doi.org/10.1039/D0CC03579C
51. Ji Y, Ma Y, Ma Y, Asenbauer J, Passerini S, Streb C (2018) Water decontamination by polyoxometalate-functionalized 3D-printed hierarchical porous devices. Chem Commun 54(24):3018–3021. https://doi.org/10.1039/C8CC00821C
52. Schneider M, Fritzsche N, Puciul-Malinowska A, Baliś A, Mostafa A, Bald I, Zapotoczny S, Taubert A (2020) Surface etching of 3D printed poly(lactic acid) with NaOH: a systematic approach. Polymers 12(8):1711

53. Kosik-Kozioł A, Graham E, Jaroszewicz J, Chlanda A, Kumar PTS, Ivanovski S, Święszkowski W, Vaquette C (2019) Surface modification of 3D printed Polycaprolactone constructs via a solvent treatment: impact on physical and osteogenic properties. ACS Biomater Sci Eng 5(1):318–328. https://doi.org/10.1021/acsbiomaterials.8b01018
54. Martin JH, Ashby DS, Schaedler TA (2017) Thin-walled high temperature alloy structures fabricated from additively manufactured polymer templates. Mater Des 120:291–297. https://doi.org/10.1016/j.matdes.2017.02.023
55. Moll A, Blandin JJ, Dendievel R, Gicquel E, Pons M, Jimenez C, Blanquet E, Mercier F (2021) Coupling powder bed additive manufacturing and vapor phase deposition methods for elaboration of coated 3D Ti-6Al-4V architectures with enhanced surface properties. Surf Coat Technol 415:127130. https://doi.org/10.1016/j.surfcoat.2021.127130
56. Cheng C, Gupta M (2017) Surface functionalization of 3D-printed plastics via initiated chemical vapor deposition. Beilstein J Nanotechnol 8(1):1629–1636
57. Yang Z, Yan C, Liu J, Chabi S, Xia Y, Zhu Y (2015) Designing 3D graphene networks via a 3D-printed Ni template. RSC Adv 5(37):29397–29400. https://doi.org/10.1039/c5ra03454j
58. Hansen A, Renner M, Griesbeck AG, Büsgen T (2020) From 3D to 4D printing: a reactor for photochemical experiments using hybrid polyurethane acrylates for vat-based polymerization and surface functionalization. Chem Commun 56(96):15161–15164. https://doi.org/10.1039/D0CC06512A
59. Peris E, Okafor O, Kulcinskaja E, Goodridge R, Luis SV, Garcia-Verdugo E, O'Reilly E, Sans V (2017) Tuneable 3D printed bioreactors for transaminations under continuous-flow. Green Chem 19(22):5345–5349. https://doi.org/10.1039/C7GC02421E
60. Amin Yavari S, Loozen L, Paganelli FL, Bakhshandeh S, Lietaert K, Groot JA, Fluit AC, Boel CHE, Alblas J, Vogely HC, Weinans H, Zadpoor AA (2016) Antibacterial behavior of additively manufactured porous titanium with nanotubular surfaces releasing silver ions. ACS Appl Mater Interfaces 8(27):17080–17089. https://doi.org/10.1021/acsami.6b03152
61. Rifai A, Tran N, Lau DW, Elbourne A, Zhan H, Stacey AD, Mayes ELH, Sarker A, Ivanova EP, Crawford RJ, Tran PA, Gibson BC, Greentree AD, Pirogova E, Fox K (2018) Polycrystalline diamond coating of additively manufactured titanium for biomedical applications. ACS Appl Mater Interfaces 10(10):8474–8484. https://doi.org/10.1021/acsami.7b18596
62. van Hengel IAJ, Gelderman FSA, Athanasiadis S, Minneboo M, Weinans H, Fluit AC, van der Eerden BCJ, Fratila-Apachitei LE, Apachitei I, Zadpoor AA (2020) Functionality-packed additively manufactured porous titanium implants. Mater Today Bio 7:100060. https://doi.org/10.1016/j.mtbio.2020.100060
63. Guo Q, Cai X, Wang X, Yang J (2013) "Paintable" 3D printed structures via a post-ATRP process with antimicrobial function for biomedical applications. J Mater Chem B 1(48):6644–6649. https://doi.org/10.1039/C3TB21415J

Surface Modification of Additively Manufactured Materials: Adding Functionality as Fourth Dimension

Materials, Surface Treatment, Chemical Functionalization, Electrochemical Functionalization, Bio-functionalization, Surface Etching, Anisotropy, Surface-initiated Polymerization

Roxanne Hubesch, Uzma Malik, PR. Selvakannan, Lakshmi Kantam Mannepalli, Suresh K. Bhargava

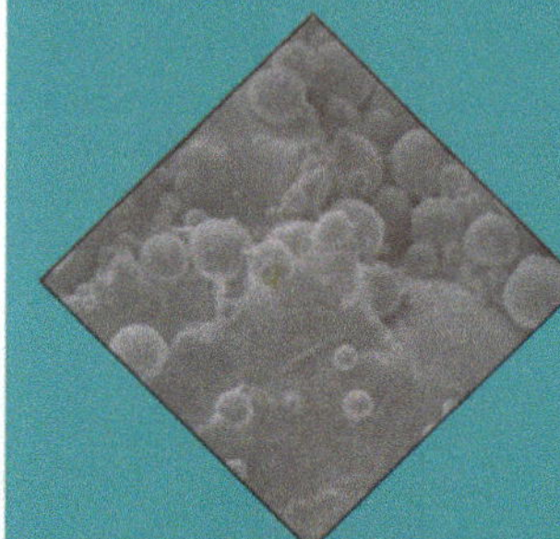

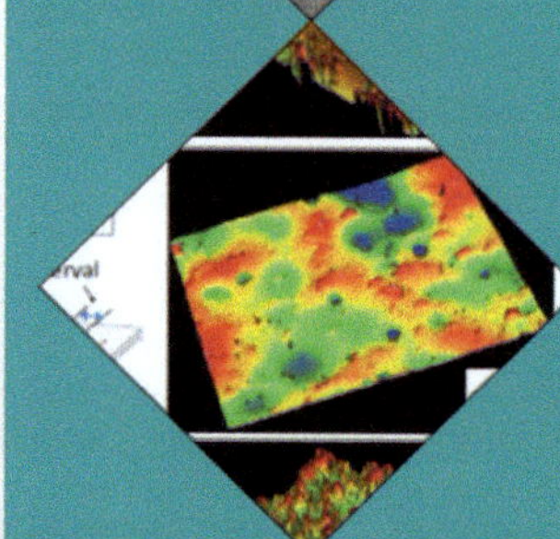

6

Chapter 7
Emerging Technological Applications of Additive Manufacturing

Sunil Mehla, PR. Selvakannan, Maciej Mazur, and Suresh K. Bhargava

Contents

Abstract Once considered to be a field specific to mechanical sciences, additive manufacturing has now proliferated into all streams of science and swiftly becoming a true global phenomenon. Moving well beyond printing of customized prototypes and trinkets, many enterprises now manufacture using 3D printing at moderate to large scale. Massive improvements in precision, quality and reliability of additive manufacturing have triggered rapid uptake of this technology in the research and development sector especially in fields such as chemical engineering, electronic engineering, materials engineering, biochemistry, optics, analytical sciences, industrial chemistry, and environmental sciences. This chapter highlights some interesting applications of additive manufacturing in chemical processes such as catalysis, separation, and high throughout experimentation, sensing devices such as microfluidics, electrochemical, optical, optoelectronic, and electrical sensors, and energy systems such as batteries and capacitors. The advantages of AM porous catalysts and adsorbents materialize from their high catalytic and separation efficiencies, hierarchical porosity, suitable flow properties, superior mass and energy transfer, novel composite formulations, enhanced product selectivity and high throughput processing of reactants. On the other hand, additively manufactured sensor and energy systems gain the benefits of high performance, better cycling performance (charging/discharging),

S. Mehla · PR. Selvakannan · S. K. Bhargava (✉)
Centre for Advanced Materials and Industrial Chemistry, RMIT University, Melbourne, VIC, Australia
e-mail: suresh.bhargava@rmit.edu.au

M. Mazur
Centre for Additive Manufacturing, RMIT University, Melbourne, VIC, Australia

S. K. Bhargava et al. (eds.), *Additive Manufacturing for Chemical Sciences and Engineering*, https://doi.org/10.1007/978-981-19-2293-0_7

multifunctionality, geometric shape complexity, customized design, shaping of amorphous materials, better integration of device components and in three dimensions, portability, device flexibility, self-powering capability, and automatic operation. In all such applications the chemical reactivity of the 3D printed construct governs its primary functionality in addition to the shape derived basic function. Clearly this is an ascension from the simple use of printed constructs as 3D objects of complex shapes and geometry. Starting with a brief discussion on the rise of additive manufacturing in chemical sciences, this chapter mainly focusses on the applications of additive manufacturing while building on the knowledge gained in the previous chapters. The applications have been classified as surface sensitive chemical processes which are confined to the first few hundred microns of the surface of a 3D printed construct, bulk sensitive chemical processes which depend on the bulk properties of 3D constructs and high throughput experimentation applications. A summary and outlook section conclude the chapter with a perspective and viewpoint on the future frontiers for additive manufacturing in chemical processes and a knowledge test has been provided for the young learners in the last section.

Keywords Additive manufacturing · Devices · Sensing · Energy storage · Catalysis

Abbreviations

3D	Three-dimensional
4D	Four-dimensional
AAO	Andic Aluminium Oxide
ABS	Acrylonitrile butadiene styrene
AJP	Aerosol Jet Printing
AM	Additive Manufacturing
APTES	(3-Aminopropyl)triethoxysilane
As	Arsenic
ATP	Adenosine Triphosphate
ATM	Ammonium Thiomolybdate
Au	Gold
BHA	Bariumhexaaluminate
BL	Bioluminescence
$BMIM^+BF_4$	1-Butyl-3-methylimidazoliumtetrafluoroborate
BST	$Bi_{0.4}Sb_{1.6}Te_3$
BTC	Benzene Tricarboxylic Acid
CAD	Computer-Aided Design
CMOS	Complementary Metal-Oxide Semiconductor
CNT	Carbon Nano Tubes
CV	Cyclic Voltammetry
DIW	Direct Ink Writing

DLP SLA	Dynamic Laser Projection Stereolithography
DMSO	Dimethylsulfoxide
DME	Dimethylether
EES	Electrical Energy Storage
EPAM	Electric polling assisted Additive Manufacturing
FDM	Fused Deposition Modelling
Ga	Gallium
GNS-GO	Graphene Nanosheets-Graphene Oxide
HTE	High Throughput Equipment
HER	Hydrogen Evolution Reaction
IC	Integrated Circuit
ICPMS	Inductively Coupled Plasma Mass Spectrometer
IJP	Ink Jet Printing
In	Indium
ITO/PEDOT:PSS	Indium tin Oxide/Poly(3,4-ethylenedioxythiophene) poly(styrenesulfonate) blend
LTPMS	Low Temperature Plasma-based Mass Spectrometry
LIFT	Laser Induced Forward Transfer
LTO	$Li_4Ti_5O_{12}$
LFP	$LiFePO_4$
LAGP	$Li_{1.5}Al_{0.5}Ge_{1.5}P_3O_{12}$
LCD	Liquid Crystal display
Li	Lithium
MAPLE DW	Matrix Assisted Pulsed Laser Evaporation Direct Write
MEH-PPV	Poly 2-methoxy, 5-(2ethylhexyloxy)-1,4-phenylene vinylene
MEMS	Microelectromechanical Systems
MOF	Metal Organic Framework
MSI	Mass Spectrometry Imaging
MTO	Methanol to Olefins
NAND	NOT-AND
NMP	N-methyl 2-pyrrolidone
OER	Oxygen Evolution Reaction
PCL	Poly-ε-caprolactone
Pd	Palladium
PDMS	Polydimethoxysilane
PE	Piezoelectric
PEC	Photo-electro Catalytic
PEDOT	Poly(3,4-ethylenedioxy-thiophene)
PEI	Polyethylenimine
PEGDA	Poly(ethylene glycol) diacrylate
PLA	Polylactic Acid
PP	Polypropylene
Pt	Platinum
PTFE	Polytetrafluoroethylene
PVDF	Polyvinylidene Fluoride

P(VDF-TrFE)	Poly(vinylidene fluoride-trifluoroethlene)
QD-LED	Quantum Dot Light Emitting Diode
RFID	Radio Frequency Identification
rGO	Reduced Graphene Oxide
Sb	Antimony
SEIRA	Surface-enhanced Infra-red Absorption
SEM	Scanning Electron Microscopy
SERRS	Surface-enhanced Resonance Raman Scattering
SERS	Surface-enhanced Raman Scattering
SiC	Silicon Carbide
SLA	Stereolithography
TE	Thermoelectric
TFT	Thin Film Transistors
TPCFL	Two Photon Continuous Flow Lithography
TPL	Two-photon Lithography
TRGO	Thermally Reduced Graphene Oxide
UHF	Ultra-High Frequency
UTAM	Ultra-thin Alumina Membranes
WHSV	Weight Hourly Space Velocity
YIG	Yttrium Iron Garnet

7.1 Introduction

The three-dimensional (3D) printing or additive manufacturing (AM) revolution originated from a simple yet powerful concept—physical realisation of designs and ideas. My own design, my own solution to my problem, solved using constructs made by me using materials chosen by me. The marvels of modern technology and advances have beautifully enabled this simple idea and the product is additive manufacturing. It is no wonder that rapid and seamless manufacturing of 3D objects from a user defined computer-generated code finds applications in … well, almost everything.

Additive manufacturing (AM) refers to manufacturing technologies and processes that produce 3D objects by combining material additively using its' precursors rather than subtracting/machining material from a larger bulk. Thus AM has a unique advantage over other manufacturing techniques i.e., a large reduction in material wastage which makes AM affordable and sustainable in the long run. Most AM technologies add material on a layer-by-layer basis, using an energy input to facilitate fusion of the printable material in a form precisely governed by the actual design. The material in the desired final form is specifically manufactured at the exact coordinates provided by a computer-generated code that controls (usually automated) the positioning of the object/nozzle in the printing process. Along with reduction in material wastage, the capability to produce geometrically complex shapes and customised as well as localised production has astoundingly enabled the 3D printing revolution.

Originally, AM was mainly employed to produce concept models and prototypes using plastic materials. After several years of development, however, focus has shifted towards fabrication of multifunctional and end-use products using AM. Recent technological advances have extended AM to a variety of materials ranging from hard metals, and ceramics to soft hydrogels and biomaterials. Beyond the rapid expansion in availability of 3D printable materials, AM processes have evolved over the years to become a new paradigm in advanced manufacturing of simple to complex shaped but most importantly customized products. Compared with other conventional manufacturing or advanced manufacturing methods, AM demonstrates advantages of freedom to design, improved material utilization, tool-less fabrication, and shortened lead time, paving the way for promising applications in aerospace, medical, transportation, construction, energy, and electronics industries. AM technologies have the potential to facilitate the integration of multiple devices and components for any custom-made application, on-site manufacturing of products, on-demand, and distributed manufacturing, which eventually reduces the need of large-scale production, their transport and associated carbon footprints.

Since the first additive manufacturing process developed by Francois Willeme in the 1860s for creating sculptures of real like objects, AM has become a global revolution fueled by widespread online communities dedicated to sharing of their object designs and pointers to manage complicated prints. Not getting left behind, chemical scientists also started to print components for their research and daily use equipment such as substrate holders, vessels, tanks, cells, and stirrers. The use of additively manufactured education aids such as solid models, unit cell of crystalline solids, body organs and organisms is phenomenal and has greatly boosted fundamental understanding at the education and research levels. While the additive manufacturing of a bionic ear using biological tissue in a hydrogel matrix with electronic components exhibiting capability to listen stereophonic music and enhanced auditory sensing of radio frequency may sound fascinating and futuristic to many, the feat was already achieved in 2013 [1]. Today, many commercial enterprises have developed a business model reliant solely on additive manufacturing such as printing using living human tissue, a 3D printer which is capable of printing majority of its own components, most automotive and aerospace industries, private and government defense industries, dental implants, prosthetics, food industries and fashion [2].

The most prominent early examples of the adoption of additive manufacturing in chemical processes have been the design and use of 3D printed reaction wares, hierarchically designed catalyst shapes, 3D printing of electrochemical materials, microfluidic sensors, and novel multifunctional composites. The need for functionalization of surfaces of printed objects was obvious since not all printing materials were chemically resistant and suitable for a reactive chemical environment. In analytical sciences, 3D printing has been extensively used in separation and sensing applications in microfluidics, for fabricating immunosensors and biosensors, and even for ex vivo analyses of neurotransmitters. 3D printed devices have also found use in point-of-care applications, in the form of all-printed analytical devices, which are low-cost and integrate both separation and sensing, including electrochemical and optical techniques.

Additive manufacturing has the unique capability for direct or indirect transformation of functional materials into useful geometries, shapes, configurations, and devices. Thus, it is no wonder that such a flexible and adaptable technology has stimulated wide interest in various fields to provide custom-made design solutions such as tissue engineering, scaffolds for cellular growth, bio-printing, microfluidics, fine chemical synthesis, drug delivery, electrochemistry, electronic devices, microreactors, mixers and heat exchangers, food sciences, robotics and so on. The uptake and scope of AM in these application areas is still in the nascent stage but rapidly increasing. The main objective of this chapter is to introduce the reader to the applications of advanced manufacturing in different chemical processes. Building on the knowledge developed in earlier chapters of the book, this chapter aims to familiarize the readers with the requirements and performance parameters that are critical for different applications and how specific AM technologies suit to some applications more than others. In this chapter different applications have been broadly classified into three categories as (1) Surface sensitive applications where the surface of AM constructs plays the most important role in governing its functionality such as functional surfaces, chemical and biosensing, optical devices and electronic devices, (2) Bulk sensitive applications where the key functionality of the AM construct derives from how the material is organized in its bulk such as in microfluidics, catalysis, batteries, capacitors and separation applications, and; (3) High throughput applications where the AM constructs are either used to design equipment enabling high throughput experimentation or AM technologies are used to generate arrays of materials enabling the testing of a wide permutation and combination of starting materials. Figure 7.1 depicts a schematic covering the scope of applications briefed in this chapter with images of related AM constructs. Finally, a summary and outlook Sect. 7.3 has been provided which takes the reader to the future of AM in chemical processes anticipating areas of further development and breakthrough innovations. For the benefit of students, a knowledge test has been provided in Sect. 7.4 where the learning outcomes can be ascertained.

7.2 Applications of Additively Manufactured Constructs

7.2.1 Surface Sensitive Applications

A wide variety of important chemical processes are surface sensitive and insensitive to the bulk composition of tested material such as molecular recognition, sensing, conduction of electricity, reflectance and plasmon excitation. Such processes have been classified as surface sensitive because they are limited to only the first few hundred microns of the surface referred to as the skin depth in optical processes. A simple strategy to fabricate highly functional 3D structured materials for surface

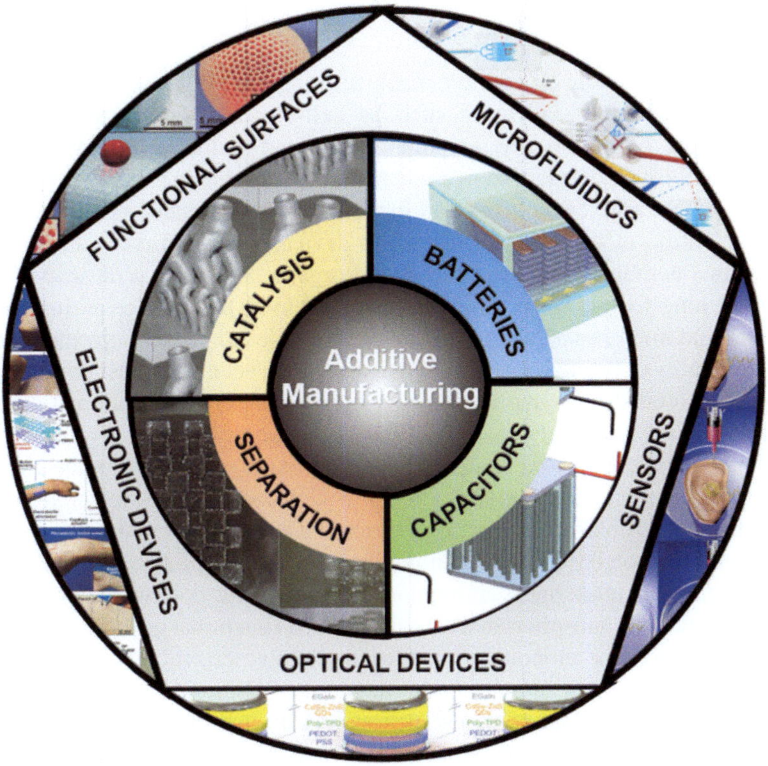

Fig. 7.1 A schematic depicting the scope of applications in this chapter and some examples of the use of additive manufacturing technologies in chemical processes, sensing and energy systems

sensitive applications is to functionalize the surface of additive manufactured 3D objects as elaborated in Chap. 6. For the discussion in this chapter surface sensitive applications have been categorized as (1) functional surfaces such as antimicrobial, hydrophobic, hydrophilic, amphiphobic, self-cleaning and shape morphing surfaces; (2) Sensing applications concerning detection of important chemicals such as pollutants, pesticides, explosives, dyes, and pharmaceuticals or biosensing applications concerning detection of important biomolecules, disease biomarkers, metabolites, bacteria, viruses and oligonucleotides or mechanophysical sensors concerned with detection of stress, strain, deformation, motion, vibration, sound and so on; (3) Optical devices dealing with manipulation of light in some form in devices such as light emitting diodes, lasers, waveguides, optical sensors, photonic crystals, optical metamaterials, and micro-optics components, and; (4) Electronic devices dealing with manipulation of current such as transistors, diodes, resistors, thermistors, integrated chips, circuits, photovoltaics, RFID tags, thermoelectric devices and piezoelectric devices.

7.2.1.1 Functional Surfaces

Many useful applications of additive manufacturing can be derived from a strategic functionalization of the surfaces such as self-cleaning, antimicrobial activity, hydrophobicity, hydrophilicity, amphiphobicity and shape morphing. Combined with the capabilities of additive manufacturing for production of 3D objects in an economic, low waste, customized and localized manner, functional surfaces allow better integration of these 3D surfaces into devices and equipment. For example, antimicrobial activity or a more general anti-organic activity is a desired functionality where long-term usage or storage is involved. With the passage of time, organic molecules and microbes adhere to the surfaces of objects. These molecules can then impact the utility of printed objects by interfering with accurate measurements and in biological or food storage applications even producing cross-contamination, infections, and manifesting safety issues. Functionalization of surfaces of printed objects with photocatalytically active materials like zinc oxide, titanium dioxide and silver can provide the needed anti-organic activity to such constructs. Another approach is to use an anti-microbial resin itself for fabrication as demonstrated by Yue et al. for photopolymerization based AM technologies by developing a composite resin using positively charged quaternary ammonium groups in the polymer matrices which imparted the desired anti-microbial activity [3]. The functional group could be added as an alkyl chain to the monomers used for photopolymerization or incorporated as a linear polymeric chain in appropriate quantities for integration in photocurable resins. Other antimicrobial coatings such as based on bacteriocin nisin, chitosan, acids, salts, plant extracts and enzymes lysozyme and lactoperoxidase, typically used in food packaging industry, can also be utilized to impart the needed antimicrobial activity to additively manufactured objects [4]. The research on development of visible light-responsive photoactive materials is also expanding rapidly and such materials can also be anticipated to find application in rendering anti-organic activity to 3D constructs.

Self-cleaning surfaces refer to those surfaces which can autogenously remove contaminants or external materials sticking to the surface using inputs from the ambient environment such as light and moisture or by the aid of a simple rinse with water. Cleaning using water droplets can be achieved on hydrophobic and superhydrophobic surfaces. Such surfaces have a contact angle greater than 90° for water, leading to suspension of water droplets on their surfaces and easy rolling without wetting the surface [5]. Since most contaminants and impurities are hydrophilic such rolling droplets produce the cleaning action by picking up the contaminants in their bulk and subsequent removal by simple rinsing with water. Today self-cleaning coatings are widely employed in textiles, glass windows, plastics, steel, cables, paints, solar panels, and cement [6]. In addition, superhydrophobic surfaces also find use in niche applications such as droplet generation, corrosion resistance, friction reduction, staining resistant and blood repellent surfaces. Fihri et al. have provided a detailed account of superhydrophobic coatings for metallic surfaces which are usable for 3D printed metallic objects to provide protection against corrosion and surface degradation [7]. Such coating technologies can be readily integrated with AM technologies for manufacturing self-cleaning 3D printed objects.

Besides the use of functionalisation methods, direct patterning, or fabrication of objects in specific topologies and integration of appropriate materials into the matrices can also be used to impart hydrophobicity to printed objects. Wettability is impacted by both surface energy as well as surface roughness (topology). Methyl and fluoride functional groups have been widely used to impart adequate surface energy to superhydrophobic surfaces whereas surface patterning by lithography, etching, templating, or machining is often used to impart the adequate surface roughness. Surface with repeating pattern of grooves were reported to promote anisotropic wetting whereas repeating patterns of micropillars exhibited isotropic wetting characteristics wherein the height of the pillars were an important parameter [8]. Re-entrant structures exhibit superoleophobic properties, sharp edged surfaces exhibit pinning of water droplets and rounded edges exhibit better self-cleaning properties [8]. Graeber et al. demonstrated the use of nano 3D printing technology, two-photon lithography (TPL), for fabrication of macrotextured superhydrophobic surfaces consisting of an alternating square array of truncated cones and inversed identically shaped indentations resulting in a 3 folds decrease in contact times for viscous liquids [9]. The design, structure and water contact angle for the fabricated surface is shown in Fig. 7.2. A master was first fabricated by TPL and subsequently transferred to PDMS and a PTFE coating was applied to obtain the hydrophobic surface. Such surfaces find applications in food industry for viscous liquid removal, prevention of food staining and chemical shielding [9].

Contrary to hydrophobicity, recent research has also seen the emergence of superwettable (superhydrophillic) surfaces which find applications in heat transfer, improved bio-adhesion, efficient separation of liquid mixtures, energy conversion devices and ink-based printing [10]. The flexibility in controlling the surface texture (topology) and composition from advanced manufacturing technologies such as 3D printing have enormously enabled such research. Another type of engineered surfaces which have received wide attention are superamphiphobic surfaces i.e., which are both superhydrophobic (water repellent) and superoleophobic (oil repellent). Such surfaces find applications in bionic science, liquid transport, microfluidics, drag-reduction, anti-fouling and oil–water separation [11].

Surfaces can also be imparted exceptional shape morphing capabilities by inducing anisotropy in composition resulting in anisotropic stress and straining. The resulting stresses can be used to imbibe stimuli responsive shape morphing capabilities to these materials which form the basis of 4D printing. 4D objects refer to those objects which change their 3D shape with time or after exposure to a stimuli such as immersion in water, exposure to light or application of a mechanical force [12]. Additive manufacturing which is a layer-by-layer coordinate-controlled manufacturing method enables easy integration of 4D capabilities into 3D objects by using a composition gradient of functional materials at adequate locations in the objects such as hinges, curves and supports as shown in Fig. 7.3. Such 4D shape morphing

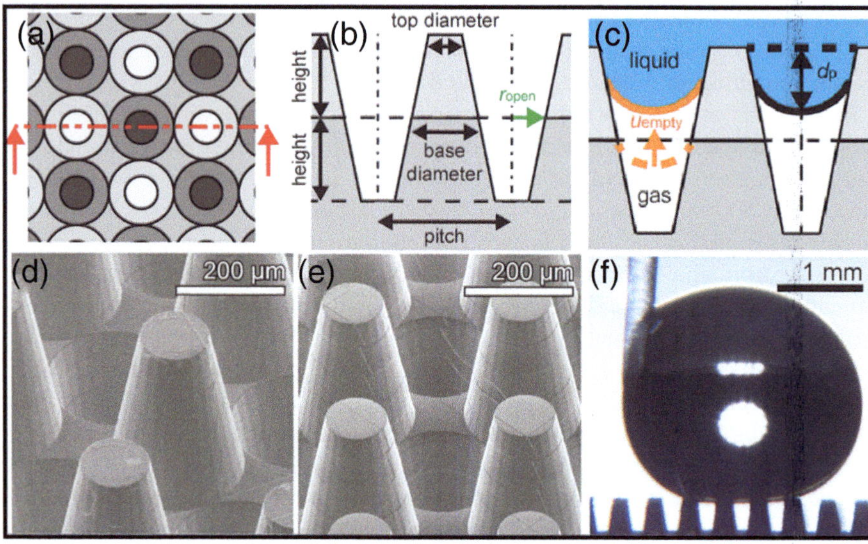

Fig. 7.2 Design and fabrication of an alternating truncated cone geometry based superhydrophobic surface using two-photon lithography; **a** Design of the structure in XY plane; **b** Vertical profile of the structure; **c** Schematic showing the fluid leaving the developed structure at an average emptying fluid velocity of U_{empty} and a penetration depth of d_p; **d** SEM micrograph of the as printed structure; **e** SEM micrograph after transferring the structure to PDMS; **f** Image of water droplet on PTFE coated PDMS structures for contact angle measurement. Reproduced from Ref. [9] with permission from American Chemical Society

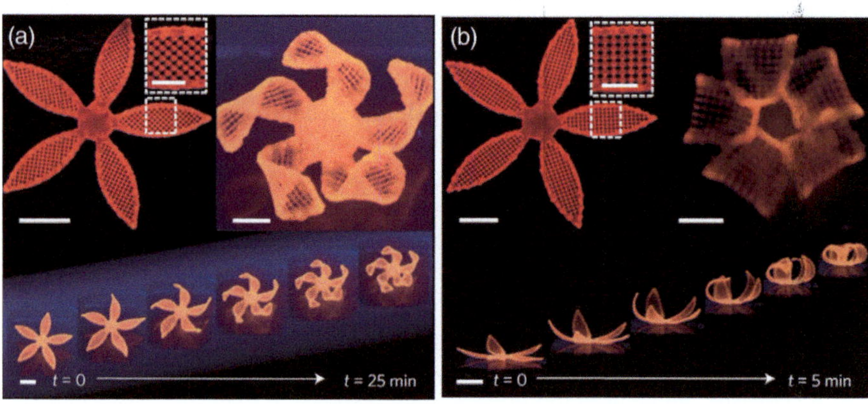

Fig. 7.3 Demonstration of complex shape morphing surfaces prepared by biomimetic 4D printing. **a** A 4D flower composed of 90∘/0∘ bilayers and **b** A 4D flower composed of 45∘/45∘ bilayers with respect to the long axis of each petal with time-lapse sequences of the flowers during the swelling process. Scale bars, 5 mm, inset = 2.5 mm. Reproduced from Ref. [13] with permission from Springer Nature

systems find applications in smart textiles, autonomous robots, bio-medical devices, drug delivery and tissue engineering [13].

7.2.1.2 Sensors

Sensing or detection refers to the quantitative estimation of the concentration of a targeted analyte or degree of completion of a chemical, biochemical, or mechanical process using an appropriate sensor technology. Sensing is an important component of analytical chemistry and finds tremendous applications in our daily life such as in detection of harmful pollutants and hazardous chemicals, quality control, biomolecules and disease biomarkers, temperature, concentration, humidity, pressure, motion, heart rate and so on. Sensing processes can be classified based on the process under investigation such as chemical sensing for detection of chemicals, biosensing for detection of biomolecules and bioprocesses, mechanical sensing for estimation of mechanical stress, strain or deformation, tactile sensing for detection of touch and contact. Another popular classification is based on the fabrication method or material of construction used for sensors such as 3D printed sensors, flexible sensors, semiconducting sensors, liquid metal sensors, resonators and so on. However, it is easiest to understand the functioning of sensors by their classification based on the mechanism of detection. In the subsequent discussion different additively manufactured sensors have been described based on their sensing mechanisms and related application examples. A list of different types of sensors fabricated using additive manufacturing is given as Table 7.1.

Optical sensors are based on comparison of differences in optical response exhibited by the sensor with and without the analyte based on different optical process such as reflectance, absorbance, transmission, scattering, emission, fluorescence, colour change, luminescence (chemiluminescence and electroluminescence) and plasmon excitation (SERS, SEIRA, SERRS) [80]. As can be seen from Table 1, AM sensors for biosensing applications have most commonly relied on optical sensing using luminescence, fluorescence and colorimetric detection such as glucose and lactate (blood and sweat), imaging (cell, bacteria and DNA), cell characteristics (viability, density, growth, aging and calcium), immunosensing and biomechanism sensing [81]. On the other hand, chemical sensing using AM sensors has mostly used measurements of optical spectrum, density, absorbance, transmission, plasmon resonane frequency changes and colorimetry for pH sensors, chemical reaction monitoring, liquid sensing, and liquid characterization.

Electrochemical sensors are based on electrochemical sensing mechanisms such as potentiometric sensors and voltametric sensors. Potentiometric sensors rely on the measurement of potential difference between working and reference electrode when no current is present. In the absence of current, potential difference corresponds to the analyte concentration (generally ions) in electrolyte. The electrodes in potentiometric sensors are often called ion selective electrodes since selective detection of a single ion from a mixture is generally observed in such sensors [82]. A chronoamperometry method uses the application of a large potential and observance of a large

Table 7.1 A compilation of different types of additively manufactured sensors based on the sensing mechanism employed, AM technology and materials used and demonstrated applications

Classification	Detected analyte/process	Sensing mechanism	Additive manufacturing technology (material)	References
Electrochemical sensors	Glucose and lactate	Amperometric	IJP (polymer)	[14]
	Cell dopamine and ATP	Amperometric	IJP (polymer)	[15]
	Cell imaging	Amperometric	IJP (polymer)	[16, 17]
	Liquid sensor	Amperometric	FDM (PET)	[18]
	Glucose and lactate	Amperometric	FDM (ABS)	[14]
	Cell imaging	Amperometric	FDM (ABS)	[18]
	Sweat glucose	Amperometric	FDM (ABS)	[19]
	Chemical reaction monitoring	Amperometric	DLP SLA (Photocurable resin)	[20]
	Gas sensor	Potentiometric	DLP SLA (Acrylic resin)	[21]
	Cellular metabolic sensor	Potentiometric	DLP SLA (Photopolymer)	[22]
	Cell density sensor	Potentiometric	PlasClear resin	[23]
	pH sensor	Potentiometric	SLM (stainless steel)	[24]
	Droplet length measurement	Potentiometric	SLA (resin)	[25]
	Sweat glucose	Conductometric	FDM (ABS)	[19]
	Influenza virus sensor	Electrochemical	FDM (PLA)	[26]
Optical sensors	Cell ATP	Chemiluminescence	IJP (polymer)	[27]
	Chemical reaction monitoring	Chemiluminescence	IJP (polymer)	[28]
	Sweat lactate	Chemiluminescence	FDM (ABS)	[29]
	DNA imaging	Electro-Chemiluminescence	SLA (clear methacrylate-based resin)	[30]
	Immunosensor	Electro-chemiluminescence	FDM (PLA)	[31]
	Bacteria sensor	Bioluminescence and absorbance	SLA (resin)	[32, 33]
	Cell toxicity	Bioluminescence	FDM (ABS)	[34]
	Cellular calcium imaging	Fluorescence	SLA (resin)	[35]

(continued)

Table 7.1 (continued)

Classification	Detected analyte/process	Sensing mechanism	Additive manufacturing technology (material)	References
	Cell imaging	Fluorescence	IJP (polymer)	[36]
	Cell viability	Fluorescence	IJP (polymer)	[37]
	Immunosensor	Fluorescence	IJP (polymer)	[38]
	pH sensor	Fluorescence	FDM (PET)	[39]
	Immunosensor	Fluorescence	FDM (ABS)	[40]
	DNA damage sensor	Fluorescence	FDM (ABS)	[41]
	Cellular aging sensor	Fluorescence	TPL (PEGDA)	[42]
	pH sensor	Colorimetric	IJP (photoresin)	[43]
	Aptamer based biosensing	Colorimetric	DLP SLA (photopolymer)	[44]
	Urine protein sensor	Colorimetric	DLP SLA (resin)	[45]
	Liquid concentration sensor	Colorimetric	DLPA SLA (colourless resin)	[46]
	Glucose sensor	Colorimetric	DLP SLA (photoresin)	[47]
	Chemical reaction monitoring	Optical absorbance	IJP (polymer)	[48]
	Biomechanism sensing	Optical absorbance	TPL (Foturan photostructurable glass)	[49]
	Bacteria sensor	Optical absorbance	FDM (ABS)	[50]
	Cell density sensor	Optical density	SLS (polyamide)	[51]
	Immunosensor	Optical density	FDM (ABS)	[52]
	pH sensor	Plasmonics	FDM (ABS)	[53]
	Chemical reaction monitoring	Optical spectrum	FDM (PP)	[54, 55]
	Chemical reaction monitoring	Optical spectrum	DIW (Acetoxy silicone polymer)	[54, 56]
	Liquid viscosity sensor	Optical imaging	DLP SLA (Transparent resin)	[57]
	Liquid characterisation	Permittivity	FDM (ABS)	[58]

(continued)

Table 7.1 (continued)

Classification	Detected analyte/process	Sensing mechanism	Additive manufacturing technology (material)	References
Electrical sensors	Radio frequency harvesting	Patch antenna	IJP (polymer)	[59]
	Wireless sensor network	Folded slot antenna	IJP (polymer)	[60]
	Bionic ear	Radio frequency reception	DIW (Cell-seeded alginate hydrogel)	[1]
	Particulate matter sensor	Quartz crystal microbalance	IJP (polymer)	[61]
	Bionic ear	Capacitance	FDM (PVDF)	[62]
	Liquid sensor	Capacitance	FDM (polyphneylsulfonate)	[63]
	Acoustic sensor	Capacitance	IJP (polymer)	[64]
	Strain sensor	Capacitance	IJP (polymer)	[65]
	Strain sensor	Capacitance and resistance	IJP (polymer)	[66]
	Strain and tactile sensor	Resistance and capacitance	FDM (Conductive composite and PLA)	[67]
	Strain sensor	Resistance	IJP (polymer)	[68]
	Temperature sensor	Resistance	FDM (FilaFlex)	[69]
	Strain sensor	Resistance	DIW (Viscoelastic carbon conductive ink)	[70]
	Tactile sensor	Resistance	DIW (CNT ink)	[71]
	Helical liquid sensor	Resistance	DIW (PLA—MWCNT composites)	[72]
	Gas sensor	Resistance	DIW (PEDOT:PSS)	[73]
	Tactile sensor	Piezoelectric resistance	IJP (polymer)	[74]
	Flow sensor	Hall effect (thermoelectric)	FDM (Magnetite nanoparticles in thermoplastic)	[75]
	Physiological sensor	Bioelectric signal	IJP (polymer)	[76]
Integrated sensors	Cell imaging sensor	Integrated sensors	FDM (ABS)	[77]
	Environmental sensor	Integrated sensors	FDM (PLA)	[78]
	Multifunctional biomembrane	Integrated sensors	Binder Jetting (polymer)	[79]

dip in current. An impedance analysis method is based on impedance differences before and after contacting with the solution. Potentiometric sensors often employ ion selective surface layers such as membranes, polymers, functional groups, and liquid crystals which can selectively capture the targeted ions [82]. Potentiometric gas sensors rely on the use of materials which are composed of an ion of the gas to be detected such as oxygen ions for O_2 detection using yttria sulphated zirconia or use of semiconductor crystals for potentiometric titrations such as InAs, InSb and GaAs. Voltametric sensors rely on the measurement of current as the potential difference between reference and working electrodes is varied and recorded current vs potential curves are called voltammograms [83]. Different types of voltametric methods are available and forms the basis of many chemical and biosensors. Additively manufactured electrochemical sensors have been frequently used and most sensors relied on the use of amperometric detection for biological and chemical processes such as chemical reaction monitoring, cell imaging, glucose, lactate, dopamine, and ATP detection while the use of potentiometric sensing has also been demonstrated for applications such as gas sensors, pH sensors, cell metabolism, cell density and droplet length detection.

Electrical or electronic sensors are based on quantification of the extent to which an electrical process has been completed in relation to the targeted analyte or investigated process such as resistive sensors measuring resistance changes, capacitive sensors measuring capacitance changes, conductometric sensors measuring conductivity changes, thermoelectric sensors measuring thermoelectric current, piezoelectric sensors measuring piezoelectric resistance or current and electromagnetic radiation reception (using antenna). The measurement of capacitance and resistance has most commonly been employed for development of AM electrical sensors which have found applications in wearable electronics and tactile sensors by detection of strain or touch [74]. Besides detection of different physical processes such as flow, temperature, and liquid levels detection of human physiological parameters has also been demonstrated using piezoelectric and thermoelectric mechanisms in addition to capacitance and resistance [84]. The use of antenna designs also forms the basis of some AM sensors such as radio frequency detectors (RFID tags) used in various applications such as proximity detection, security, wireless communication, and the bionic ear. The forming of functional materials and multimaterial integration capabilities of additive manufacturing are evident from the design of a bionic ear by Mannoor et al. which exhibits enhanced auditory sensing and complementary hearing from the designed left and right ears for radio frequency reception [1]. An intertwined conducting polymer composed of silver nanoparticles was printed in the form of a coil antennae and a cell-seeded hydrogel matrix provided the physical form of the ears as shown in Fig. 7.4. In-vitro culturing of cartilage tissue around the antennae was used to finish the hybrid bioelectronic 3D construct. The most sophisticated sensors are fully integrated sensors which rely on the use of multiple sensors communicating with each other and recording data simultaneously. The real potential of additive manufacturing is evident from such examples where the superior integration capabilities, rapid prototyping and free form fabrication capabilities are put into good use. For example, Xu et al. demonstrated the use of binder jetting and standard lithographic

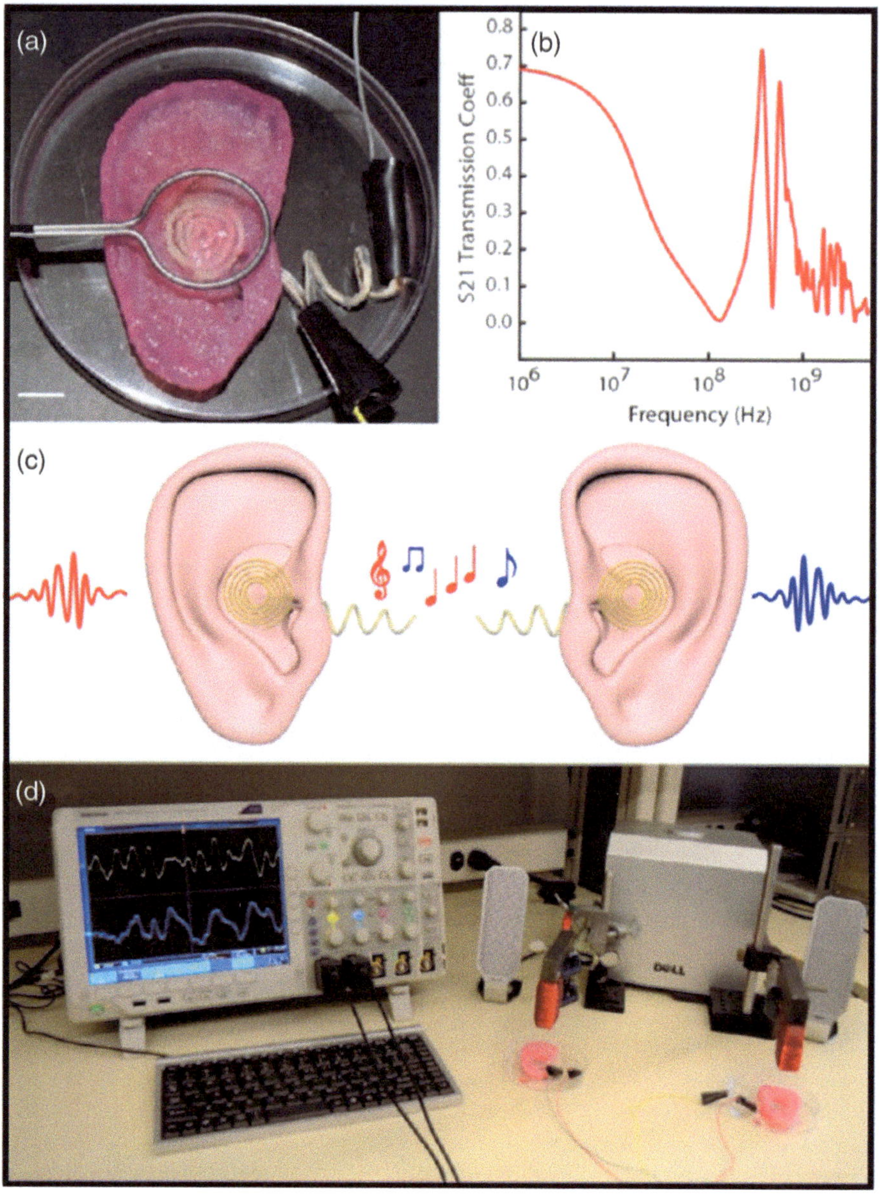

Fig. 7.4 Design, principle and electrical characterization of bionic ear; **a** Experimental setup used for electrical characterization wherein a loop antenna transmits the input signal and output signal is collected from the electrodes on cochlea; **b** Response of the bionic ear to input frequencies as seen from the plot of forward power transmission coefficient S21 variation with frequencies; **c** Schematic representation of signal reception from two bionic ears; **d** Image depicting complementary bionic ears listening to stereophonic audio music and collection of output signal. Reproduced from Ref. [1] with permission from American Chemical Society

techniques for integration of a multisensing platform in the form of 3D integumentary (skin-like covering) membranes which were used for cardiac measurements as well as stimulation across the entire epicardium (membrane attached to the walls of heart) [79]. The developed elastic devices completely enveloped the heart and provided a mechanically stable biotic/abiotic interface with a range of functionalities such as measurement and mapping of cardiological parameters, pH, temperature, mechanical strain, and actuation using electrical, thermal, and optical stimulation. Such devices are a great tool for recording the spatial and temporal effects of insults, diseases, and therapies by collecting information about the metabolic, excitable, ionic, contractile, and thermal states. This technology can be used to identify the origin of pathophysiological conditions such as arrhythmias, ischaemia, or heart failure [79].

Additively manufacturing has alleviated many issues with conventional sensor fabrication such as multistep serial fabrication methods and large chemical wastage resulting in comparatively lower economic viability, restriction to 2D or 2.5D devices and device integration limitations. AM can easily replace the existing device packaging and integration processes which have low turnover times and dependence on externals vendors and facilities. New processes based on AM can provide highly efficient and reliable multimaterial, free-form and localized fabrication. At the same time the sensing performance of AM sensors are comparable or even better then conventionally fabricated sensors because both types of sensors fundamentally use the same sensing materials. AM also enables fast and easy integration of sensors with end use equipment like network analyzers, analytical equipment, and smartphones. For example, Cevenini et al. demonstrated the 3D printing of a low-cost smartphone adaptor with a mini-dark box, cartridge with assay reagents and aligned optical interface with phone camera as a smartphone-based bioluminescence (BL) whole-cell toxicity biosensor [34]. They used genetically engineered human embryonic kidney cells which express a green-emitting luciferase mutant and used them as sentinel cells. The developed standalone device which also provided two droppers and a dedicated app for analysis allowed the user to perform a toxicity assay within 30 min as depicted in Fig. 7.5.

7.2.1.3 Optical Devices

Optical devices manipulate light using fundamental optical processes such as transmission, reflection, absorption, scattering, emission, luminescence, fluorescence, interference, and diffraction. Optical transducers which form the basis of optical sensors record changes in the magnitude of these optical processes as they interact with analyte under study. A basic criterion for designing optically active materials is the precise control in composition and 3D architecture of the used materials since most optical processes are affected even by minute changes in composition or geometry. Thus, additive manufacturing has enormous potential in photonics, plasmonics, wearable electronics and optoelectronics owing to its capability to manufacture combination of different materials in complex geometrical forms [84, 85]. In addition,

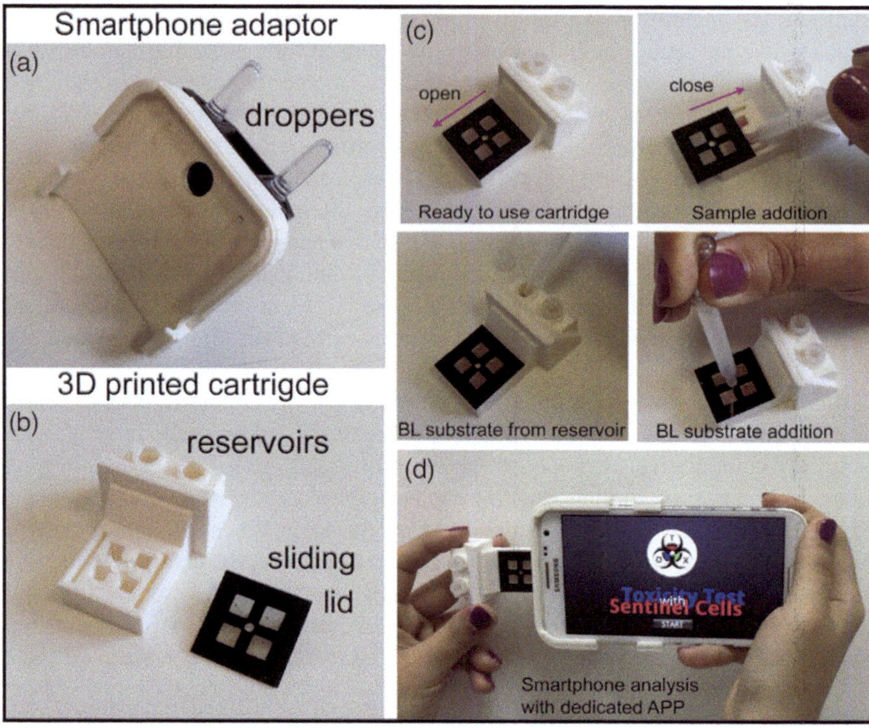

Fig. 7.5 Design and operation of a 3D printed smartphone sensor interface printed with FDM of ABS for a Samsung Galaxy Note smartphone; **a** 3D printed smartphone adaptor with slots for droppers; **b** 3D printed ABS cartridge comprising 4 wells, bioluminescence (BL) substrate reservoirs and a black sliding lid with transparent ABS windows; **c** Demonstration of steps for using the toxicity sensor including loading, sample addition, BL substrate addition and transfer to the phone cartridge for analysis. Reproduced from Ref. [34] with permission from Elsevier

compared to conventional thin film fabrication techniques such as photolithography and electron beam lithography, additive manufacturing allows direct printing on non-planar surfaces, curvilinear substrates, and tortuous 3D patterns allowing integration capabilities for devices in three dimensions [86]. This 3D integration is of paramount importance since it allows a more efficient and dense integration of device components such as light sources, lasers, lenses, waveguides, and photodetectors. The recent progress in optical devices mostly focusses on the development of novel fabrication methods, expansion of materials that can be used for fabrication, improvement in resolution of manufacturing techniques and improvement in accessibility of manufactured geometries [85, 87]. Some interesting examples of optical devices fabricated by additive manufacturing are briefed below. Kong et al. used direct ink writing of custom ink formulations for the unprecedented 3D printing of semiconducting quantum dot light-emitting diodes (QD-LEDs) exhibiting pure and tunable colour emission properties [88]. Moreover, the QD-LEDs could be conformally printed

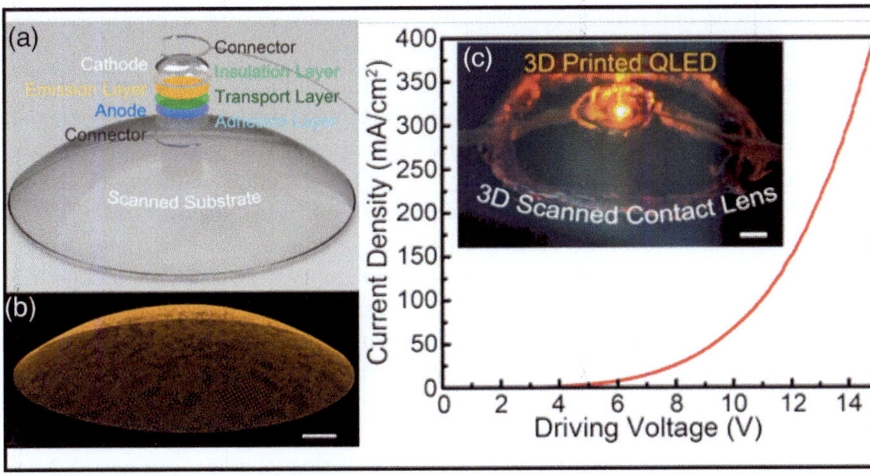

Fig. 7.6 Design of the 3D printed QD-LED for a curvilinear contact lens substrate; **a** CAD model showing QD-LED components and conformal integration on a lens; **b** 3D model of a contact lens acquired by 3D structured-light scanning; **c** Current density vs voltage characteristics of the 3D printed QD-LEDs atop the contact lens. The inset shows the electroluminescence output from the printed QD-LED. Scale bars are 1 mm. Reproduced from Ref. [88] with permission from American Chemical Society

directly on curvilinear substrates such as contact lenses enabling smart glasses technology as shown in Fig. 7.6. Direct ink writing allowed seamless integration of five different materials including emissive semiconducting inorganic nanoparticles, elastomer matrix, organic polymers as charge transport layers, solid and liquid metal conductors, and UV-adhesive transparent substrate layer. 3D printing of a complete $2 \times 2 \times 2$ cube matrix was demonstrated as an example. Authors anticipated further application of QD-LEDs array with integrated active components, interconnects and embedded methodology to micro electromechanical system-based devices, transistors, solar cells and photodiodes [88]. Ota et al. demonstrated the fabrication of smart objects and devices with integrated sensors, integrated circuit (IC) chips, antennas and interconnects using direct ink writing of a stretchable 3D material (Filaflex, Recreus), fused deposition modelling of polylactic acid, PDMS and liquid metals [69]. The method involved an initial printing of a base layer with pre-designed slots for subsequent integration of electronic components and microchannels. In the next stage, the microchannels were filled with liquid metals to complete the interconnect circuitry. In the third stage, integrated chips, resistors, capacitors, and other electronic components were added with their terminals connected to the liquid metal interconnects. A PDMS layer or transparent resin layer was finally applied to complete one layer of the smart object. In subsequent stages more layers were added in a similar manner to impart addition functionalities to the smart object. Such 3D integration is not possible using conventional thin film based electronic device integration technologies. Using the developed integration method, authors demonstrated the fabrication of a form fitting glove with embedded programmable heater, temperature sensor, associated control electronics for thermotherapeutic treatment [69].

In near future, photonic circuits are projected to outperform conventional electronic circuits which are unable to cope with the current demand of large-scale data storage, high speed computations and parallel computing due to their inherent size-derived (dimensions ~ 2 nm are already in use) performance limitation. Optical waveguides are like interconnects in electronic circuitry and used to link different nanoscale photonic elements. Ideal waveguides should exhibit low losses, enable high integration densities, multiple functions, and ability to be manufactured at high throughputs. Pyo et al. used meniscus guided direct ink writing for fabrication of free-standing 3D polystyrene waveguides and demonstrated its capability to connect randomly distributed nanophotonic elements with low substrate coupling leakage losses [89]. Polystyrene was used for waveguide fabrication because of its optical transparency, high refractive index (~1.59), and high integration capability. Demonstrated stacked connections without direct contact, Y-shaped joints, and stretchability of connections due to shape, enabled high integration capabilities for the fabricated waveguides even for stretchable components. Similarly, Weidenbach et al. used fused deposition modelling of polystyrene to fabricate 1.5 mm thick rectangular single mode waveguides for 120 GHz and demonstrated its use for fabrication of 1×2 splitters, multimode-interference 1×3 splitter and a planar waveguide coupler with tunable coupling response [90].

Luo et al. succeeded in additive manufacturing of fused quartz by feeding fused quartz filaments and subsequent deposition while melted using a CO_2 laser which couples to the phonon modes of quartz and produces rapid localised heating. Fused quartz has transparency in a large optical window and better thermal as well as mechanical resistance compared to glass. Fused quartz is widely used in optical components, communications, electronics, hermetic seals, chemistry labware and chemical and photochemical reactors. The technology once matured can potentially eliminate glass blowing or at the least complement it in manufacturing of custom designed and geometrically complex quartz components.

Two photo lithography (TPL) is one of the most precise additive manufacturing technologies with true 3D fabrication capabilities (even overhanging parts can be printed without support structures) and sub-micron resolution (<1 μm) capabilities. The superior fabrication capabilities of TPL are now being used to explore novel applications such as quantum computing and optical circuit boards. Boson sampling refers to the computational complexity in simulation of output distribution of bosons propagating in linear optical interferometers. Crespi et al. demonstrated the use of TPL to fabricate five mode integrated interferometer to study three photon interference and verified the quantum mechanical simulations [91]. It is anticipated that linear-optical quantum computers will enable quantum computation using adaptive measurements soon. Similarly, Spagnolo et al. used ultra-fast laser writing in borosilicate glass to fabricate a three-port beam splitter called tritter used to study bosonic coalescence of three photons [92]. Laza et al. demonstrated an improvisation to TPL called the two photon continuous flow lithography (TPCFL) wherein photoresists were exposed by the two photon absorption process in microfluidics channels for fabrication of 3D polymer particles with arbitrary shapes such as fibres, helices and bow-tie particles with sub-micron resolution and surface roughness as low as 10 nm [93]. Shutter opening time and fluid flow was used to control the length of the polymer particles and throughput of the process was increased by using multiple beams for exposure. Such 3D polymer particles find applications in photonics, micro-rheology, tissue engineering, drug delivery and biomolecule analysis. Thiele et al. demonstrated the fabrication of eagle eye inspired highly miniaturised foveated camera by 3D printing of different multi-lens objectives directly on a complementary metal-oxide semiconductor (CMOS) image sensor using two photon lithography [94]. The developed system combined four doublet lenses with different focal lengths in a 2 × 2 arrangement and a 70° full field of view and increasing angular resolution of up to 2 cycles/deg field of view in the centre of the image was achieved. The entire footprint of the device was less than 300 × 300 μm and height less than 200 μm. Developed miniaturized multiaperture imaging system have applications in endoscopy, optical metrology, optical sensing, surveillance drones and security.

AM technologies are rapidly changing the technological landscape for design and production of passive and active optical components, photonic and optoelectronic devices, and high-density 3D integration of device components. Furthermore, the controlled manipulation of light by custom designed 3D optical devices is expected to enhance optical resolution and reduce aberrations in vision components for robots, drones, and autonomous movable systems. Rapid progress in the field of wearable smart devices and sensors utilising integration of optical devices with gloves, bioimplants, endoscopes, catheters, contact lenses, glasses, and optical fibres are anticipated for high level functions such as real time monitoring of body fluids and recovery, augmented reality, improved vision and so on.

7.2.1.4 Electronic Devices

Today electronic devices are ubiquitous and commonly seen in everyday use devices such as pen drives, printed circuit boards of all programmable home tools and mother boards of computers and laptops. Additively manufactured or printed electronics refer to those electronic devices which have been manufactured using 3D printing techniques imbibing superior properties in such devices because of flexibility of design, free forming capabilities, low wastage of materials improving economic viability, better conductivity of 3D interconnects, flexible structure, and many others. As a result, research, development, and applications of printed electronics are rapidly increasing. In addition, the rise of wearable electronics in recent research demands electronic devices to be stretchable, foldable, portable, and mechanically durable while retaining performance of electronic components. The market for printed and flexible electronics was estimated to reach $73 billion by 2027 [95]. Additive manufacturing has profoundly enabled the flexible electronics revolutions by providing a wide variety of materials that can be printed, facilitating development of conducting ink formulations, integrated printing of all electronic components (interconnects, sensors, devices, packaging) and emergence of novel methods for AM of flexible electronics. Compared to well proven and conventional methods of electronic device fabrication such as photolithography, metal evaporation and wet chemical etching, AM methods provide much higher economic viability, low wastage, comparable or better performance and superior functionality. Some examples of recent developments in additive manufacturing methods for printed electronic devices and their applications are briefed below.

Direct write methods of AM are widely used in fabrication of printed electronics owing to the well-understood and simple conversion of CAD design to devices. Direct ink writing, inkjet printing and aerosol jet printing are some examples of direct write methods often used for electronic devices. In addition, other direct write methods for specific use in electronics research have also been reported such as matrix assisted pulsed laser evaporation direct write (MAPLE DW) and laser induced forward transfer (LIFT) which rely on the precise exposure of electronically active materials to be deposited with a laser. Laser based direct writing methods were one of the earliest demonstrations of AM at the microscales. Different strategies can be used for micromanufacturing using precise motion of lasers such as laser induced vaporisation of materials and controlled transfer, micromachining of unwanted components and photopolymerization based additive manufacturing [96]. In LIFT a laser evaporates a dense coating of atoms, ions or small molecules which is transferred to a substrate in a controlled manner allowing direct write capabilities. However, the process is very aggressive, thereby, limiting the use for materials which decompose from heat or vaporisation. Shaw-Stewart et al. demonstrated fabrication of polymer light emitting diodes by LIFT methodology using a sacrificial triazene polymer layer coated with aluminium as cathode material, poly-(ethylene oxide) to improve electron injection and poly[2-methoxy, 5-(2ethylhexyloxy)-1,4-phenylene vinylene] (MEH-PPV) as the light emitting polymer [97]. Al-PEO-MEH-PPV laminates were successfully transferred to bilayer ITO/PEDOT:PSS (poly(3,4-ethylenedioxythiophene) poly(styrenesulfonate) blend) electrodes and a 100-fold increase of luminous efficiency was obtained for a maximum of 0.45 cd/A at a brightness of 400 cd/m^2. Chrisey et al. innovated the MAPLE DW methodology as a variation to LIFT wherein a transparent ribbon is used to coat the material to be deposited and a pulsed laser contacts the transparent ribbon rather than the coated material itself [98]. This makes the process much softer and allows the deposition of most materials like metals, metal oxides, ceramics, ferrites, polymers, chemical precursors and even living cells of E. Coli bacteria. MAPLE DW deposits high quality conformal coatings in a contactless manner as per CAD design with high write speeds (200 mm/s) and resolution as high as 2 μm. The use of MAPLE DW was demonstrated in fabrication of capacitors, interconnects, phosphor displays, yttrium iron garnet (YIG) core inductors and coplanar resistors [98].

Aerosol jet printing (AJP) has also been widely used for fabrication of a wide variety of 3D electronic devices owing to its maskless and contactless direct write capabilities with high resolution (μm) and throughput [99–102]. Marinov et al. deposited interdigitated Ag electrodes using AJP which were 20–50 μm wide and 8–10 μm thick with good adhesion to the substrate and electrical conductivity of 4–12 μΩ.cm directly on printed circuit boards [103]. The Ag electrode were further functionalised with a chemoselective polymer/carbon black composite for binding nerve gas stimulant dimethylmethyl phosphonate to demonstrate its chemiresistive detection. Conductors operating at radio frequency with low signal loss were also demonstrated. Horteis et al. demonstrated fabrication of front contacts for solar cells with AJP and observed efficiencies higher than 20% [104]. AJP was used to direct write a silver seed layer which was subsequently plated using light induced plating to result in 45 μm contact lines with contact resistivities as low as $2m\Omega cm^2$, open-circuit voltage of 661 mV and short circuit current of 38.4 mA/cm^2 on a $110\Omega/sq$ emitter and cell area of 4 cm^2. Cost-effective materials, high throughout, high resolution deposition techniques and low temperature processing have enormous potential for photovoltaic devices with even higher efficiencies [105].

Radio frequency identification (RFID) tags have found wide implementation in our modern technological society and used in electronic devices such as proximity cards, passports, transportation payments, animal tracking, security applications and asset management. Passive RFID tags rely on the principle of an interrogating device emitted radio frequency reception by an antenna integrated on the RFID tag. The antenna generates a rectified direct current in response to the input frequency which is used to power subsequent communication back to the receiver of interrogating device through an IC chip also integrated on the RFID tag. RFID antennas are fabricated using cheap metals such as copper, silver, or aluminium. Hoey et al. investigated the efficacy of MAPLE DW and AJP DW for fabrication of RFID tags compared to commercial ultra-high frequency (UHF) etching-derived copper-based RFID tags [106]. Direct write-derived RFID tags showed equal and even higher performances compared to commercial tags without any decrease in communication distances. Interestingly, antenna thickness of 0.3–0.5 μm, surface roughness and edge definition derived from direct write methods did not make any measurable interference in RFID communication.

Ha et al. demonstrated the fabrication of thin film transistors (TFT), NAND gates, ring oscillators and inverters for flexible electronics on polyimide and silicon substrates using AJP of aqueous semiconducting CNT-based ink, high-capacitance ion gel gate dielectric and a conducting polymer gate electrode using PEDOT-PSS. Use of liquid inks for direct printing of electronic devices on flexible plastic foils for devices like flexible displays and large-area sensors are highly desired from economics viewpoint because of ease in manufacturing. However, slow switching frequencies and associated high supply voltages hamper the development of such flexible electronics. TFTs of 50 μm channel length developed by Ha et al. showed substantial improvement in printing of electronic circuits using AJP and exhibited ambipolar transport with electron and hole mobilities greater than 20 $cm^2/V.s$ [107].

Thermoelectric (TE) devices are solid state devices that convert heat to electrical power or electrical power to heat. A variety of thermoelectric electronic devices are available today such as Peltier coolers, body-heat energy harvesters, automotive heat recovery, high-power industrial heat harvesters, solar-thermal systems, and temperature sensors [108]. However, large scale and commercial application of this technology is not feasible with current devices because of several practical issues such as lack of suitable materials, manufacturing challenges, low conversion efficiencies and economic viability [109]. Bismuth-telluride (Bi_2Te_3) and lead-telluride (Pb_2Te_3) show the highest performance for TE conversion at room temperature but are costly, scarce, toxic, and brittle. The toxicity and brittleness of these materials are especially a big hurdle for large scale manufacturing and commercial applications and more so because of their low utility for curvilinear and flexible substrates. Other commonly used inorganic TE materials are semimetals and doped semiconductors such as chalcogenides, silicides, clathrates, skutterudites, half-Heusler alloys and oxide and tetrahedrites [109]. Recent research has seen the emergence of organic and inorganic–organic hybrid TE devices owing to their ease of processing and manufacturing, low cost, low density, flexibility, stretchability, non-toxicity, and inherently low conductivities. Some examples of organic TE materials are polyaniline, polypyrrole and polythiophene which suffer from relatively low electrical conductivity and Poly(3,4-ethylenedioxy-thiophene) (PEDOT) composites with poly(styrenesulfonate) referred to as PEDOT:PSS [109]. These organic materials can also be readily integrated with inorganic TE materials with the objective to retain high efficiency of inorganic components at reduced loading and retaining the high processability and flexibility of the polymer components. Because of its unique advantages of rapid fabrication of complex geometric structures, waste minimization, reduced consumption of raw materials and time, freedom of design, and automation and capability to integrate different materials into a coherent designed form, additive manufacturing of such materials make them highly suitable for TE applications.

TE materials are usually compared in terms of their efficiencies which is related to the dimensionless figure of merit ZT ($= S^2\sigma T/\kappa$, where S is the Seebeck coefficient, σ is the electrical conductivity, κ is the thermal conductivity, and T is the absolute temperature) or the power factor ($S^2\sigma$) [110]. Thus, TE materials which show a high-power factor, but low thermal conductivity are desired for a high ZT value or TE conversion efficiency. Bi-Te exhibits a ZT value around 1 at room temperature. He et al. used stereolithography with custom resin formulations wherein a bismuth, antimony, and tellurium alloy ($Bi_{0.5}Sb_{1.5}Te_3$, BST) obtained from dry milling were homogeneously integrated to fabricate circular and square TE patches [111]. TE materials with 60% loading of BST alloy exhibited a ZT of 0.12 at 43 °C. Moreover, authors concluded that the developed material had desired exceptionally low thermal conductivity (0.2 W m^{-1} K^{-1}) but the ZT value was reduced because of the inferior electrical conductivity. Oztan et al. successfully demonstrated fused deposition modelling of acrylonitrile butadiene styrene (ABS, 20 wt%) and bismuth telluride (Bi_2Te_3, 80 wt%) composites into concentric cylinders and 5-point stars exhibiting an exceptional ZT value of 0.54 [108]. ABS was used as a sacrificial printing template and removed post print by annealing. Authors concluded that the high porosity of resulting formed matrices reduced electrical conductivity and was responsible for the relatively lower ZT values compared to pure Bi_2Te_3. Shi et al. investigated the use of selective laser sintering for AM of $Bi_{0.5}Sb_{1.5}Te_3$ alloys and observed significant porosity in the resulting structures with mixed benefits. Due to the porous structure, the electrical conductivity was observed to be 1.8 folds lower than bulk BST alloy and thermal conductivity 18% lower than bulk BST alloy. Slightly higher Seebeck coefficient resulted in a ZT value higher than bulk material (1.29) but with some loss of mechanical strength due to imbibed porosity.

Qiu et al. demonstrated the use of selective laser melting of $Bi_{0.4}Sb_{1.6}Te_3$ (BST) alloys for fabrication of a sub-millimetre (0.2 mm) TE device exhibiting a ZT value of 1.1 [112]. Single crystal BST alloys show the highest conversion efficiencies but relatively lower compressive strengths (37 MPa) which is not desired since bulk device manufacturing using such brittle materials requires good compressive strength. Authors demonstrated that SLM of BST alloys produces a polycrystalline material with exceptional compressive strength (91 MPa) with an orientation factor of 0.9 i.e. a structure very similar to single crystal. High ZT values of 1.1 with high compressive strengths allowed the fabricated devices to reach overall conversion efficiencies of 4% and maximum cooling temperature difference of 62 °C [112]. Kee et al. demonstrated fabrication of self-healing and stretchable organic thermoelectric devices using direct ink writing of DMSO, PEDOT-PSS and polymeric surfactant based ink formulations [113]. Stable TE performance was observed up to 35% tensile strain and TE performance was retained to higher than 85% of initial capacity even after repetitive cutting and healing cycles with a healing response time of just 1 s. A power factor of 2.5 $\mu W\ m^{-1}\ K^{-2}$ was exhibited in the temperature range of 25–106 °C. Du et al. demonstrated an innovative bilayer structure using FDM of polylactic acid (PLA) composites with Bi_2Te_3 and silver to obtain power factors of 875 $\mu W\ m^{-1}\ K^{-2}$ at 87 °C [110]. Their strategy was to fuse printed PLA-Bi_2Te_3 (Bi_2Te_3 39.6 wt%) layers with PLA-Ag (Ag 41.7 wt%) layers with high electrical conductivity to boost the power factor by using the improved carrier transport and interfacial effects in the developed junctions. The field of thermoelectric electronic devices is still in a nascent stage but rapidly emerging. Flexible TE devices could power wearable electronics and enable low-grade waste heat recovery from non-planar surfaces such as human body or even industrial pipelines. Most importantly, additive manufacturing is gainfully and positively enabling the thermoelectric research with multiple advantage.

Piezoelectric (PE) materials generate electrical charge on application of mechanical stress. The PE effect is also reversible such that an application of an electric field generates an internal mechanical strain. Piezoelectricity originates from the electromechanical interactions between the electrical and mechanical state of PE materials. A variety of natural and synthetic piezoelectric materials are currently available such as ceramics including quartz, topaz, Rochelle salt, bone, tourmaline group minerals, perovskites such as barium titanate ($BaTiO_3$) lead titanate ($PbTiO_3$), lead zirconate titanate ($PbZr_xTi_{1-x}O_3$, PZT), potassium niobate ($KNbO_3$), lithium tantalate ($LiTaO_3$), lithium niobate ($LiNbO_3$) and organics such as silk, wood, enamel, hair, dentin, rubber, paper, polyvinylidine fluoride (PVDF), Poly (vinylidene difluoride- trifluoro ethylene) P(VDF-TrFE), and polypropylene (PP) [114, 115]. These materials find applications in devices such as self-powered sensors, energy harvesting, energy storage, actuators, transducers, acoustics, damping, biodiagnostics and implants [116]. With the rise of internet of things technology, the number of sensors and data acquisition and communication devices are expected to increase exponentially. Similarly, future wearable sensors will require a reliable and flexible power source. Since such miniaturised sensors have low power requirements from milliwatt to microwatts, it is ideal for these devices to be powered by piezoelectric devices while harvesting the waste energy and without the need of any external power sources. Common natural and synthetic motions can be easily utilised by piezoelectric materials such as vibrations, pressures, strain, and sound for subsequent conversion to electricity. Similarly, supply of external energy for actuation and transduction also forms the basis of some piezoelectric devices.

Conventional PE devices manufactured using substractive manufacturing (machining) are not competitive in their performance because of relatively poor resolution and integration densities of machining processes and not economically competitive because of long production runs and low utilization of materials. Mechanical stress can also cause grain loss, strength degradation and near-surface area depolarization of piezoelectric current in absence of adequate device integration and packaging [117]. Comparatively, additive manufacturing provides design flexibility, capability to fabricate complex geometrical features, efficient device integration, assembly and packaging work flow, superior integration capabilities with other electronics, high resolution and excellent electrochemical response from 3D design, improved piezoelectric response and device sensitivity [117]. Some examples of additively manufactured PE devices and their fabrication methods are briefed below:

PE devices require poling which is the process of alignment of the dipoles of the crystalline phase in the PE materials before their functional use. Poling performed using a strong electric field is called contact poling whereas poling using a corona needle to induce ionisation of the air molecules above the PE material is called corona poling. PVDF is a semicrystalline PE material which exists in its α, β, γ and δ crystalline phases depending on the orientation of polymer molecules. While the α crystalline form of PVDF is nonpolar because of random orientation of dipoles, its β crystalline form is polar with all dipoles pointing in the same direction in an all-trans zigzag conformation and primarily responsible for the piezoelectric properties of PVDF [118]. α-PVDF can be transformed into β-PVDF at elevated temperatures by mechanical stretching, contact poling, corona poling or electrospinning. Lee et al. innovated an additive manufacturing method called the electric poling-assisted additive manufacturing (EPAM) wherein PVDF filaments and electric field can be used for continuous and simultaneous poling as well as free form fabrication of PE devices [118]. During the FDM like process, molten polymer at elevated temperatures experiences mechanical stress from the leading nozzle and poled by the strong electric field. The poled PVDF PE devices exhibited enhanced PE performance whereas in the absence of electric field dipoles were not aligned. Bodhke et al. also used the EPAM method for PVDF-$BaTiO_3$ hybrids with 10 wt% $BaTiO_3$ loading to fabricate a 40-layer circular spiral, 40-layer square spiral and mesh scaffolds and demonstrated its use in fabrication of gait sensors [119]. 300% increase in piezoelectric charge output was achieved by poling of developed hybrids compared to pure PVDF using an electric field of 1 MV m^{-1} beyond which deposited films underwent breakdown. Cu tape and Ag paint electrodes were used for current collection and develop gait sensors successfully differentiated between walking and stomping using PE elements in the shoe sole front and hind as shown in Fig. 7.7.

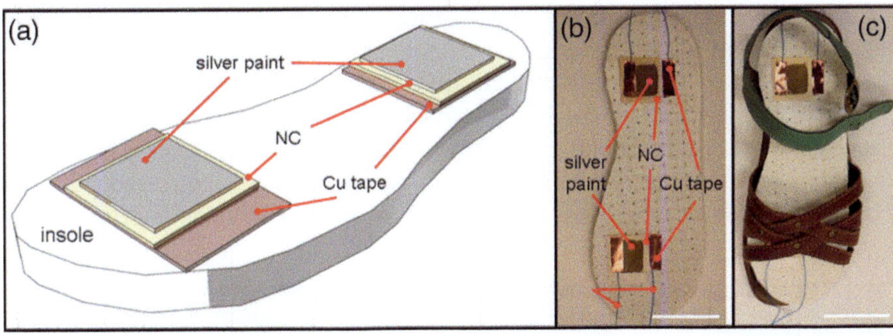

Fig. 7.7 **a** Schematic depiction of the construction of 3D printed piezoelectric sensor for gait monitoring integrated with copper tape and silver current collectors sandwiching piezoelectric material (NC); **b** Integration of the sensor on the shoe sole; **c** Re-installation of the shoe-sole in shoe base. Scale bars are 2 cm. Reproduced from Ref. [119] with permission from American Chemical Society

Kim et al. used DLP SLA using acrylate functional groups decorated barium titanate nanocrystals integrated in photoliable polymers such as polyethylene glycol diacrylate for fabrication of organic–inorganic hybrid PE devices in complex 3D shapes. $BaTiO_3$ loading up to 10 wt% could be achieved and covalent bonding of nanocrystals with the polymer matrix enabled better transfer of mechanical stress leading to piezoelectric coefficients ~ 40 pC/N, which is over 10 times larger than composites prepared using unfunctionalized nanocrystals. Fast fabrication of (<2 s).

complex arrays such as dot array, square array with variable pore sizes, honeycomb arrays, cross arrays, 3D mushroom arrays, tapered cantilever array and honeycomb nanotube arrays were demonstrated using the DLP SLA additive manufacturing technology [120]. Chen et al. demonstrated the use of DLP SLA using 70 wt% loading of $BaTiO_3$ nanoparticles in photocurable resin for the fabrication of ultrasonic PE sensors for ultrasonic imaging [121]. Ball milling of $BaTiO_3$ with surfactant and custom solvent mixtures allowed the preparation of dispersed $BaTiO_3$ nanoparticles of 100 nm size. Ball milling was also used for preparation of homogeneous photo resin composites with 70 wt% $BaTiO_3$ nanoparticles loading. Post fabrication annealing treatments at high temperature and in presence of air allowed the removal of all organic material and preparation of dense $BaTiO_3$ bodies with a density of 5.64 g/cm^3 which is 93.7% of bulk $BaTiO_3$. Piezoelectric constant of 160 pCN^{-1} was exhibited by the developed PE material which was close to pure $BaTiO_3$ (190 pCN^{-1}). Lim et al. [121] demonstrated inkjet printing of a flexible PE device using by using 55 vol% $BaTiO_3$ in ink formulation, silver bottom and copper top electrode on a polyimide substrate with a SU-8 epoxy packaging [123]. Annealing treatments were not required, and contact poling was done using electric fields from 30 – 200 kV cm^{-1} for 3 h. Developed devices exhibited an impressive open circuit voltage of 7 V and a short circuit current of 2.5 μA leading to an effective power output of 5 μW. Similarly, Chen et al. investigated the performance of alkaline niobate $K_{0.48}Na_{0.52}NbO_3$ based PED devices fabricated by micro stereolithography and subsequent annealing [124].

Piezoelectric coefficient value of 170 pCN^{-1} was exhibited which was close to that of bulk material. Lim et al. developed a novel laminar piezoelectric composite based on sheets of polylactic acid, single-walled carbon nanotubes and silver nanowires that exhibited high transparency, mechanical stretchability, deformability and low power consumption [122]. The composite heterostructure was used to develop an interactive human–machine interface (iHMI) which enables hardware control (robotic arm) by hand and feedback collection as shown in Fig. 7.8. Such human interfaces when

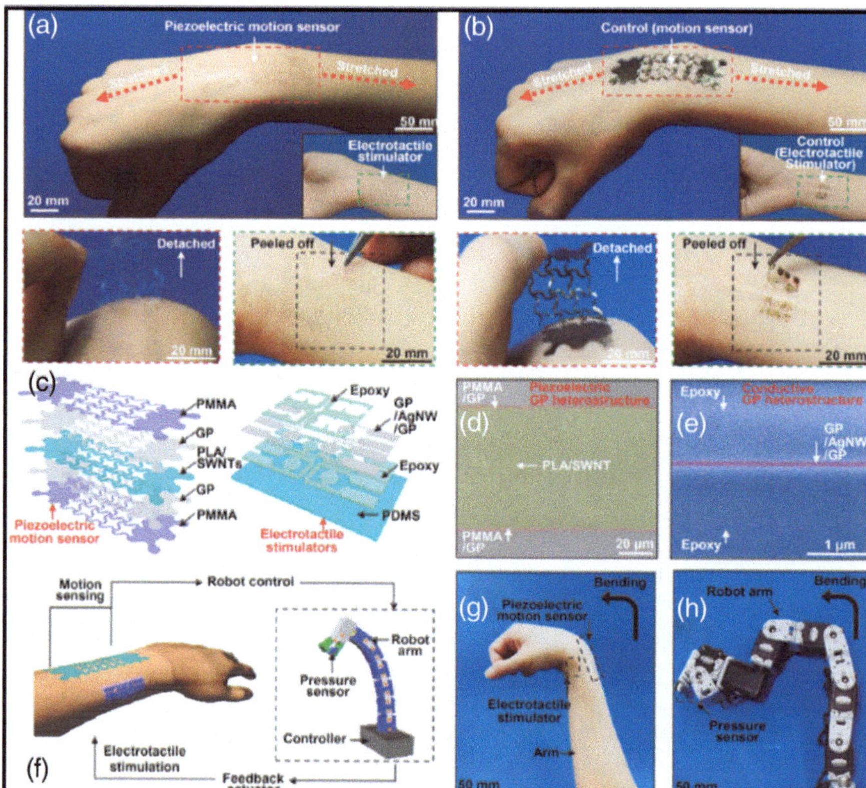

Fig. 7.8 Design of a transparent piezoelectric motion sensor and electrotactile stimulator for interactive human robot interface; **a** Transparent piezoelectric motion sensor in conformal contact with the top and bottom of a human wrist. Partially detached images of the motion sensor (bottom left) and the electrotactile stimulator (bottom right) are also shown; **b** Images of the control specimens of the motion sensor and the stimulator; **c** Exploded schematics of the fabricated heterostructures used in the motion sensor and the electrotactile stimulator; **d** Colorized cross-sectional scanning electron microscopy (SEM) image of the heterostructures used in the motion sensor (PMMA/GP: red, PLA: green); **e** Colorized cross-sectional SEM image of the electrotactile stimulator (Epoxy: blue, GP/AgNW/GP: pink); **f** Schematics of the closed-loop system showing flow of data and resultant feedback stimulation; **g** Representative image showing human motion to control the robot arm; **h** Representative image of the corresponding position of the robot arm, which is controlled by the human motion. Reproduced from [122] with permission from Wiley

developed using conventional methods show limitations of poor wearability, user discomfort, low signal-to-noise ratios due to conformability and undesired aesthetics [122]. Manufacturing of PE devices with AM still poses some issues such as precision and the weak strength of the printed parts arising from layer-to-layer bonding, high porosities undesired for PE applications, large print time for complicated geometries and overhanging parts due to requirement of support structures and need for even better printable photoelectric, current collector and packaging materials. The continuous development of AM technologies for PE device fabrication is expected to alleviate these issues and facilitate even better performances, higher energy storage density, electromechanical conversion efficiencies and wide applications in different areas.

7.2.2 Bulk Sensitive Applications

Unlike surface sensitive chemical processes which are confined to the first few hundred microns on the surface of a functional object, bulk sensitive chemical processes depend on the composition and arrangement of bulk material because of the bulk nature of mechanisms involved in such processes such as absorption, adsorption, catalysis, separation, batteries, capacitors, and microfluidics. These processes typically involve the diffusion of molecules or charge carriers in the bulk of the material and thus heavily benefit from a porous network, high specific surface areas and adequate functional composition of the bulk material. Physical phenomenon such as mass transfer, heat transfer, maldistribution, temperature gradients and composition homogeneity begin to play an increasingly important role as the size of objects increase [125]. Thus, forming and shaping is widely implemented in bulk sensitive applications to retain the functional activity of the bulk materials while improving their performance characteristics. As a bulk processing technology, additive manufacturing outperforms other manufacturing techniques due to reduced wastage of materials, decentralized and customized production, capacity to manufacture complex shapes and patterns and availability of wide range of materials that can be printed. On the other hand, conventional manufacturing technologies can still outperform AM at large scales due to the related economy of scaling. Thus, AM is finding great acceptance for rapid prototyping, lab to pilot scale experimentation, and small-medium scale production of reactors, microreactors, microfluidics, mixers, and heat exchangers.

7.2.2.1 Microfluidics and Flow Devices

Microfluidics refer to microfabricated devices consisting of a network of interconnected microchannels (channels with diameter in microns) designed to handle small quantities of fluids. Use of microfluidics leads to reduction in consumption of reactants, solvents or analytes, increased throughput, improved safety and favorable economics of production and operation [126]. These advantages have promoted wide interest from analytical and medical diagnostic communities where sample size reduction to picolitre or single cell quantities has obvious benefits [127]. Microfluidics are thus often called "lab on a chip" or micro total analysis systems (microTAS/μTAS) [128, 129]. Active microfluidic devices are based on the actuation of pumps and valves to control fluid flow using an energy input whereas passive microfluidic devices consisting of separators, mixers and distributors use external actuation or capillary forces to control the fluid flow. Using such active and passive components microfluidics can integrate several functionalities into a single device such as enzymatic processing, cell/tissue incubation controls, biochemical processing, optoelectronic readouts, automated and computerized control, and wireless capabilities [126, 130].

Manufacturing of microfluidic devices is an emerging field with enormous potential due to widespread point-of-care applications. Conventionally microfluidic manufacturing has relied on the use of microfabrication capabilities in cleanroom which are tedious and time consuming. Initial microfluidic devices were mostly PDMS (polydimethoxysilane) based and manufactured by soft lithography on glass or silicon substrates in planar geometries due to restraints of the fabrication process [131]. Moreover, such processes led to wastage of high volume of hazardous chemicals such as photoresists and etchants. Thus, the microfluidics research community is rapidly shifting towards additive manufacturing for manufacturing of microfluidic devices. These devices can be directly printed with plastics, metals or ceramics using AM, however, PDMS still remains the material of choice owing to its desirable properties such as optical transparency, gas permeability, water permeability, biocompatibility, low Young's modulus (1000 times lower than plastics) and economic viability [126, 128]. However, a few issues have hindered the rapid commercialization and scale-up in the microfluidic industry such as lack of standardized user interfaces hampering integration with existing equipment, [132] operation and maintenance of complex microfluidic control systems need trained personnel and low commercial viability of the PDMS process itself [128, 133]. Thus, recent research has seen the shift towards manufacturing of microfluidics using AM which presents great incentives of customized and decentralized production.

A great deal of recent innovation in microfluidics is currently focused on development of novel materials which can replace PDMS but also be additively manufactured and demonstration of additively manufactured PDMS-free microfluidics with available materials. In this regard, materials are desired which demonstrate biocompatibility, optical transparency, cost, and ease of manufacturing [126]. Optical transparency allows integration of microfluidics with optical readout functionality and interesting applications in optoelectronic and microelectromechanical systems (MEMS) industry [134]. Consequently, photopolymerization based AM technologies with capability to produce transparent components with high resolution are being actively pursued. However, the resins used in SLA are not biocompatible with limited gas and liquid permeability. Bonding of the microfluidics becomes a persistent issue in FDM derived microfluidics. Another important criterion behind experimenting with different AM technologies for production of microfluidics is the manufacturing of blockage free channels with uniform dimensions. Fusion and jetting based AM technologies often leave undesired residue which can block the microfluidic channels. Similarly, in SLA prolonged solvent is often required to remove uncured resins or material used to provide support in microfluidics channels.

Many reports of exciting progress in 3D microfluidics are available in the literature and a few representative examples are briefed below. Lee et al. demonstrated fabrication of modular passive microfluidic elements using stereolithography with a desktop 3D printer having 30 μm resolution for mixing, gradient generation, reaction and multiple input/output generation [38]. The modules could be simply assembled in horizontal or vertical orientations using metal pins, O-rings and uniquely designed connecting elements such as cone-shaped alignment features, O-ring hole and horseshoe-shaped metal pin holes as shown in Fig. 7.9. Different passive microfluidic elements of the device were used to demonstrate detection of alpha-fetoprotein biomarker [38]. Rogers et al. also relied on stereolithography to print integrated membrane based valve systems for pressure based actuation and demonstrated its use up to 800 actuation cycles [135]. Carrell et al. demonstrated simple sealing of additively manufactured open face microfluidic channels with PDMS using plasma treatment and either a silicone spray or (3-Aminopropyl)triethoxysilane (APTES) functionalization [136]. Au et al. demonstrated the use of stereolithography for fabrication of biocompatible and optically transparent valves and pumps for microfluidic automation including automatic routing, dispensing, mixing or separation of fluids

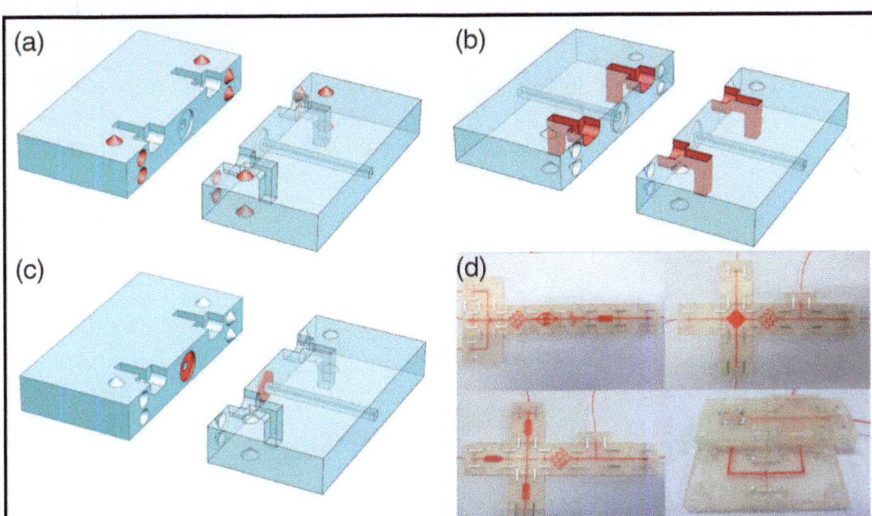

Fig. 7.9 Schematic showing the design of modular microfluidics chambers and the stereolithography fabricated transparent microfluidic assemblies; **a** Convex and concave cone-shaped alignment features; **b** Horse-shoe shaped metal pin hole alignment features; **c** O-ring hole-based alignment features; **d** Different possible assembly combinations used as integrated microfluidic device for biosensing. Reproduced from [38] with permission from Royal Society of Chemistry

through microchannels [35]. The research work systematically describes the development of active microfluidic components such as single valve switch, two-valve switch, peristatic pump and a four-valve switch as a cell perfusion system using fluid circuit diagrams and actuation mechanisms as shown in Fig. 7.10 [35].

Another important aspect of microfluidics research is the synthesis of micro and nanoparticles by microfluidic constructs using multistep synthesis chemistry such as polymerization, precipitation, and synthesis using sol–gel, hydrothermal and thermolysis methods [137]. The ultrasmall channels and miniscule liquid handling capabilities in microfluidics allow excellent control over shape and size distribution of

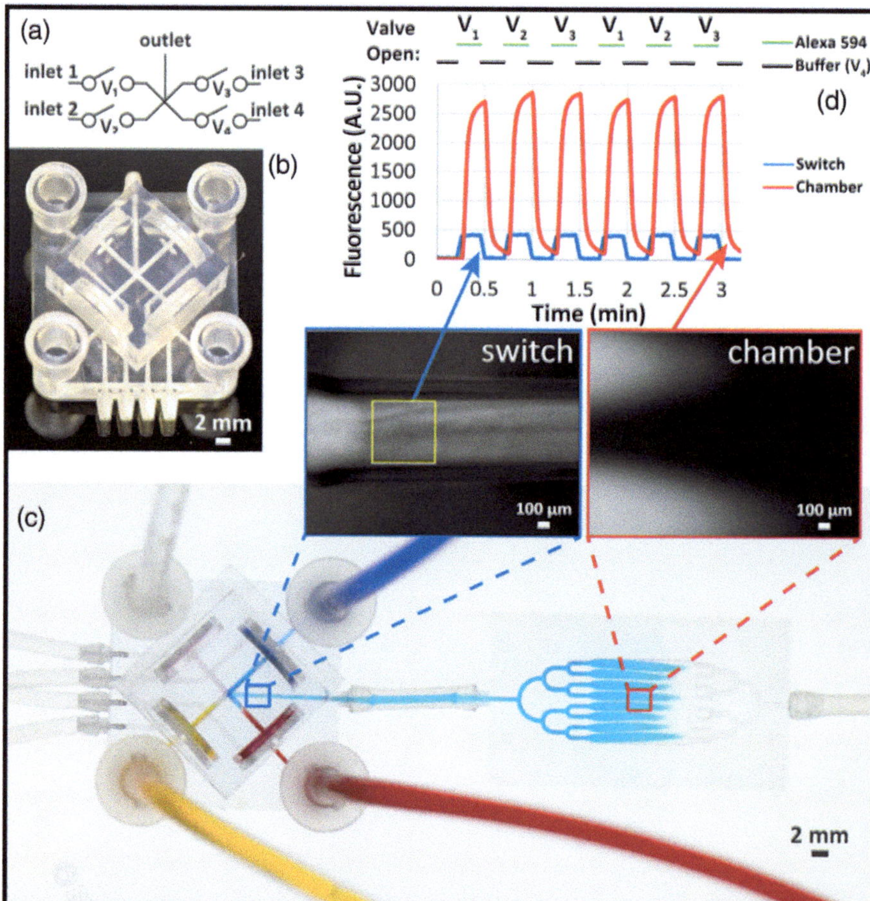

Fig. 7.10 An automated microfluidic 3D-printed cell perfusion system; **a** Flow circuit diagram; **b** Image of the fabricated four valve switch; **c** Dye filled switch connected to a cell culture chamber (Insets indicate fluorescent micrograph of the indicated locations; **d** Fluorescent plots indicating concentration profile during in the switch and the chamber. Reproduced from Ref. [35] with permission from Royal Society of Chemistry

synthesized particles. Moreover, microchannels can safely sustain higher pressures and safely handle hazardous reagents in minimal quantities allowing experimentation with novel synthesis systems. The additive manufacturing revolution is anticipated to play an increasingly important role in microfluidics research. AM promotes modular and team-based CAD design approach while minimizing waste products, tooling requirements and prototyping times. Considering the multi-disciplinary nature of microfluidics research, digital content design and sharing of useful valve and distributor architectures through internet is a niche aspect with great potential to rapidly change the technological landscape.

7.2.2.2 Catalysis

A catalyst is a substance which speeds up the rate of a chemical reaction without itself getting consumed. Catalysts are the backbone of industrial manufacturing and without them it would be impractical and uneconomical to manufacture important chemicals which are commonly used in our daily life such as active pharmaceutical ingredients, fuels, plastics, solvents, chemicals, fertilizers, and paints. Ideally a catalyst should have an infinite lifetime since it does not get consumed in the reaction. However, in practice lifetimes of catalyst can vary from seconds to years depending upon the composition of catalyst and reaction being catalyzed. This happens because the active sites of catalysts get damaged during the reaction due to poisoning from contaminants, blockage of active sites, reduction in accessibility and other damaging processes. Thus, catalysis is an important branch of chemical science dealing with design of active catalysts, their characterization, performance evaluation, regeneration, and re-use. When the phase of the catalyst is different from the reaction phase the process is called heterogenous catalysis (gas–solid, liquid–solid, slurry(gas–liquid)-solid) whereas when the catalyst is present in the same phase as the reaction medium itself the process is called homogeneous catalysis. Homogenous catalysts show the advantages of high selectivity, good control over reaction rates but separation of the catalyst from the reaction medium is a major issue. Heterogeneous catalysis is easy to manage since the catalyst can be easily separated from the reaction medium and even regenerated for reuse. Generally, heterogeneous catalysts are used in their solid form and shaped into packing material such as pellets, granules, extrudates of different shapes (cylindrical, bi-lobe, tri-lobe, quadra-lobe) and spheres to allow high flow processing of reactants and products. The shaping of catalyst bodies is performed to minimize the pressure drop across reactor bed. Catalysis is also commonly classified based on the nature of chemical process involved as thermocatalysis where energy for reaction is provided in the form of heat, electrocatalysis where energy for reaction is provided in the form of electrical potential and photocatalysis where energy for reaction is provided in the form of electromagnetic radiation (light). Recent research has seen the emergence of hybrid processes such as electrothermal, photothermal and photoelectrochemical catalysis. Additive manufacturing is revolutionizing the field of catalysis by providing a means to fabricate catalyst bodies with functionally optimized intricate shapes and novel catalytic reactors with enhanced performance and process intensification capabilities [138–141]. Chemical reactor design and engineering aspects of catalysis are discussed in more detail in Chap. 8. Some examples of additively manufactured catalyst bodies and derived benefits are briefed below.

The use of robocasted (direct ink writing) frameworks in woodpile configuration functionalized by catalyst coatings was pioneered by Stucker et al. as early as 2004 for catalytic combustion of methane [142]. The flexibility of robocasting technology enabled fabrication of woodpile lattices with desired dimensions allowing a useful comparison with extruded cordierite honeycombs of same dimensions. A bariumhexaaluminate (BHA) catalyst prepared by co-precipitation was wash coated onto the supports using an aqueous adhesive slurry containing an ammonium salt of poly (acrylic acid). The woodpile structure generated a highly turbulent flow by the virtue of no line-of-sight pathways and the highly porous structure with high surface area catalyst coating facilitated high mass and heat transfers. Interestingly, functionalized robocasted woodpile lattices were observed to convert at least six times more methane than extruded honeycomb supports at 600 °C. By enhancing the catalyst loading conversion could be increased by a factor of 2 to 45% methane conversion at 600 °C. Using the true advantage of robocasting, authors also showed that a woodpile lattice printed directly with BHA catalyst demonstrated 89% conversion at 600 °C and a noteworthy 100% conversion of methane at 700 °C [142].

Lefevere et al. investigated the importance of support structure in methanol to olefins (MTO) reaction using robocasted stainless steel (SS316L) fibers in woodpile configuration coated with zeolite ZSM-5 [143]. ZSM-5 was deposited on the steel surface using wash coating method (dip coating) by using a customized slurry containing milled zeolite powder, methylcellulose, and a colloidal silica solution. Coating with an optimized slurry allowed deposition of homogeneous catalyst layer with good adhesion. Interestingly, at low temperatures a framework structure with straight channels favored high selectivity towards dimethylether (DME). Whereas at high temperatures and high mass flow rates (WHSV) or low contact times, a framework structure with tortuous channels and a 1–3 configuration of the woodpile architecture exhibited highest yield of light olefins. The optimised catalyst was observed to be 20% more active than a packed bed of same catalyst and as active as cordierite honeycomb monolith. Clearly, the utilisation of a catalyst coated AM framework structure yielded a hierarchical catalyst system with macropores arising from the framework and micropores contributed from the zeolite structure and resulting high efficiency of heat and mass transfer enabled conversion of methanol to olefins in high yields [143]. Noyen et al. studied the activity of zeolite mordenite coated woodpile structures composed of porous Titanium grade 5 alloy (TI6Al4V) and manufactured using robocasting for decomposition of nitrous oxide [144]. A stable and well adhered coating was obtained by an alkali treatment of woodpile framework structures and repeated cycles of direct hydrothermal crystallization by immersion in synthesis gel. Efficient mass and heat transfer in hierarchically porous framework structure again demonstrated enhanced catalytic activities. The successful demonstration of loading, activation, and catalysis with additively manufactured bodies functionalized with conventional acidic catalysts such as zeolites opens the pathways for commercial applications of 3D printed catalysts into the petroleum refining and petrochemical sectors for applications such as gasoil and crude cracking, naphtha isomerization, lubricant dewaxing, and heavy naphtha cracking [145–148].

Danaci et al. investigated hydrogenation of CO_2 to methane (Sabatier reaction) using a Ni-Al_2O_3 catalyst coated on the surface of robocasted stainless teel (SS316L) woodpile framework [149]. The slurry composition was optimized to dip coat a milled Ni/Al_2O_3 catalyst prepared by impregnation of Ni salt in boehmite with 12% Ni loading. An aqueous slurry with 1% acetic acid, 3% poly (vinyl alcohol), 2% colloidal silica and 20% catalyst loading allowed homogeneous deposition of catalyst coating with good control over thickness and adhesion to framework surface. The hierarchical catalyst system composed of functionalized SS316L in woodpile configuration demonstrated enhanced CO_2 conversion at high temperatures and improved catalyst stability over time (up to 200 h) [149].

Applications of additively manufactured catalyst structures have also been demonstrated in photocatalysis such as the use of Au-TiO_2 robocasted woodpile structures developed by Elkoro et al. for hydrogen photoproduction from water/ethanol gaseous mixtures [150]. Au nanoparticles could be integrated in the titania paste used for extrusion or decorated post fabrication by impregnation. A reduction in size of titania filaments in woodpile configuration and impregnation method of gold decoration was identified to be optimum and yields up to 0.24 mol H_2 min^{-1} g^{-1}Au could be obtained [150]. Arin et al. integrated TiO_2 nanoparticles in their aqueous suspension inks prepared using microwave-assisted hydrothermal treatment of titaniumisopropoxide in ethylenediamintetraaceticacid and triethanolamine for inkjet printing of transparent photocatalytically active TiO_2 films [151, 152]. The microwave synthesis chemistry rapidly reduced synthesis time and the low temperature drying properties of such inks made these formulations very attractive for printing and functionalisation of soft materials such as wood, polymer, paper, and fabric while rendering photocatalytic active to these materials [151, 152]. Similarly, Cerna et al. developed transparent photoactive TiO_2 thin films using inkjet printing of their customised inks for decomposition of 2,6-dichloroindophenol under UV irradiation [153]. Polyethylene glycol was used as an anti-cracking agent and allowed direct printing on soda lime glass using titanium alkoxide solutions. A sponge-like structure was printed using an FDM printer loaded with a polymer containing TiO_2 powder dispersion. The TiO_2 inclusion remained active after printing and could demonstrate photocatalyzed degradation of Rhodamine 6G in solution. Maximum printable loadings of 10% TiO_2 could be achieved but exhibited comparatively lower activity than higher loadings.

Tubio et al. designed a copper nitrate loaded alumina ink for direct extrusion of the Cu-Al_2O_3 catalyst into woodpile structures for Ullmann reactions [154]. An aqueous slurry composed of ceramic alumina powder with hydroxypropyl methylcellulose as a viscosity modifying agent and poly(ethylenimine) as a cationic polyelectrolyte was used as the ink. The ink was ready for extrusion after controlled evaporation of water using a nozzle with an aperture of 410 μm attached to a syringe pump with print speeds up to 2 mm/s. A high temperature annealing treatment decomposed the copper salt along with other polymeric components of the ink and enabled integration of copper onto a porous alumina matrix in woodpile configuration with high mechanical strength and surface to volume ratios. The developed solid woodpile catalyst frameworks demonstrated high efficiency in catalysis for Ullman reactions, good recyclability and no leaching of copper was reported [154]. The disclosed method of formulating a catalytic ink using ceramic alumina exhibits superior advantages of high catalytic reactivity, ease of preparation, recyclability, and stability. Zhu et al. used robocasting of bimetallic alloy clays and subsequent dealloying for fabrication of metal monoliths exhibiting high specific areas, catalytic activities, mass transfer, low pressure drops and supercapacitance characteristics [155]. A 70:30 Ag:Au clay prepared by mixing Ag clay (~90% Ag powder, organic binder, and water) and Au clay (~92% Au powder, organic binder, water) with organic binder and water was robocasted using a tapered nozzle with 200 μm orifice and a 3-axis positioning stage. 70 wt% nitric acid was used to dealloy the structures for 48 h. The dealloyed monoliths exhibit a bicontinuous solid-void structure unique to dealloying.

It evident from the examples given above that robocasting (DIW) is the most popular additive manufacturing technology used in shaping of catalyst bodies. Woodpile configuration has been widely investigated because of its ease of fabrication by direct extrusion writing. Catalyst integration into printable materials as well as use of novel ink formulations are important functionalization methods for catalysis applications because they allow control over the bulk structure of the catalyst. However, surface functionalization strategy also has enormous potential in catalysis since it can be used to deposit active catalytic coatings in AM microreactors and complex shaped catalyst bodies [156]. Catalysis research can gain significantly from additively manufactured reactors, microreactors and shaping of catalysis bodies in designs derived from in-depth hydrodynamic studies. These aspects of catalysis are elaborated in Chap. 8.

7.2.2.3 Batteries

The rapid development and extensive use of portable electronics has led to an enormous demand for light weight, safe, durable, reliable, and cost-effective energy storage devices to achieve the super objective of power-independent electronics systems. Batteries are the frontrunners among the most important and widely used electrochemical energy storage devices and extensively studied for energy storage and delivery in a wide range of forms, capacities, and power densities. AM technologies because of their unique advantages such as localised production, customisation, process flexibility, reduced waste, and freedom in shaping have great potential for integration in the electrochemical storage industry. However, application of AM in energy storage industry heavily relies on functionally active electrochemical materials and composites with superior performance characteristics in terms of electrochemical response, stability, durability, cyclability and integration with end use devices. Large-scale production of such ideal micro batteries is still a challenge since printed battery electrodes can have poor mechanical properties and other issues such as need for adequate air and moisture proof packaging.

The focus of current and emerging research in energy storage devices is development of new electrode materials, electrolytes, cell structures, and novel fabrication approaches to improve electrochemical performance of batteries while reducing the associated costs [157]. The most well-known battery is the lithium-ion (Li-ion) battery, popular for availability in a variety of shapes and forms and proven high energy density, low self-discharging rate, and high current charge/discharge cycling capability. However, further improvement in charge/discharge rate and power density is marred by inferiority of available electrode materials exhibiting low electronic conductivities and inadequate diffusion of lithium ions. The high cost of Li-ion batteries especially utilising nickel and cobalt is also a major hurdle for rapid uptake of battery technology in transportation sector. Alternatives such as lithium–air and lithium–sulphur batteries are also receiving increased attention due to advantages of lower cost and higher energy capacity. In addition, replacement of lithium by abundant metals is ideal for sustainability in the long run employing zinc, calcium, aluminium, and sodium-based batteries. Energy density and areal capacitance can also be increased by utilising thicker electrodes with higher material loading while preserving rapid ion diffusion. However, electron transport distances and overall electrical impedance of the electrode also increase inevitably resulting in power density and rate capability reduction [158]. Compared to 2D planar structures, porous 3D structures yield shorter diffusion pathways and lower resistance to the ion-transport process [159]. Moreover, energy density increases from the improved electrode reaction and ion transfer in the highly porous structure with large specific surface areas.

Various AM technologies have been innovatively used for the fabrication of batteries as briefed in the below sections. In addition, template assisted electrodeposition has also been briefly described because of its high relevance for fabrication of electrodes for energy storage devices. The templates are usually derived from 2D lithography techniques but when combined with electrodeposition and electroplating techniques allow good control over morphology of nanostructures along with an overall resolution of sub 10 μm structures at high throughput deriving from the used lithography. Andic aluminium oxide (AAO) and ultra-thin alumina membranes (UTAM) have also been extensively used as templates for template assisted electrodeposition of 3D structures [160, 161].

Direct ink writing (DIW) is the most widely used AM technology for battery fabrication due its advantages of ease of fabrication, wide availability of ink formulations, mask-free process, and low cost of fabrication [162]. Sun et al. developed custom ink formulations for DIW of $Li_4Ti_5O_{12}$ (LTO) and $LiFePO_4$ (LFP) high aspect ratio filaments for fabrication of a Li-ion battery on gold interdigitated electrodes using glycerol, ethylene glycol, hydroxy propyl cellulose, ethyl cellulose and water [163]. A three-axis stage was used for movement in three dimensions whereas air powered syringes with glass nozzles of 30 μm diameter were used for extrusion writing at 250 $\mu m\ s^{-1}$ as shown in Fig. 7.11. The 3D microbattery was packaged and sealed with polymethylmethacrylate and PDMS and exhibited high areal energy density of 9.7 J cm^{-2} at a power density of 2.7 mW cm^{-2} [163]. Wei et al. demonstrated the fabrication of all battery components by DIW using four different inks

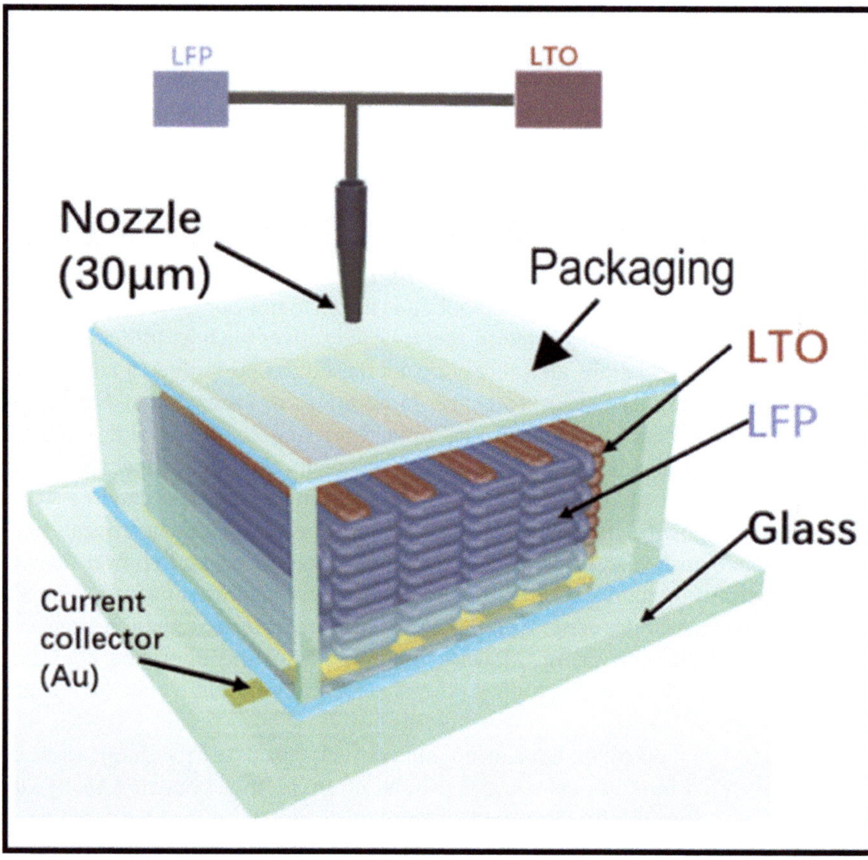

Fig. 7.11 Schematic representation of a 3D interdigitated microbattery fabricated by direct ink writing. Reproduced from Ref. [163] with permission from Wiley

including packaging, anode, separator, and cathode. The design flexibility of DIW was used to study the effect of LFP/(LTO) electrode thickness from 50 μm to 1 mm by comparing areal energy and power densities [164]. Thick electrodes retained the high areal power densities characteristics of thinner electrodes but exhibited much higher area energy densities. Zhang et al. employed DIW of monodisperse SnO_2 quantum dots with graphene oxide and water and demonstrated fabrication of complex geometries with the filaments such as zigzag lines, spiral rectangles, periodic micro lattices and mosquito coils [165]. Printed electrodes exhibited ultrahigh charging capacity, cycling stability and superior performance compared to SnO_2 quantum dots.

Lyu et al. employed the pyrolysis of a cobalt based metal organic framework to obtain a hierarchical nitrogen doped mesoporous carbon material with embedded cobalt nanoparticles which when loaded with Li_2O_2 allowed manufacturing of Li-O_2 battery cathodes with specific energy density of 798 Wh kg^{-1} much higher than specific energy densities of well-established Li–S batteries (<500 Wh kg^{-1}) and Li-ion batteries (<300 Wh kg^{-1}) [166]. Sub 10 μm resolution can be obtained using DIW by optimization of nozzle parameters but weak mechanical bonding between subsequent layers remains a problem for DIW based battery fabrication. Like DIW, fused deposition modelling (FDM) mostly relies on the integration of electrode or electrolyte materials with thermoplastics to obtain a conducting filament which can be 3D printed and used for subsequent preparation of electrical energy storage (EES) devices. Wei et al. integrated graphene oxide sheets in n-methyl 2-pyrrolidone (NMP) solvent into ABS using a homogeniser to obtain a conducting filament [167]. Maurel et al. integrated graphite into PLA matrix using PEGDME500 plasticizer up to 40 wt% to print negative disk electrodes by FDM [168]. Reyes et al. demonstrated the printing of a fully customizable Li-ion battery and fuels cells with integrated anodes, cathode, separators, and current collectors by relying on an integrated PLA filament with improved ionic conductivity [169]. The successful integration of batteries to wearable electronics such as LCD glasses and LCD bangle highlights capability of such printed batteries to enable end use applications. Compared to DIW, FDM has lower resolution and the need for integration of functionally active materials into thermoplastic matrices for electrochemical applications constraints its application.

Inkjet printing is emerging as a battery fabrication technology as increasing number of ink formulations with excellent droplet formation and rheological properties are becoming available. Lawes et al. used an ink formulation comprising of silicon nanoparticles, carbon black and a polymer binder (poly(3,4-ethylenedioxythiophene):polystyrene sulfonate, PEDOT:PSS) in water and used to print anodes for Li ion batteries with a desktop inkjet printer [170]. The battery showed excellent electrochemical performance and under limited depth of discharge measurement showed excellent stability up to 1000 cycles. Brown et al. developed porous Na-ion battery anodes by using ammonium thiomolybdate (ATM) and reduced graphene oxide (rGO) based ink formulation in water and used freeze drying post inkjet printing to immobilise it on nickel foams and subsequent annealing to obtain the MoS_2-rGO aerogel on nickel as anode material. Reasonable electrical storage capacity was demonstrated but irreversible reduction of Mo^{+2} to Mo^0 and Na^+ insertion led to drastic reduction in specific energy capacity with cycling [171]. Inkjet printing has great potential in battery fabrication because of its high resolution, capability for utilizing composite ink formulations and high material utilisation but print speeds are dominated by drying and droplet hydrodynamics. Saleh et al. demonstrated the use of aerosol jet printing (AJP) for fabrication of hierarchically porous silver electrode microlattices for battering leading to excellent mechanical stability, increase in specific capacity by 400% and areal capacity by 100% [172]. Although only limited applications have been demonstrated using AJP, this AM technology is anticipated to enable highly efficient, durable, and robust EES devices with high performance owing to its high resolution and contactless fabrication capacity.

Template assisted electrodeposition (TAE) technique relies on the use of physical templates for providing a desired structure to an electrodeposited material. Initially demonstrated for anodic aluminium oxide (AAO) membranes in the so called "on-wire lithography" the technique has now been generalised for use with lithography produced polymer templates of 2D or 3D structure. Zhang et al. demonstrated the fabrication of a bi-continuous bulk cathode by electrodeposition in a template prepared using monolayer of monodisperse polystyrene nanospheres (colloidal lithography) [173]. The porous nickel scaffold obtained after PS removal was used for further electrodeposition of electrolytically active materials such as NiOOH and lithiated MnO_2. The fabricated battery exhibited ultra-high charge/discharge capability with minimal capacity loss and fast charging times. Using the same concept Pikul et al. prepared interdigitated electrodes with colloidal lithography derived porous structure coated with lithiated manganese oxide (LMO) for Li-ion batteries exhibiting high power and energy densities [174]. Ning et al. used holographic lithography and photolithography using SU-8 and AZ9260 photoresists in conjunction with electrodeposition for fabrication of porous nickel current collectors and further electrodeposition of MnO_2 for cathode and Ni–Sn alloys for anode as shown in Fig. 7.12 [175]. The developed batteries showed high power and energy densities with good retention of capacity even after 100 cycles. Chen et al. demonstrated an

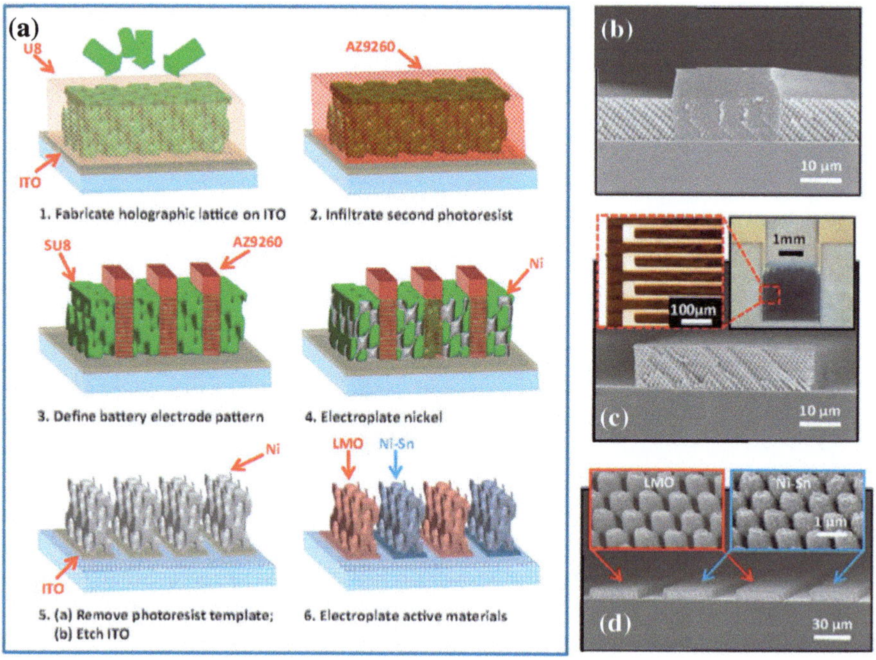

Fig. 7.12 **a** Schematic representation of the templated electrodeposition process employing combination of holographic lithography and photolithography for fabrication of a 3D microbattery; **b** SEM image depicting embedding of AZ9260 resist in holographic lattice; **c** SEM image depicting the structure of a single digit of the interdigitated Ni current collector; **d** SEM image of 3D interdigitated electrodes with LMO cathode and Ni–Sn anode. Reproduced from Ref. [175] with permission from the Proceedings of the National Academy of Sciences of the United States of America

interesting variation of this philosophy using micro-stereolithography in fabricating a 3D micro sized Li-ion battery consisting of trench shaped current collectors separated by a UV curable poly(ethylene glycol) based zigzag shaped membrane acting as the gel electrolyte [176]. The inherent porosity of the membrane due to interconnecting channels assembled on aluminium foil and scotch tape with polymer coating enabled high ionic conductivity with a charge/discharge plateau near to theoretical plateau values for $LiFePO_4$. However, poor cycling performance and capacity was obtained. Cohen et al. employed stereolithography to prepare perforated spherical, cylindrical and cubic shaped graphene filled conductive substrates and used electrophoretic deposition to fabricate a tri-layered structure of $LiTiO_2$ anode, polyethylene oxide (PEO)–$LiAlO_2$ or $Li_{1.5}Al_{0.5}Ge_{1.5}P_3O_{12}$ (LAGP)-polyethylenimine (PEI) membrane, and $LiFePO_4$ cathode on graphene-filled conducting substrates [177]. Developed 3D micro batteries exhibited excellent electrochemical performance and high areal energy densities and outperformed similar assembly on perforated silicon chips. A major limitation of the templated assisted electrodeposition using physical templates is that it is a template-based process requiring removal of the template from the fabricated material leading to material wastage and enhanced costs.

Application of additive manufacturing for fabrication of electrochemical energy storage devices is still in a nascent stage and many critical issues need to be addressed. There is an immediate requirement for novel active materials aptly designed for AM for use as electrodes, electrolytes, and current collectors. Recent research has mostly demonstrated integration of existing inorganic materials or composites into AM technologies. While the high porosity of AM constructs enables better energy and power densities, the mechanical stability remains low due to the inferior bonding strength between layers. High temperature post-processing requirements for most AM technologies as well as high oxygen/air/moisture sensitivities of electrochemically active materials impose adequate packaging restrictions presenting additional issues that need to be resolved. Finally, manufacturing of batteries is inherently a cost-intensive process which is typically made economically viable by employing large scale production. Thus, diffused, and localised production with AM technologies appears counterproductive under the current scenario. Thus, the rapid prototyping capability of AM remains the most useful incentive for such implication.

7.2.2.4 Capacitors

With the increase in awareness about global climate change, uptake, and utilisation of renewable energy resources such as solar, wind, hydro, and geothermal energy is increasing rapidly. Such non-continuous energy sources which show fluctuations in supplied energy throughout the day largely depend on storage of energy. Batteries and capacitors are the most widely implemented and researched energy storage devices. Like batteries, capacitors are two electrode energy storage devices separated by a dielectric material. Metal electrodes as current collectors can be used in different shapes such as wires, plates and microbeads as required by the form of the capacitor. Figure 7.13 depicts the conceptual differences between a battery and capacitor as

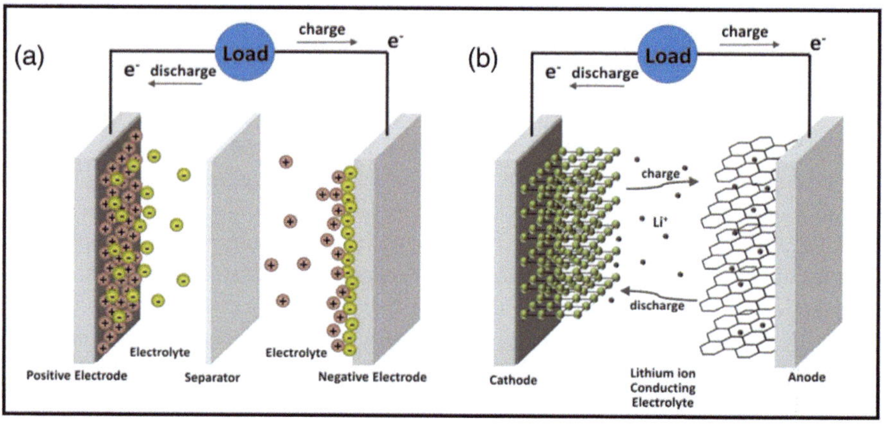

Fig. 7.13 Conceptual differences between a capacitor and batteries as energy storage materials; **a** In a capacitor the two electrodes are separated by an electrolyte and a dielectric medium. Energy is stored by induction of opposing charges on electrode–electrolyte interface during charging and release of charge carriers during discharging; **b** In a Li-ion battery the two electrodes are separated only by a lithium ion conducting electrolyte. The cathode is fabricated from lithium donor materials and anode is fabricated to have high lithium ion accepting capability with reversible transport of Li ions between the electrodes during charging and discharging. Reproduced from Ref. [159] with permission from Wiley

energy storage materials. Some examples of common capacitor dielectric materials are glass, mica, air, paper, plastics, and ceramics. Capacitors are commonly used in electronics today such as for blocking direct current and allowing alternating current, for smoothing output of power supplied in analog filter networks, for tuning of antenna to specific frequencies in resonant circuits and stabilising voltage and power flow in electrical power transmission. Capacitor stores energy in the form of opposing charges on its current collectors when disconnected from the charging circuit. Supercapacitors or ultracapacitors are capacitors that show high capacitance but with lower voltage limits. These capacitors can be charged and discharged much faster and for higher number of cycles than rechargeable batteries.

Additive manufacturing enables fabrication of 3D current collectors which minimise transport lengths for ions and electrons due to shorted diffusion paths and higher surface areas at low cost due to minimisation of metal use [178]. This led to improved device efficiencies and increase in energy density per unit volume. Use of custom ink formulations and integration of electrochemically active materials such as carbon nanotubes, graphene, conducting polymers, and electrolyte gels in ink formulations allows direct ink writing of capacitors on appropriate substrates [158]. Ho et al. demonstrated a generalised method for DIW of solid-state thick film carbon supercapacitors on electronic devices at room temperature and ambient conditions. Mesocarbon microbeads with PVDF gel binder were used as electrode materials and $BMIM^+BF_4$ (1-butyl-3-methylimidazoliumtetrafluoroborate) ionic liquid was used as the electrolyte for high ionic conductivity and good mechanical strength. At a discharge rate of 0.1 mA cm^{-2} a capacitance of 0.55 mF cm^{-2} was exhibited with maximum power density of 140 μW cm^{-2} and energy density of 0.681 μW h cm^{-2} from a single capacitor. Similarly, Nathan-Walleser et al. used DIW of surfactant-free thermally reduced graphene oxide (TRGO) inks in isopropanol with concentrations up to 15 g L^{-1} on carbon cloths as electrode material [179]. The capacitor device was completed by sealing a separator and electrolyte layer between the TRGO-carbon cloth electrodes and titanium discs for current collection. TRGO material exhibiting a surface area of 593 m^2 g^{-1} and conductivity of 16 S cm^{-1} allowed very high capacitance values of 3.55 mF cm^{-2} and charging-discharging at high voltage scan rates up to 15 V s^{-1}. Use of 1 molar tetraethyl ammonium tetrafluoroborate and acetonitrile as electrolytes exhibited high energy density of 4.43 mW h cm^{-3}, power density of 42.74 kW cm^{-3} and volumetric capacitance of 8.5 F cm^{-3} [179]. These performance results are at least one order of magnitude higher than comparable graphene-based capacitors and have application in AC line filter networks.

Sun et al. also used DIW for fabrication of graphene based interdigitated flexible micro supercapacitors by printing graphene oxide inks, subsequent reduction of graphene oxide by hydroiodic acid followed by printing of poly (vinyl alcohol)—sulfuric acid electrolytes on PET substrates [180]. Interestingly, no additional organic binders, conducting additives, metal collectors and polymeric separators were required in this configuration. Developed micro-supercapacitors with excellent superior volumetric capacitance of 41.8 F cm^{-3}, areal energy density of 7.6 mW h cm^{-2} and power density of 29.2 mW cm^{-2} compared to devices fabricated by lithography and conventional manufacturing [180]. Liu et al. described the use of an electrochemical ion intercalation treatment to improve the capacitance of graphene paper and DIW graphene aerogel supercapacitors [181]. In a three-electrode setup with Pt counter electrode, Ag/AgCl reference electrode and graphene electrode as the working electrode, lithium perchlorate solution in polycarbonate electrolyte was used to induce the Li ion intercalation by applying –2.5 V for 2 h followed by perchlorate ion intercalation by applying + 2.5 V for 2 h. The developed electrochemical treatment led to graphene exfoliation and functionalisation with oxygen moieties. Without any annealing requirements, gravimetric capacitance of 101.7 F g^{-1} at current densities of 10 A g^{-1} was exhibited with good rate capability and 1000 fold enhancement in areal capacitance compared to untreated electrodes [181].

Zhao et al. used SLS for printing 3D interdigitated titanium alloy (Ti6Al4V) electrodes consisting of hexagonally ordered pillars of 500 μm diameter and 8 mm height connected to a base plate, functionalised their surfaces by electro-polymerisation of pyrrole and used poly (vinyl alcohol)-phosphoric acid polymer electrolytes to fabricate 3D micro-supercapacitors as shown in Fig. 7.14 [182]. The developed micro-supercapacitor exhibited performances comparable to lithographically fabricated devices with a volumetric capacitance of 2.4 F cm^{-3}, maximum energy density of 213.5 W h m^{-3} and power density of 4 kW m^{-3} [182]. Yang et al. demonstrated the use of DLP SLA for fabrication of capacitors in complex 3D shapes such as honeycombs, cylindrical arrays, bowl and star shapes by employing custom resins composed of commercial Flex resins integrated with silver coated lead zirconate titanate (PZT) nanocrystals [183]. PZT crystals were initially ball milled to a size

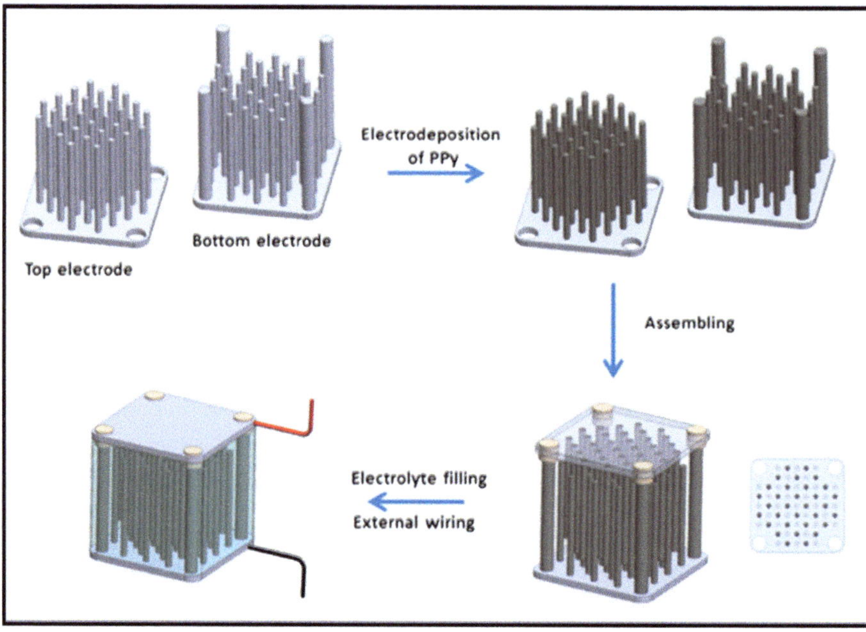

Fig. 7.14 Schematic representation of the design of a 3D interdigitated microcapacitor fabricated by selective laser sintering. Reproduced from Ref. [182] with permission from Elsevier

of 400 nm and subsequently hydroxylated and functionalised with silanes before coating with silver. Integration of PZT@Ag nanocrystals into the resin led to increase in its viscosity and reduction in curing depth but no issues with stereolithography. The developed dielectrics showed dielectric permittivity values of 120 for 18 vol% PZT@Ag nanocrystal integration which was around 30 folds higher than Flex resin alone. Gravimetric capacitance of 63 F g^{-1} at current densities of 0.5 A g^{-1} was exhibited for developed capacitor [183]. Zhu et al. employed DIW for fabrication of graphene aerogel-based supercapacitors with thick electrodes using custom ink formulations based on functionalised graphene nanosheets-graphene oxide composites (GNS-GO), fumed silica nanoparticles as a rheology control additive and resorcinol formaldehyde solution as a gelation agent [184]. After printing, gelation was performed followed by supercritical drying, carbonization and silica etching using hydrofluoric acid. An optimum ink formulation used 12.5 wt% GNP and 4.2 wt% SiO_2 and allowed the fabrication of supercapacitors with 90% capacitance retention with current density increase from 0.5–10 A g^{-1}. Even with printed 1 mm thick electrodes, energy density of 0.14 mW h cm^{-3} and power density of 2.6 kW cm^{-3} was exhibited which was equal or higher compared to supercapacitors made with electrodes than were 10–1000 times thinner [184]. Yao et al. employed unprecedented high loadings of MnO_2 on graphene aerogel-based supercapacitors fabricated by DIW of graphene nanosheets and methylcellulose based inks followed by freeze drying and annealing to obtain up to 8 mm thick electrodes with high capacitance retention as shown in Fig. 7.15 [185]. Pseudocapacitive materials like MnO_2 show excellent gravimetric capacitance values but their loading is typically limited up to 1 mg cm^{-2} beyond which drastic decay of capacitive performance is observed due to sluggish ion diffusion. Authors demonstrated MnO_2 loadings up to 182.2 mg cm^{-2} for fabricated supercapacitors with high areal capacitance of 44.13 F cm^{-2} which retained 92.9% of their initial capacitance even after 20,000 cycles [185]. Besides the above examples recent research has also explored the use of barium titanate-PVDF piezoelectric composites as high dielectric capacitive materials as covered in depth by Wang et al. [186]. Deiner and Reitz have focussed on the use of inkjet and aerosol jet printing for development of energy storage devices [159].

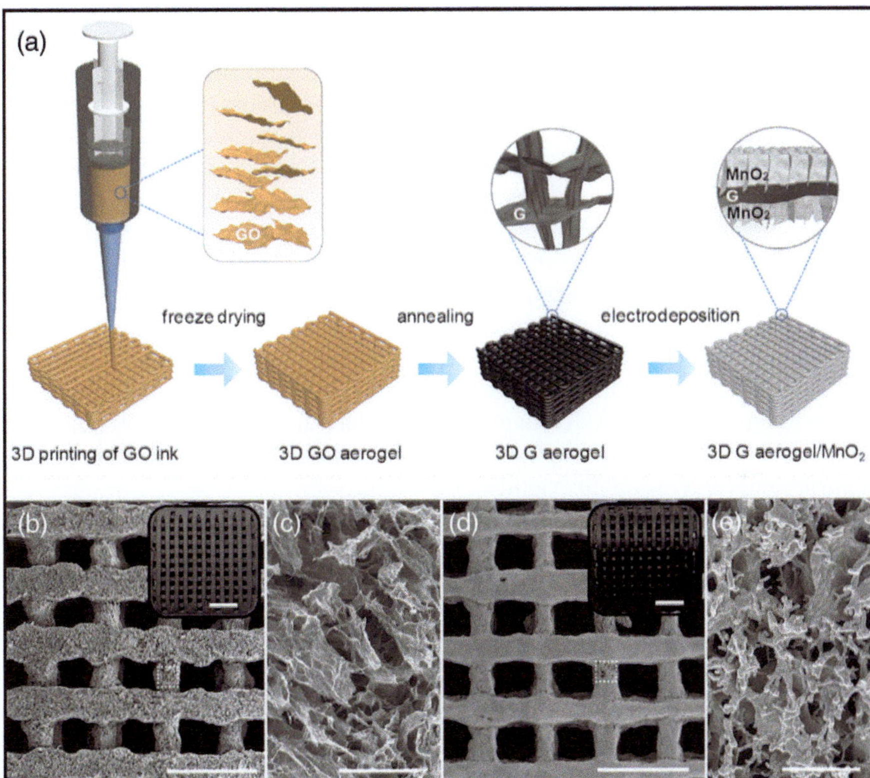

Fig. 7.15 **a** Schematic representation of the fabrication process for a micro supercapacitor involving direct ink writing using graphene oxide-based ink formulation, freeze drying, annealing and MnO_2 electrodeposition; **b** SEM image of the 3D printed graphene aerogel lattice. Scale bar, 1 mm; **c** Magnified image of a cylindrical filament highlighted in (B). Scale bar, 40 mm; **d** SEM image aerogel lattice after electrodeposition of MnO_2 for 600 s. Scale bar, 1 mm; **e** Magnified image of a cylindrical filament highlighted in (D). Scale bar, 40 mm. Insets in (A) and (C) are optical images of 3D printed graphene aerogel and graphene aerogel/MnO_2, respectively. Scale bars, 5 mm (insets). Reproduced from Ref. [185] with permission from Elsevier

The capability of additive manufacturing in fabrication of micro-supercapacitors and flexible supercapacitors with excellent retention and capacitance values is evident from the briefed examples. The one step free form fabrication such as in direct ink writing methods eliminates tedious serial fabrication steps, high wastage, and slow prototyping typical of conventional lithographic fabrication methods. Moreover, AM allows integration of several fabrication steps such as direct writing of electrodes and separation of dielectric and gel or polymer electrolytes. AM capacitive devices are anticipated to meet the future demands of light weight, flexible, high capacity, portable and miniaturised energy storage devices.

7.2.2.5 Separation

The capability of additive manufacturing for controlled fabrication of 3D constructs with a wide variety of material choices also makes them attractive for use in filtration and separation applications [139, 187]. Separation applications can be based on different mechanisms such as molecular sieving mechanism involving sieving of specific molecules from a mixture in the appropriately sized pores of a porous material as used in membrane separations, selective retention mechanism involving selective retention of specific molecules in the mixture on the surface of a solid medium as used in chromatographic separations and solid phase extraction and phase separation mechanisms relying on vapor pressure differences as used in distillation. Many important analytical methods and industrial chemical processes rely on separation phenomenon such as chromatographic separations for analysis, membrane separation for water purification or contaminants removal, distillation for purification of solvents, phase separation during industrial production of pharmaceutical and organic chemicals and oil removal during oil spills.

AM has been used in separation technology for pre-separation and post-separation interface devices, shaping of porous materials using AM for use in separation devices, and separation columns and packing material for phase separation and extraction. The flexibility in fabrication of custom designed geometrically complex shapes using thermally and chemically stable materials enables the use of additive manufacturing for pre-separation and post-separation interface devices. Sample preparation is an essential step for most analytical techniques and different preparation methods such as purification, dilution, filtration, extraction, digestion, pre-concentration, and derivatization are essential to improve the selectivity, sensitivity, or detection limit of analytical methods [188]. Recent research has focused on automation and improvisation of each of these unit analytical operations using custom-designed equipment and devices. Wang et al. fabricated a multifunctional stator (static part of a rotary system) as a front-end device to liquid chromatography by printing a mesofluidic chip device using magnetic nanoparticles for in-line sorptive microextraction in a rotary microvalve using stereolithography [189]. The 3D printed microflow assembly was used to detect organic contaminants such as methylparaben, ethylparaben, propylparaben, phenylparaben, butylparaben and triclosan in the range of 1.1 – 4.5 ng per ml in saliva and urine samples. The chip device could be used for more than 20 injections with minimal losses of moderately nonpolar compounds due to adhesion on device internal surfaces. Martinez-Jarquin et al. fabricated an ambient temperature plasma probe for mass spectrometry imaging (MSI) able to produce plasma jet of diameter ~ 200 μm with reproducible analysis results [190]. The same group also demonstrated custom printing of robotic feeding arms and plasma probes for low temperature plasma-based mass spectrometry (LTPMS) as an improvisation to direct injection electrospray ionisation-based mass spectrometry for chemical fingerprinting of tequila and mezcal Mexican spirit admixtures [192]. The high throughput analysis using LTPMS with custom 3D printed components reduced analysis time from 6 min to 20 s. Pre-concentration is often required before inductively coupled plasma mass spectrometer (ICPMS) to improve sensitivity of the method by removal of unwanted salt matrices in samples such as sea water. Su et al. fabricated a solid phase extraction pre-concentrator for quadrupole-based ICPMS using stereolithography of polyacrylate polymers as shown in Fig. 7.16. [191]. An ordered 3D array of cuboids was used to maximize mixing and extraction efficiency and allowed interference-free determination of Mn, Ni, Zn, Cu, Cd, and Pb in sea water and urine without any deterioration of detection limits. Such geometries provide highly efficient mixing of liquids with high permeability, low pressure drops, rapid column flushing and regeneration capabilities. Calderilla et al. used stereolithography to integrate analyte oxidation, disk-based solid phase extraction and analyte complexation

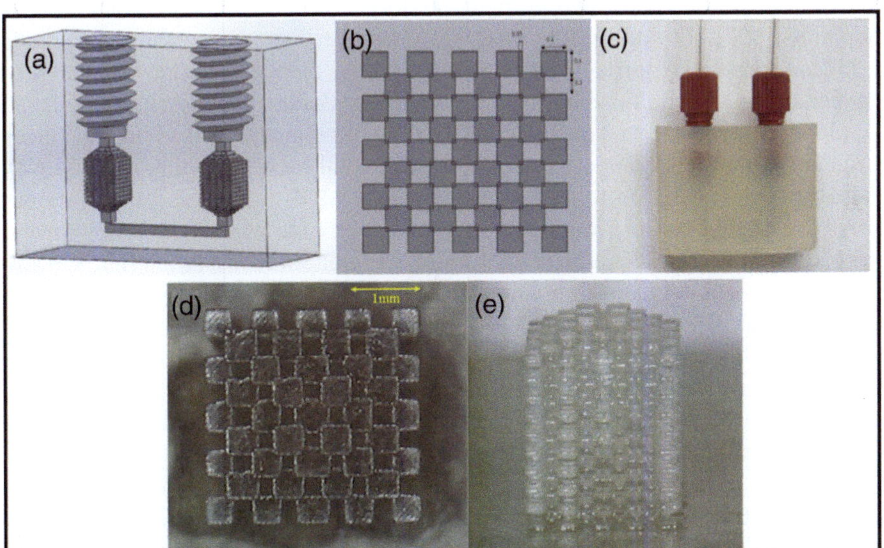

Fig. 7.16 Design and fabrication of a pre-concentrator using stereolithography for quadrupole-based ICPMS detection of metal ions in sea water; **a** Schematic depicting the overall design of the pre-concentrator; **b** Top view of the design based on 3D arrangement of cuboids; **c** Image of the fabricated pre-concentrator device; **d** Optical image of the top-view of fabricated stacked cuboid structure; **e** Front view of fabricated stacked cuboid structure. Reproduced from Ref. [191] with permission from American Chemical Society

in a single modular 3D device connected to a multisyringe flow injection analysis system [193]. Detection of iron in water was demonstrated with improved sensitivity and a detection limit of 7 ng per 3 ml sample and standard deviation of less than 4.3%. Similarly, Cr (VI) detection was demonstrated with the same device relying on complexation of Cr (VI) with 1, 5-diphenylcarbazide in acidic medium and retention of the complex in designed solid phase extraction disc [194]. A detection limit of 1 ng per 2 ml sample and standard deviation less than 4.8% was obtained.

Another class of separation application is based on the appropriate shaping of molecular sieving 3D porous solids (adsorbents) such as zeolites, mesoporous silicas, mesoporous carbon, and crystalline frameworks. Wang et al. functionalized the surface of ABS laminates and skeletons in woodpile configuration fabricated by FDM using Cu-BTC (benzene tricarboxylic acid) metal organic framework (MOF) by repeated soaking cycles in MOF precursor solutions to obtain a uniform coating of Cu-BTC nano crystals of average size 450 nm [195]. The developed Cu-BTC coated skeletons were used for selective removal of methylene blue from water and recoveries up to 98.3% were obtained for 5 mg/L MB solution within 10 min. Thakkar et al. demonstrated direct robocasting of MOF monoliths with MOF loadings as high as 85 wt% for removal of CO_2 from air [196]. MOF-74 (Ni) and UTSA-16 (Co) were used for robocasting of monoliths and CO_2 uptake capacities of 1.35 and 1.31 mmol/g which were 79% and 87% of pure MOF capacities, respectively, were observed when exposed to 0.5 wt% CO_2 in air at 25 °C. Couck et al. demonstrated the separation of CO_2 gas from methane and nitrogen mixtures using robocasted zeolite ZSM-5 monoliths [197]. Zeolite loading of 65% in the matrix could be achieved without significant drop in adsorption performance and the monolith framework could be regenerated by helium flushing for 1 h without any heating requirements. Salazar-Aguilar et al. demonstrated the direct ink writing of prismatic and cylindrical silicon carbide (SiC)—MOF composite monoliths with solid content loading up to 64 wt% and overall porosities up to 78% [198]. Fe-BTC MOF was used for its catalytic activity in phenol hydroxylation with hydrogen peroxide whereas SiC was used to provide mechanical and thermal stability to the monolith structure. High and low molecular weight polyethyleneimine and hydroxymethyl cellulose were used in ink formulation along with SiC and MOF components to adjust the rheological properties. 20% yield and 80% selectivity towards dihydroxybenzene was achieved when monoliths were used in a custom stirrer reactor.

AM has also been employed in custom design of separation equipment such as distillation columns and shaped solid phase extraction sorbents as briefed below. Mardani et al. demonstrated the development of modular distillation columns using stereolithography including AM packing material [199]. Distillation and separation of hexane from cyclohexane in full reflux and continuous mode was demonstrated. The size limitation of the printer was overcome by the modular design with module sealing achieved using flanges. A major limitation of SLA fabricated columns was the compatibility of resins used for printing with solvents which could be improved using SLM printed metallic parts or using suitable coatings [199]. Worawit et al. demonstrated automated microextraction of acidic drugs such as ketoprofen, ibuprofen, diclofenac and naproxen from urine and their detection in therapeutically relevant concentrations using nanostructured polypropylene hollow fiber membrane filled with carbon nano fibers in a stereolithography printed 3D extraction chamber [200]. The entire analytical process of acceptor handling, sample loading, regeneration of organic film, on-line dilution and injection into a liquid chromatograph was automated using 3D printed chambers and flow components. Lahtinen et al. demonstrated successful scavenging of Au, Pt and Pd precious metals from electronic waste using stereolithography printed filters composed of polypropylene and 10 wt% anion exchange resin [201]. Compared to commercial anion exchange resins, the used composite material showed comparative scavenging performance. Belka et al. demonstrated the fabrication of shaped sorbents able to fit in laboratory vials for small molecules extraction from water by using the fused deposition modelling of a PVA based composite material Layfomm 60 from which PVA can be removed post fabrication to result into a porous matrix [202]. Extraction of glimepiride in water was demonstrated with a recovery of 82% and a standard deviation of less than 5%. Using the FDM of the same PVA based Layfomm 60, Konieczna et al. demonstrated the extraction of natural steroids from human plasma such as aldosterone, cortisol, β-estradiol, testosterone, dihydrotestosterone, methyltestosterone and betamethosone and subsequent analysis using liquid chromatography-based mass spectrometry [203]. De Middeleer et al. used a molecularly imprinted polymer (MIP) on polymer functionalization strategy for functionalization of 3D scaffolds prepared by FDM of poly-ε-caprolactone (PCL) and used it as a solid phase extraction sorbent for detection of metergoline and ergot alkaloid mycotoxins. Metergoline was first imprinted into polymeric spheres obtained from methacrylic acid monomer and trimethylolpropane trimethancrylate crosslinker and subsequently removed using a methanol/triethanolamine solvent [204].

7.2.2.6 High Throughout Experimentation

Combinatorial chemistry and high throughout experimentation deal with large number of parallel experiments in synthesis/characterisation/performance evaluations and subsequent data analysis to facilitate rapid fulfilment of experiment objectives. Two critical elements of high throughout experimentation are design of experiments and high throughput equipment (HTE). A major limitation of commercial HTE is their specificity towards the experimental objective which is great for quality control and testing in industrial production environments where the experimental objective remains same for the lifetime of equipment. However, in research and development environment where experimental objectives change periodically with change of research projects, process improvements and new materials, commercial HTE lose their usefulness in the long run. Procurement of multifunctional HTE with flexibility of operation for a wide umbrella of applications may prove uneconomic and unsustainable. The unique advantages of AM namely design flexibility, capability of customization and wide choice of materials make it suitable for localised fabrication of HTE as per requirement. Rapid prototyping and low turnover times can enable highly efficient experimentation at low overall costs. As such the uptake of additive manufacturing in high throughput experimentation has been enormous as can be seen from below examples.

Perhaps the widest use of additive manufacturing in high throughput experimentation has been the research on inkjet printed metal oxide photoelectrocatalysts for water splitting. In such experiments, inkjet printing is used to quantitatively disperse aqueous metal nitrate and metal chloride salt solutions as precursors on conductive glass and subsequently converted to metal oxides by pyrolysis. The ink formulations typically employed ethanol and isopropanol as solvents and pH modifiers such as acetic acid, ammonium nitrate, hydrochloric acid and nitric acid were used along with polymeric surfactants such as Pluronic F127 or viscosity modifiers such as diethylene glycol monobutyl and diethylene glycol. Woodhouse et al. were amongst the earliest pioneers of this methodology and identified the performance evaluation of deposited metal oxide electrodes as the slowest step in the high throughput experimental process [205]. They relied on the use of a laser scanning system for performance evaluation. Using the developed high throughout methodology they identified a novel p-type semiconductor ternary oxide of cobalt, aluminium and iron formed in a spinel structure with iron and aluminium substituting into cobalt sites and a nominal stoichiometry of $Co_{3-x-y}Al_xFe_yO_4$ ($x = 0.18$ and $y = 0.3$) [206]. The metal oxide exhibited a band gap of 1.5 eV near to the ideal value for an efficient single photoelectrode for water splitting but the photocurrent was observed to be limited by slow kinetics of hydrogen evolution. Katz et al. used a commercial piezoelectric inkjet printer and used up to 8 metal nitrates to obtain 252 metal oxides and used automatic performance evaluation of these metal oxides for their open circuit potentials, short circuit photocurrent densities, and current density *vs.* applied potential behaviour under visible light irradiation [207]. He et al. identified titanium as the most effective dopant for α-Fe_2O_3 amongst silicon, aluminium and titanium dopants for its water photoelectron catalytic (PEC) activity synthesised by spray pyrolysis of metal oxides and subsequent high throughout laser scanning analysis [208].

Using a similar approach for photocatalyst evaluation, Liu et al. used inkjet printing to successfully synthesise a million formulations per hour of up to 8 components using fixed quantities of metal precursors and evaluated the formulations with multidimensional group testing strategies to identify highest H_2 evolution rates for $Mg_{0.82}Ni_{0.02}Cu_{0.5}Ti_{1.0}O_w$ ternary oxide and $Pd_{0.0003}Ti_{1.0}O_w$ binary oxide [209]. As an example of developed methodology, 5000 catalyst species of quaternary mesoporous metal oxides $Mg_xNi_yCu_zTi_{1.0}O_w$ ($0 < x, y, z < 1.0$) were printed and split into 8 groups of 625 each. Each set of 625 catalyst formulations underwent batch testing in their respective groups. The photocatalytic hydrogen production from formaldehyde water solutions was determined by gas chromatography. In their first iteration, group 4 was most active and a further 5000 compositions were subdivided into this groups' scope ($Mg_{0.5-1}Ni_{0-0.35}Cu_{0.5-0.9}Ti_{1.0}O_w$). After the 4th iteration of this method, optimal catalyst compositions were finalised.

Gregoire et al. developed a highly integrated scanning droplet cell for rapid electrochemical and photoelectrochemical evaluation of a library of quaternary metal oxides prepared by inkjet printing [210]. The droplet cell was carefully designed based on several critical design criterion such as a suspended droplet generating setup that can generate a single droplet of electrolyte for contact with the metal oxide, use of reference and counter electrodes in optimal configurations, current density measurements from 10 μA cm^{-2} – 1 A cm^{-2} with standard potentiostat, 1 mm horizontal resolution in movement, efficient mass transfer of reactants and products in liquid and gas phase, low settlement time and rapid movement between measurements, optimum integration of light sources and avoidance of cross contamination. A 0.6 V full-cycle CV measurement could be completed in 4 s and 5456 quaternary metal oxides of iron, cobalt, nickel, and titanium were evaluated [210]. The same group also derived an expression for the efficiency of oxygen evolution catalyst based on its optical and catalytic properties [211]. Optimal oxygen evolution catalytic activity was observed near $Fe_{0.3}Ni_{0.7}O_x$, and the optimal catalyst efficiency was observed near $Fe_{0.23}Co_{0.13}Ni_{0.07}Ti_{0.57}O_x$.

Seley et al. used a pH sensitive fluorescence indicator (3-(2′-pyridyl)-[1, 2, 3]-triazolo[1,5- a]pyridine, PTP) sensitive to pH changes because of release of protons from water oxidation for evaluation of inkjet printed mixed oxide electrocatalysts for water electrolysis. Ternary and quaternary mixtures of metal oxides of 5 noble metals Ir, Ru, Pt, Rh, Pd and other metal oxides of Mn, Al, Nb and Sn were evaluated. Promising formulations were evaluated further by measuring polarisation curves and Tafel slopes and compared with an IrO_2 standard [212]. In order to eliminate the difficulties in measurement of small localised pH changes, Xiang et al. developed a bubble screening method for evaluation of the performance of inkjet printed metal oxide catalyst libraries for hydrogen evolution reaction (HER) and oxygen evolution reaction (OER) [213]. A customised electrochemical cell was employed, and the transient of bubble intensity was quantitatively analysed to yield a figure of merit that directly corresponded to the reaction rate. Active formulations from a library of 231 metal oxide formulations in a ternary mixture of nickel, iron and cobalt oxides were successfully identified in less than a minute [213]. Serial screening of the same library with conventional methods yielded excellent agreement with the bubble screening method. It is evident that inkjet printing enables rapid synthesis of compositionally controlled catalyst libraries for high throughput evaluation. However, the performance evaluation of such libraries remains the bottleneck in rapid identification of promising compositions. As such, a great deal of innovation has been employed in development of high throughput catalyst screening methods relying on custom built evaluation cells. Additive manufacturing is anticipated to advance this field even further by facilitating development of novel and well-integrated performance evaluation cells utilising microfluidics, automated valves and light manipulation capabilities using advanced optoelectronic circuitry.

Kitson et al. demonstrated the discovery of two new coordination polymers by using an array of single use hydrothermal synthesis reactors fabricated by fused deposition modelling of polypropylene [55]. The reactants were filled in the voids during fabrication and the vessels were directly sealed by printing of a solid body eliminating the need of any sealing components like gaskets and flanges. Such single use labware is most suited for localised production, high throughput experiments and recycle of polymeric material for reuse.

7.3 Summary and Outlook

The key focus of this chapter was to appreciate the demonstrated applications of 3D printed constructs and understand the strategies and methods used for imparting additional functionalities to additively manufactured constructs for their use in chemical processes. Justification was provided for selection of a specific AM technology for a targeted application based on usable printable materials, build resolution in XY and Z directions, type of required features, mechanical strength of finished object, chemical, thermal, and mechanical stability, flexibility, elasticity, electrical conductivity, optical transparency, capability for integration of functional materials, fabrication time and post-processing requirements. In-fact, careful consideration should be given to these important parameters even before selecting or buying a specific 3D printer which is also the theme of a critical analysis by Tully and Meloni in "A Scientist's Guide to Buying a 3D Printer: How to Choose the Right Printer for Your Laboratory" [214]. However, even after such careful selection of print materials and AM technology, in most scenarios the printed object does not exhibit the desired performance in its as-printed form. Thus, different functionalisation strategies as described earlier in Chap. 6 can be used for imbibing additional functionalities to 3D printed constructs. The applications of AM in chemical processes classified as surface sensitive applications, bulk sensitive applications and high throughout experimentation were critically evaluated and the underlying common themes and rationale behind the demonstrated strategies were elucidated.

It is anticipated that the reader will take advantage of such a large compilation of application examples and contemplate upon the revealed trends. A careful observation clearly highlights why some applications are more inclined towards the use of specific AM technologies and why others are rejected. More importantly, gaps can be identified by careful analysis to discern AM technologies which have not yet been explored or less explored for specific applications with potential benefits. For example, direct ink writing has been widely used for printing of 3D graphene-based electrodes for batteries and capacitors. The same method and material while also advantageous for catalysis applications due to high specific surface area, ordered porosity, capacity for subsequent functionalisation and high chemical reactivity has not been widely employed by the catalysis research community. Similarly, it can also be noted that high throughput experimentation has mostly focussed on the rapid performance evaluation of inkjet-printed electrocatalysts and photoelectrocatalysts for water splitting but not used at the same levels for evaluation of catalysts in other reactions. The possibility of rapid performance measurements in such systems using innovative methods such as bubble screening or use of a droplet scanning cell enables such high throughput evaluation. However, researchers should be able to find similar innovative solutions for rapid evaluation in other important reactions such as carbon dioxide conversion, methane oxidation and ammonia splitting whether catalysed thermally, electrochemically or photoelectrochemically. In addition, the thematic classification of application areas in this chapter is anticipated to induce cross fertilisation as well as introduce the younger readers to the fundamentals of different research fields.

Additive manufacturing technologies demonstrate unique advantages of design flexibility, capability for fabrication of geometrically complex structures, rapid prototyping, low wastage of materials, high resolution, mask-free and contactless (aerosol jet printing) fabrication, manufacturing capability in variety of material choices, multi material printing, economic viability, and rapid integration capabilities for device components. With the online sharing of design architectures and localised production capabilities facilitated by AM, 3D printing can greatly empower the fourth industrial revolution. It is rare to find a single manufacturing philosophy that integrates and enables such wide-reaching research areas. AM is cultivating a culture of multidisciplinary learning, objective oriented localised design optimization and inducing rapid broad-spectrum advancement of chemical sciences.

For surface sensitive applications, surface functionalisation strategies are highly effective in meeting the research objectives at low cost. While surface functionalisation can be used for all AM technologies, photopolymerization-based and fusion-based technologies can significantly benefit from these methods. For example, despite the high print speeds and high resolution of DLP SLA, the biocompatibility of used resins is a big problem for its bio-applications where inert gold coatings or coatings of bio-compatible materials can easily resolve the issue. Similarly, the high mechanical and thermal stability of SLM fabricated metal reactors can be used to impart additional catalytic functionalities allowing operation at high gas and liquid flow rates. Using surface functionalisation methods, the advantages of complex 3D shapes in effective mixing, mass transfer and heat transfer can be retained while additional reactive, chemical, and optical functionalities can be imparted. Increased uptake of surface functionalisation in 4D printing and active origami is also anticipated to impart novel stimuli responsive capabilities beyond simple immersion in water or a solvent. Research in sensing, optical and electronic devices can gain the most by using AM microfluidics, sample holders, integration of needed electronic components (LED, photodetectors, valves, pumps, heating coils and sensors) and generalised system automation. Photopolymerization-based (SLA, DLP SLA and TPL) and fusion-based (SLS, SLM and LOM) technologies are most suited for facilitating such advances in sensing, electronic and optical devices.

Despite rapid progress in the field of optical devices some challenges still hamper the commercial viability of demonstrated applications. Most AM techniques suitable for photonics and optoelectronics such as two photon lithography and direct laser writing are still prototyping tools for research rather than production tools. High throughout fabrication capabilities are essential for commercial production. TPL despite its advantages in free form 3D fabrication and high resolution are limited by throughout, surface finish and limited choice of photoresists that can be used. Emerging research on development of more efficient photoinitiators, high power lasers and faster motion stages can potentially eliminate some of the throughout issues. Use of multiple beams and holographic projection systems are also anticipated to make favourable throughput gains at least for some applications. Other critical issues that need to be resolved are multiscale printing and multimaterial processing.

Since most objects of real time use are macroscopic in size, combination of different AM technologies to print non-functional bodies and functional surfaces or front ends are desired.

For bulk sensitive applications integration and use of novel printable materials are the most effective methods for imbibing additional functionalities in AM constructs to meet research objectives at low cost. Extrusion-based (DIW and FDM) and material jetting-based (IJP and AJP) technologies are most suited in this regard because of a great flexibility in use of customised ink formulations using functionally active materials and composites. The use of DIW and FDM has been widely demonstrated in printing of metals, alloys, metal oxides, mixed metal oxides, ceramics, active carbons, hydrogels, and polymers. Multimaterial printing capabilities can make both technologies even more competent where they lack in comparison to inkjet and aerosol jet printing technologies. The uptake of aerosol jet printing technologies is widely increasing for design of electronic and optoelectronic devices owing to its mask-less, contact-less, and high-resolution printing capabilities. However, further advances are desired to enable printing of sharp-edged lines in AJP. Additive manufacturing is anticipated to fuel an enormous increase in research for process intensification and development of efficient microreactors, battery electrodes and fuel cells. Three-way converters for automobile exhaust treatment is a good example of the use of monolith reactors for gas conversion at high throughput. With the rapid progress and uptake of AM in catalysis technology more sophisticated and efficient microchannel reactors, monoliths, and geometrically complex catalyst bodies are anticipated to enable high throughput gas and liquid processing in important reactions such as carbon dioxide utilisation [215] methane oxidation [216] hydrogen production [217] and Fischer–Tropsch slurry cracking [218, 219].

Finally, additive manufacturing is anticipated to penetrate further into niche markets like emergency response in disaster management, high throughput experimentation, advanced labware and advanced design solutions. Disasters are often accompanied by localised shortage of useful equipment, pharmaceutical ingredients, chemicals, and antimicrobial bandages. When such demand cannot be met from the international market due to border closure such as in the COVID-19 pandemic, AM can be used to establish short term production facilities in research and government laboratories to meet such requirements. AM is also a great enabler for high throughput experimentation. Most commercial and fixed-objective high throughput research equipment (HTE) lose their effectiveness and utility in the research and development environment where experimentation objectives tend to change with time as new research projects, themes and grants are obtained. Thus, custom designing and localised production of HTE as per experimentation needs caters perfectly to this demand. As the use of AM for localised production of labware and glassware increases it is also anticipated that in near future local suppliers will be replaced by suppliers of AM goods and objects. Similarly, open sharing and even selling of useful advanced labware, reactor and device designs can also be foreseen.

7.4 Test Your Knowledge

1. Which additive manufacturing technology has the highest resolution in XY and Z direction as well as allows unsupported fabrication of overhanging parts?
2. Which additive manufacturing technology allows the highest print speeds for fabrication?
3. How can the fabrication times be drastically reduced for photopolymerization based AM technologies?
4. Identify the only correct statement from the following:
   (a) Direct ink writing (robocasting) provides highest resolution and lowest fabrication times.
   (b) Selective laser sintering can be used to fuse the powdered form of any material
   (c) Resolution in the Z direction and fabrication time are inversely proportional for any additive manufacturing technology.
   (d) Additive manufacturing cannot be used for printing of elastic materials
5. Identify the only correct statement from the following:
   (a) Microreactors are used because they increase the rate of reactions.
   (b) High pressure reactions cannot be performed in microreactor because of safety issues.
   (c) Microreactors are only suitable for lab testing and not commercial production as they involve small volumes of reactants and products.
   (d) Physical processes like mass transfer, heat transfer, temperature gradients, maldistribution and channelling are reduced at small scales involved in microreactors.
6. Identify the only incorrect statement from the following:
   (a) Nano sized powders are most suited for chemical reactors owing to their high specific surface areas and high chemical reactivities.
   (b) Catalyst pellets and extrudates also serve the dual purpose of improvement in mixing for continuous flow reactors.
   (c) Robocasted catalyst bodies can potentially replace the use of catalyst pellets and extrudates.
   (d) Foam packed reactors are use in diffusion limited reactions to enhance mass and heat transfer.
7. Identify the only incorrect statement from the following:
   (a) Additive manufacturing enables better integration of optoelectronic device components in three dimensions compared to conventional assembly techniques.
   (b) Additive manufacturing can be used to fabricate flexible and elastic 3D objects suited for biotic/abiotic interfaces with body organs

(c) DLP SLA is the most promising AM technology for applications of optoelectronic devices in biosensing, tissue engineering, cell culturing and organ printing because its high resolution, high print speeds, smooth finish to objects and capability to print optically transparent objects.
(d) Optoelectronic devices can greatly benefit by integration with microfluidics because of reduced consumption of analytes arising from the small scale.

8. Identify the only correct statement from the following:

(a) Batteries and capacitors are identical energy storage devices composed of two electrodes and an electrolyte and exhibiting same charging/discharging characteristics.
(b) Lithium-ion batteries are the most promising energy storage solutions because of the use of low-weight and low-cost lithium which is abundantly found in earth crust.
(c) Supercapacitors are going to completely replace batteries soon because of their capabilities for fast charging and discharging.
(d) Two metal plates separated from each other can be used as a capacitor.

9. Identify the only incorrect statement from the following:

(a) Extraction of organics from an aqueous mixture can be performed using 3D printed mixers and phase separators.
(b) Membrane filtration is based on sieving as well as retention by chemical interactions.
(c) Separation using zeolites and crystalline frameworks is based on their chemical interaction with molecules to be retained.
(d) Throughput in membrane filtration can be improved by using 3D printed formed bodies without significant deterioration of separation performance.

10. Match the following:

S. no	AM technology	Mechanism of fabrication	Application	Classification of process
1	Inkjet printing	Extrusion of ink from a nozzle in three dimensions as per design coordinates	(a) Contact-less fabrication of conducting interconnects on 3D objects	(a) Form and shape sensitive process
2	Selective laser sintering	(b) Deposition of atomised liquid particles carried with carrier gas through a nozzle	(b) Rapid prototyping using material sheets	(b) Bulk sensitive process

(continued)

(continued)

S. no	AM technology	Mechanism of fabrication	Application	Classification of process
3	Laminate object manufacturing	(c) Melting and extrusion of a thermoplastic filament through a nozzle	(c) Cheap 3D objects using ABS	(c) Fusion-based additive manufacturing
4	(d) Aerosol jet printing	(d) On-demand ink droplet generation and deposition	(d) Batteries and capacitors	(d) High throughput experimentation
5	(e) Direct ink writing	(e) Selective and localised sintering of powders using an IR laser	(e) Rapid preparation of composite materials	(e) Surface sensitive process
6	(f) Stereolithography	(f) Photopolymerization of photocurable resin in a 3D voxel using an IR laser	(f) 3D objects with sub-micron features	(f) Energy storage materials
7	(g) Two-photon lithography	(g) Photopolymerization of photocurable resin using a UV light source	(g) Rapid prototyping, single use molds and microfluidics	(g) Photopolymerization-based additive manufacturing
8	(h) Fused deposition modelling	(h) Laser machining of a material sheet in 2D as per CAD design and binding of cut laminates	(h) Chemical Reactors	(h) Extrusion-based additive manufacturing

References

1. Mannoor MS, Jiang Z, James T, Kong YL, Malatesta KA, Soboyejo WO, Verma N, Gracias DH, McAlpine MC (2013) 3D printed bionic ears. Nano Lett 13(6):2634–2639. https://doi.org/10.1021/nl4007744
2. Gross BC, Erkal JL, Lockwood SY, Chen C, Spence DM (2014) Evaluation of 3D printing and its potential impact on biotechnology and the chemical sciences. Anal Chem 86(7):3240–3253. https://doi.org/10.1021/ac403397r
3. Yue J, Zhao P, Gerasimov JY, van de Lagemaat M, Grotenhuis A, Rustema-Abbing M, van der Mei HC, Busscher HJ, Herrmann A, Ren Y (2015) 3D-printable antimicrobial composite resins. Adv Func Mater 25(43):6756–6767. https://doi.org/10.1002/adfm.201502384
4. Joerger RD (2007) Antimicrobial films for food applications: a quantitative analysis of their effectiveness. Packag Technol Sci 20(4):231–273. https://doi.org/10.1002/pts.774
5. Wang X, Cai X, Guo Q, Zhang T, Kobe B, Yang J (2013) i3DP, a robust 3D printing approach enabling genetic post-printing surface modification. Chem Commun (Camb) 49(86):10064–10066. https://doi.org/10.1039/c3cc45817b

6. Ganesh VA, Raut HK, Nair AS, Ramakrishna S (2011) A review on self-cleaning coatings. J Mater Chem 21(41). https://doi.org/10.1039/c1jm12523k
7. Fihri A, Bovero E, Al-Shahrani A, Al-Ghamdi A, Alabedi G (2017) Recent progress in superhydrophobic coatings used for steel protection: a review. Colloids Surf, A 520:378–390. https://doi.org/10.1016/j.colsurfa.2016.12.057
8. Jafari R, Cloutier C, Allahdini A, Momen G (2019) Recent progress and challenges with 3D printing of patterned hydrophobic and superhydrophobic surfaces. Int J Adv Manufact Technol 103(1–4):1225–1238. https://doi.org/10.1007/s00170-019-03630-4
9. Graeber G, Martin Kieliger OB, Schutzius TM, Poulikakos D (2018) 3D-printed surface architecture enhancing superhydrophobicity and viscous droplet repellency. ACS Appl Mater Interfaces 10(49):43275–43281. https://doi.org/10.1021/acsami.8b16893
10. Liu M, Wang S, Jiang L (2017) Nature-inspired superwettability systems. Nat. Rev. Mater. 2:1–17. https://doi.org/10.1038/natrevmats.2017.36
11. Yan C, Jiang P, Jia X, Wang X (2020) 3D printing of bioinspired textured surfaces with superamphiphobicity. Nanoscale 12(5):2924–2938. https://doi.org/10.1039/c9nr09620e
12. Ge Q, Dunn CK, Qi HJ, Dunn ML (2014) Active origami by 4D printing. Smart Mater Struct 23(9):094007. https://doi.org/10.1088/0964-1726/23/9/094007
13. Gladman AS, Matsumoto EA, Nuzzo RG, Mahadevan L, Lewis JA (2016) Biomimetic 4D printing. Nat Mater 15(4):413–418. https://doi.org/10.1038/nmat4544
14. Gowers SA, Curto VF, Seneci CA, Wang C, Anastasova S, Vadgama P, Yang GZ, Boutelle MG (2015) 3D printed microfluidic device with integrated biosensors for online analysis of subcutaneous human microdialysate. Anal Chem 87(15):7763–7770. https://doi.org/10.1021/acs.analchem.5b01353
15. Erkal JL, Selimovic A, Gross BC, Lockwood SY, Walton EL, McNamara S, Martin RS, Spence DM (2014) 3D printed microfluidic devices with integrated versatile and reusable electrodes. Lab Chip 14(12):2023–2032. https://doi.org/10.1039/c4lc00171k
16. Ragones H, Schreiber D, Inberg A, Berkh O, Kósa G, Freeman A, Shacham-Diamand Y (2015) Disposable electrochemical sensor prepared using 3D printing for cell and tissue diagnostics. Sens Actuators, B Chem 216:434–442. https://doi.org/10.1016/j.snb.2015.04.065
17. Ragones H, Schreiber D, Inberg A, Berkh O, Kósa G, Shacham-Diamand Y (2015) Processing issues and the characterization of soft electrochemical 3D sensor. Electrochim Acta 183:125–129. https://doi.org/10.1016/j.electacta.2015.04.109
18. Bishop GW, Satterwhite JE, Bhakta S, Kadimisetty K, Gillette KM, Chen E, Rusling JF (2015) 3D-printed fluidic devices for nanoparticle preparation and flow-injection amperometry using integrated prussian blue nanoparticle-modified electrodes. Anal Chem 87(10):5437–5443. https://doi.org/10.1021/acs.analchem.5b00903
19. Dias AA, Cardoso TMG, Cardoso RM, Duarte LC, Muñoz RAA, Richter EM, Coltro WKT (2016) Paper-based enzymatic reactors for batch injection analysis of glucose on 3D printed cell coupled with amperometric detection. Sens Actuators, B Chem 226:196–203. https://doi.org/10.1016/j.snb.2015.11.040
20. Snowden ME, King PH, Covington JA, Macpherson JV, Unwin PR (2010) Fabrication of versatile channel flow cells for quantitative electroanalysis using prototyping. Anal Chem 82:3124–3131
21. Bauer R, Stewart G, Johnstone W, Boyd E, Lengden M (2014) 3D-printed miniature gas cell for photoacoustic spectroscopy of trace gases. Opt Lett 39(16):4796–4799. https://doi.org/10.1364/OL.39.004796
22. Dantism S, Takenaga S, Wagner P, Wagner T, Schöning MJ (2016) Determination of the extracellular acidification of Escherichia coliK12 with a multi-chamber-based LAPS system. Phys Status Solidi (a) 213(6):1479–1485. https://doi.org/10.1002/pssa.201533043
23. Takenaga S, Schneider B, Erbay E, Biselli M, Schnitzler T, Schöning MJ, Wagner T (2015) Fabrication of biocompatible lab-on-chip devices for biomedical applications by means of a 3D-printing process. Phys Status Solidi (a) 212(6):1347–1352. https://doi.org/10.1002/pssa.201532053

24. Ambrosi A, Moo JGS, Pumera M (2016) Helical 3D-printed metal electrodes as custom-shaped 3D platform for electrochemical devices. Adv Func Mater 26(5):698–703. https://doi.org/10.1002/adfm.201503902
25. Bhargava KC, Thompson B, Malmstadt N (2014) Discrete elements for 3D microfluidics. Proc Natl Acad Sci USA 111(42):15013–15018. https://doi.org/10.1073/pnas.1414764111
26. Krejcova L, Nejdl L, Rodrigo MA, Zurek M, Matousek M, Hynek D, Zitka O, Kopel P, Adam V, Kizek R (2014) 3D printed chip for electrochemical detection of influenza virus labeled with CdS quantum dots. Biosens Bioelectron 54:421–427. https://doi.org/10.1016/j.bios.2013.10.031
27. Chen C, Wang Y, Lockwood SY, Spence DM (2014) 3D-printed fluidic devices enable quantitative evaluation of blood components in modified storage solutions for use in transfusion medicine. Analyst 139(13):3219–3226. https://doi.org/10.1039/c3an02357e
28. Spilstead KB, Learey JJ, Doeven EH, Barbante GJ, Mohr S, Barnett NW, Terry JM, Hall RM, Francis PS (2014) 3D-printed and CNC milled flow-cells for chemiluminescence detection. Talanta 126:110–115. https://doi.org/10.1016/j.talanta.2014.03.047
29. Roda A, Guardigli M, Calabria D, Calabretta MM, Cevenini L, Michelini E (2014) A 3D-printed device for a smartphone-based chemiluminescence biosensor for lactate in oral fluid and sweat. Analyst 139(24):6494–6501. https://doi.org/10.1039/c4an01612b
30. Bishop GW, Satterwhite-Warden JE, Bist I, Chen E, Rusling JF (2016) Electrochemiluminescence at bare and DNA-coated graphite electrodes in 3D-printed fluidic devices. ACS Sens 1(2):197–202. https://doi.org/10.1021/acssensors.5b00156
31. Kadimisetty K, Mosa IM, Malla S, Satterwhite-Warden JE, Kuhns TM, Faria RC, Lee NH, Rusling JF (2016) 3D-printed supercapacitor-powered electrochemiluminescent protein immunoarray. Biosens Bioelectron 77:188–193. https://doi.org/10.1016/j.bios.2015.09.017
32. Lee W, Kwon D, Chung B, Jung GY, Au A, Folch A, Jeon S (2014) Ultrarapid detection of pathogenic bacteria using a 3D immunomagnetic flow assay. Anal Chem 86(13):6683–6688. https://doi.org/10.1021/ac501436d
33. Lee W, Kwon D, Choi W, Jung GY, Jeon S (2015) 3D-printed microfluidic device for the detection of pathogenic bacteria using size-based separation in helical channel with trapezoid cross-section. Sci Rep 5:7717. https://doi.org/10.1038/srep07717
34. Cevenini L, Calabretta MM, Tarantino G, Michelini E, Roda A (2016) Smartphone-interfaced 3D printed toxicity biosensor integrating bioluminescent "sentinel cells." Sens Actuators, B Chem 225:249–257. https://doi.org/10.1016/j.snb.2015.11.017
35. Au AK, Bhattacharjee N, Horowitz LF, Chang TC, Folch A (2015) 3D-printed microfluidic automation. Lab Chip 15(8):1934–1941. https://doi.org/10.1039/c5lc00126a
36. Alapan Y, Hasan MN, Shen R, Gurkan UA (2015) Three-dimensional printing based hybrid manufacturing of microfluidic devices. J Nanotechnol Eng Med 6(2). https://doi.org/10.1115/1.4031231
37. Anderson KB, Lockwood SY, Martin RS, Spence DM (2013) A 3D printed fluidic device that enables integrated features. Anal Chem 85(12):5622–5626. https://doi.org/10.1021/ac4009594
38. Lee KG, Park KJ, Seok S, Shin S, Kim DH, Park JY, Heo YS, Lee SJ, Lee TJ (2014) 3D printed modules for integrated microfluidic devices. RSC Adv 4(62):32876–32880. https://doi.org/10.1039/c4ra05072j
39. Hossain A, Canning J, Ast S, Rutledge PJ, Teh Li Y, Jamalipour A (2015) Lab-in-a-phone: smartphone-based portable fluorometer for pH measurements of environmental water. IEEE Sens J 15(9):5095–5102. https://doi.org/10.1109/jsen.2014.2361651
40. Heger Z, Zitka J, Cernei N, Krizkova S, Sztalmachova M, Kopel P, Masarik M, Hodek P, Zitka O, Adam V, Kizek R (2015) 3D-printed biosensor with poly(dimethylsiloxane) reservoir for magnetic separation and quantum dots-based immunolabeling of metallothionein. Electrophoresis 36(11–12):1256–1264. https://doi.org/10.1002/elps.201400559
41. Heger Z, Zitka J, Nejdl L, Moulick A, Milosavljevic V, Kopel P, Zavodsky O, Kapus J, Lenza L, Rezka M, Adam V, Kizek R (2016) 3D printed stratospheric probe as a platform for determination of DNA damage based on carbon quantum dots/DNA complex fluorescence

increase. Monatshefte für Chemie - Chem Mon 147(5):873–880. https://doi.org/10.1007/s00706-016-1705-y

42. Spivey EC, Xhemalce B, Shear JB, Finkelstein IJ (2014) 3D-printed microfluidic microdissector for high-throughput studies of cellular aging. Anal Chem 86(15):7406–7412. https://doi.org/10.1021/ac500893a
43. Yafia M, Ahmadi A, Hoorfar M, Najjaran H (2015) Ultra-portable smartphone controlled integrated digital microfluidic system in a 3D-printed modular assembly. Micromachines 6(9):1289–1305. https://doi.org/10.3390/mi6091289
44. Dirkzwager RM, Liang S, Tanner JA (2016) Development of aptamer-based point-of-care diagnostic devices for malaria using three-dimensional printing rapid prototyping. ACS Sens 1(4):420–426. https://doi.org/10.1021/acssensors.5b00175
45. Chan HN, Shu Y, Xiong B, Chen Y, Chen Y, Tian Q, Michael SA, Shen B, Wu H (2015) Simple, cost-effective 3D printed microfluidic components for disposable, point-of-care colorimetric analysis. . ACS Sens 1(3):227–234. https://doi.org/10.1021/acssensors.5b00100
46. Shallan AI, Smejkal P, Corban M, Guijt RM, Breadmore MC (2014) Cost-effective three-dimensional printing of visibly transparent microchips within minutes. Anal Chem 86(6):3124–3130. https://doi.org/10.1021/ac4041857
47. Comina G, Suska A, Filippini D (2015) Autonomous chemical sensing interface for universal cell phone readout. Angew Chem Int Ed Engl 54(30):8708–8712. https://doi.org/10.1002/anie.201503727
48. Achilli E, Minguzzi A, Visibile A, Locatelli C, Vertova A, Naldoni A, Rondinini S, Auricchio F, Marconi S, Fracchia M, Ghigna P (2016) 3D-printed photo-spectroelectrochemical devices for in situ and in operando X-ray absorption spectroscopy investigation. J Synchrotron Radiat 23(2):622–628. https://doi.org/10.1107/S1600577515024480
49. Hanada Y, Sugioka K, Shihira-Ishikawa I, Kawano H, Miyawaki A, Midorikawa K (2011) 3D microfluidic chips with integrated functional microelements fabricated by a femtosecond laser for studying the gliding mechanism of cyanobacteria. Lab Chip 11(12):2109–2115. https://doi.org/10.1039/c1lc20101h
50. Tsuda S, Jaffery H, Doran D, Hezwani M, Robbins PJ, Yoshida M, Cronin L (2015) Customizable 3D printed "plug and play" millifluidic devices for programmable fluidics. PLoS ONE 10(11):e0141640. https://doi.org/10.1371/journal.pone.0141640
51. Ude C, Hentrop T, Lindner P, Lücking TH, Scheper T, Beutel S (2015) New perspectives in shake flask pH control using a 3D-printed control unit based on pH online measurement. Sens Actuat, B Chem 221:1035–1043. https://doi.org/10.1016/j.snb.2015.07.017
52. Singh H, Shimojima M, Shiratori T, le An V, Sugamata M, Yang M (2015) Application of 3D printing technology in increasing the diagnostic performance of enzyme-linked immunosorbent assay (ELISA) for infectious diseases. Sensors (Basel) 15(7):16503–16515. https://doi.org/10.3390/s150716503
53. Hill RT, Kozek KM, Hucknall A, Smith DR, Chilkoti A (2014) Nanoparticle-film plasmon ruler interrogated with transmission visible spectroscopy. ACS Photon 1(10):974–984. https://doi.org/10.1021/ph500190q
54. Kitson PJ, Symes MD, Dragone V, Cronin L (2013) Combining 3D printing and liquid handling to produce user-friendly reactionware for chemical synthesis and purification. Chem Sci 4(8):3099–3103. https://doi.org/10.1039/c3sc51253c
55. Kitson PJ, Marshall RJ, Long D, Forgan RS, Cronin L (2014) 3D printed high-throughput hydrothermal reactionware for discovery, optimization, and scale-up. Angew Chem Int Ed Engl 53(47):12723–12728. https://doi.org/10.1002/anie.201402654
56. Symes MD, Kitson PJ, Yan J, Richmond CJ, Cooper GJ, Bowman RW, Vilbrandt T, Cronin L (2012) Integrated 3D-printed reactionware for chemical synthesis and analysis. Nat Chem 4(5):349–354. https://doi.org/10.1038/nchem.1313
57. Venkateswaran PS, Sharma A, Dubey S, Agarwal A, Goel S (2016) Rapid and automated measurement of milk adulteration using a 3D printed optofluidic microviscometer (OMV). IEEE Sens J 16(9):3000–3007. https://doi.org/10.1109/jsen.2016.2527921

58. Chahadih A, Cresson PY, Hamouda Z, Gu S, Mismer C, Lasri T (2015) Microwave/microfluidic sensor fabricated on a flexible kapton substrate for complex permittivity characterization of liquids. Sens Actuat, A 229:128–135. https://doi.org/10.1016/j.sna.2015.03.027
59. Kimionis J, Isakov M, Koh BS, Georgiadis A, Tentzeris MM (2015) 3D-printed origami packaging with inkjet-printed antennas for RF harvesting sensors. IEEE Trans Microw Theory Tech 63(12):4521–4532. https://doi.org/10.1109/tmtt.2015.2494580
60. Shamsinejad S, De Flaviis F, Mousavi P (2016) Microstrip-fed 3-D folded slot antenna on cubic structure. IEEE Anten Wirel Propag Lett 15:1081–1084. https://doi.org/10.1109/lawp.2015.2493146
61. Zhao J, Liu M, Liang L, Wang W, Xie J (2016) Airborne particulate matter classification and concentration detection based on 3D printed virtual impactor and quartz crystal microbalance sensor. Sens Actuat, A 238:379–388. https://doi.org/10.1016/j.sna.2015.12.029
62. Suaste-Gomez E, Rodriguez-Roldan G, Reyes-Cruz H, Teran-Jimenez O (2016) Developing an ear prosthesis fabricated in polyvinylidene fluoride by a 3D printer with sensory intrinsic properties of pressure and temperature. Sensors (Basel) 16(3). https://doi.org/10.3390/s16030332
63. Shemelya C, Cedillos F, Aguilera E, Espalin D, Muse D, Wicker R, MacDonald E (2015) Encapsulated copper wire and copper mesh capacitive sensing for 3-D printing applications. IEEE Sens J 15(2):1280–1286. https://doi.org/10.1109/jsen.2014.2356973
64. Haque RI, Ogam E, Loussert C, Benaben P, Boddaert X (2015) Fabrication of capacitive acoustic resonators combining 3D printing and 2D inkjet printing techniques. Sens (Basel) 15(10):26018–26038. https://doi.org/10.3390/s151026018
65. Laszczak P, Jiang L, Bader DL, Moser D, Zahedi S (2015) Development and validation of a 3D-printed interfacial stress sensor for prosthetic applications. Med Eng Phys 37(1):132–137. https://doi.org/10.1016/j.medengphy.2014.10.002
66. Borghetti M, Serpelloni M, Sardini E, Pandini S (2016) Mechanical behavior of strain sensors based on PEDOT:PSS and silver nanoparticles inks deposited on polymer substrate by inkjet printing. Sens Actuat, A 243:71–80. https://doi.org/10.1016/j.sna.2016.03.021
67. Hong J, Leigh SJ, Bradley RJ, Purssell CP, Billson DR, Hutchins DA (2012) A simple, low-cost conductive composite material for 3D printing of electronic sensors. PLoS ONE 7(11). https://doi.org/10.1371/journal.pone.0049365
68. Chang Y-H, Wang K, Wu C, Chen Y, Zhang C, Wang B (2015) A facile method for integrating direct-write devices into three-dimensional printed parts. Smart Mater Struct 24(6). https://doi.org/10.1088/0964-1726/24/6/065008
69. Ota H, Emaminejad S, Gao Y, Zhao A, Wu E, Challa S, Chen K, Fahad HM, Jha AK, Kiriya D, Gao W, Shiraki H, Morioka K, Ferguson AR, Healy KE, Davis RW, Javey A (2016) Application of 3D printing for smart objects with embedded electronic sensors and systems. Adv Mater Technol 1(1). https://doi.org/10.1002/admt.201600013
70. Cui X, Gao G, Qiu Y (2013) Accelerated myotube formation using bioprinting technology for biosensor applications. Biotechnol Lett 35(3):315–321. https://doi.org/10.1007/s10529-012-1087-0
71. Yun H-y, Kim H-c, Lee I-h (2015) Research for improved flexible tactile sensor sensitivity. J Mech Sci Technol 29(12):5133–5138. https://doi.org/10.1007/s12206-015-1112-z
72. Guo SZ, Yang X, Heuzey MC, Therriault D (2015) 3D printing of a multifunctional nanocomposite helical liquid sensor. Nanoscale 7(15):6451–6456. https://doi.org/10.1039/c5nr00278h
73. Chang WS, Kim JH, Kim D, Cho SH, Kwon Seol S (2014) Individually addressable suspended conducting-polymer wires in a chemiresistive gas sensor. Macromol Chem Phys 215(17):1633–1638. https://doi.org/10.1002/macp.201400220
74. Vatani M, Lu Y, Engeberg ED, Choi J-W (2015) Combined 3D printing technologies and material for fabrication of tactile sensors. Int J Precis Eng Manuf 16(7):1375–1383. https://doi.org/10.1007/s12541-015-0181-3

75. Leigh SJ, Purssell CP, Billson DR, Hutchins DA (2014) Using a magnetite/thermoplastic composite in 3D printing of direct replacements for commercially available flow sensors. Smart Mater Struct 23(9). https://doi.org/10.1088/0964-1726/23/9/095039
76. Salvo P, Raedt R, Carrette E, Schaubroeck D, Vanfleteren J, Cardon L (2012) A 3D printed dry electrode for ECG/EEG recording. Sens Actuators, A 174:96–102. https://doi.org/10.1016/j.sna.2011.12.017
77. Walzik MP, Vollmar V, Lachnit T, Dietz H, Haug S, Bachmann H, Fath M, Aschenbrenner D, Abolpour Mofrad S, Friedrich O, Gilbert DF (2015) A portable low-cost long-term live-cell imaging platform for biomedical research and education. Biosens Bioelectron 64:639–649. https://doi.org/10.1016/j.bios.2014.09.061
78. Salamone F, Belussi L, Danza L, Ghellere M, Meroni I (2015) Design and development of nEMoS, an all-in-one, low-cost, web-connected and 3D-printed device for environmental analysis. Sensors (Basel) 15(6):13012–13027. https://doi.org/10.3390/s150613012
79. Xu L, Gutbrod SR, Bonifas AP, Su Y, Sulkin MS, Lu N, Chung HJ, Jang KI, Liu Z, Ying M, Lu C, Webb RC, Kim JS, Laughner JI, Cheng H, Liu Y, Ameen A, Jeong JW, Kim GT, Huang Y, Efimov IR, Rogers JA (2014) 3D multifunctional integumentary membranes for spatiotemporal cardiac measurements and stimulation across the entire epicardium. Nat Commun 5:3329. https://doi.org/10.1038/ncomms4329
80. Mehla S, Kandjani A, Coyle V, Harrison CJ, Low MX, Kaner RB, Sabri Y, Bhargava SK (2022) Gold sunflower microelectrode arrays with dendritic nanostructures on the lateral surfaces for antireflection and surface-enhanced raman scattering. ACS Appl Nano Mater 5(2):1873–1890 https://doi.org/10.1021/acsanm.1c03501
81. Ni Y, Ji R, Long K, Bu T, Chen K, Zhuang S (2017) A review of 3D-printed sensors. Appl Spectrosc Rev 52(7):623–652. https://doi.org/10.1080/05704928.2017.1287082
82. Bakker E, Telting-Diaz M (2002) Electrochemical sensors. Anal Chem 74:2781–2800
83. Privett BJ, Shin JH, Schoenfisch MH (2010) Electrochemical sensors. Anal Chem 82:4723–4741
84. Yang T, Xie D, Li Z, Zhu H (2017) Recent advances in wearable tactile sensors: materials, sensing mechanisms, and device performance. Mater Sci Eng R Rep 115:1–37. https://doi.org/10.1016/j.mser.2017.02.001
85. Camposeo A, Persano L, Farsari M, Pisignano D (2019) Additive manufacturing: applications and directions in photonics and optoelectronics. Adv Opt Mater 7(1):1800419. https://doi.org/10.1002/adom.201800419
86. Conformal Robotic Stereolithography (2016). 3 (4):226–235. doi:https://doi.org/10.1089/3dp.2016.0042
87. Alaman J, Alicante R, Pena JI, Sanchez-Somolinos C (2016) Inkjet printing of functional materials for optical and photonic applications. Materials (Basel) 9(11). https://doi.org/10.3390/ma9110910
88. Kong YL, Tamargo IA, Kim H, Johnson BN, Gupta MK, Koh TW, Chin HA, Steingart DA, Rand BP, McAlpine MC (2014) 3D printed quantum dot light-emitting diodes. Nano Lett 14(12):7017–7023. https://doi.org/10.1021/nl5033292
89. Pyo J, Kim JT, Lee J, Yoo J, Je JH (2016) 3D printed nanophotonic waveguides. Adv Opt Mater 4(8):1190–1195. https://doi.org/10.1002/adom.201600220
90. Weidenbach M, Jahn D, Rehn A, Busch SF, Beltran-Mejia F, Balzer JC, Koch M (2016) 3D printed dielectric rectangular waveguides, splitters and couplers for 120 GHz. Opt Express 24(25):28968–28976. https://doi.org/10.1364/OE.24.028968
91. Crespi A, Osellame R, Ramponi R, Brod DJ, Galvão EF, Spagnolo N, Vitelli C, Maiorino E, Mataloni P, Sciarrino F (2013) Integrated multimode interferometers with arbitrary designs for photonic boson sampling. Nat Photon 7(7):545–549. https://doi.org/10.1038/nphoton.2013.112
92. Spagnolo N, Vitelli C, Aparo L, Mataloni P, Sciarrino F, Crespi A, Ramponi R, Osellame R (2013) Three-photon bosonic coalescence in an integrated tritter. Nat Commun 4:1606. https://doi.org/10.1038/ncomms2616

93. Laza SC, Polo M, Neves AA, Cingolani R, Camposeo A, Pisignano D (2012) Two-photon continuous flow lithography. Adv Mater 24(10):1304–1308. https://doi.org/10.1002/adma.201103357
94. Thiele S, Arzenbacher K, Gissibl T, Giessen H, Herkommer AM (2017) 3D-printed eagle eye: Compound microlens system for foveated imaging. Sci Adv 3(2):e1602655
95. Kamyshny A, Magdassi S (2019) Conductive nanomaterials for 2D and 3D printed flexible electronics. Chem Soc Rev 48(6):1712–1740. https://doi.org/10.1039/c8cs00738a
96. Piqué A, Auyeung RCY, Kim H, Charipar NA, Mathews SA (2016) Laser 3D micro-manufacturing. J Phys D: Appl Phys 49(22). https://doi.org/10.1088/0022-3727/49/22/223001
97. Shaw-Stewart J, Lippert T, Nagel M, Nuesch F, Wokaun A (2011) Laser-induced forward transfer of polymer light-emitting diode pixels with increased charge injection. ACS Appl Mater Interfaces 3(2):309–316. https://doi.org/10.1021/am100943f
98. Chrisey DB, Pique A, Modi R, Wu HD, Auyeung RCY, Young HD (2000) Direct writing of conformal mesoscopic electronic devices by MAPLE DW. Appl Surf Sci 168:345–352
99. Wilkinson NJ, Smith MAA, Kay RW, Harris RA (2019) A review of aerosol jet printing—a non-traditional hybrid process for micro-manufacturing. Int J Adv Manufact Technol 105(11):4599–4619. https://doi.org/10.1007/s00170-019-03438-2
100. Secor EB (2018) Principles of aerosol jet printing. Flexible Print Electron 3(3). https://doi.org/10.1088/2058-8585/aace28
101. Hoey JM, Lutfurakhmanov A, Schulz DL, Akhatov IS (2012) A review on aerosol-based direct-write and its applications for microelectronics. J Nanotechnol 2012:1–22. https://doi.org/10.1155/2012/324380
102. Folgar CE, Suchicital C, Priya S (2011) Solution-based aerosol deposition process for synthesis of multilayer structures. Mater Lett 65(9):1302–1307. https://doi.org/10.1016/j.matlet.2011.01.069
103. Marinov VR, Atanasov YA, Khan A, Vaselaar D, Halvorsen A, Schulz DL, Chrisey DB (2007) Direct-write vapor sensors on FR4 plastic substrates. IEEE Sens J 7(6):937–944. https://doi.org/10.1109/jsen.2007.895964
104. Hörteis M, Glunz SW (2008) Fine line printed silicon solar cells exceeding 20% efficiency. Prog Photovoltaics Res Appl 16(7):555–560. https://doi.org/10.1002/pip.850
105. Habas SE, Platt HAS, Hest MFAMv, Ginley DS, (2010) Low-cost inorganic solar cells: from ink to printed device. Chem Rev 110:6571–6594
106. Hoey JM, Reich MT, Halvorsen A, Vaselaar D, Braaten K, Maassel M, Akhatov IS, Ghandour O, Drzaic P, Schulz DL (2009) Rapid prototyping RFID antennas using direct-write. IEEE Trans Adv Packag 32(4):809–815. https://doi.org/10.1109/tadvp.2009.2021768
107. Ha M, Xia Y, Green AA, Zhang W, Renn MJ, Kim CH, Hersam MC, Frisbie CD (2010) Printed, sub-3V digital circuits on plastic from aqueous carbon nanotube inks. ACS Nano 4(8):4388–4395
108. Oztan C, Ballikaya S, Ozgun U, Karkkainen R, Celik E (2019) Additive manufacturing of thermoelectric materials via fused filament fabrication. Appl Mater Today 15:77–82. https://doi.org/10.1016/j.apmt.2019.01.001
109. Boudouris BW, Yee S (2017) Structure, properties and applications of thermoelectric polymers. J Appl Polymer Sci 134(3). https://doi.org/10.1002/app.44456
110. Du Y, Chen J, Meng Q, Xu J, Lu J, Paul B, Eklund P (2020) Flexible thermoelectric double-layer inorganic/organic composites synthesized by additive manufacturing. Adv Electron Mater 6(8). https://doi.org/10.1002/aelm.202000214
111. He M, Zhao Y, Wang B, Xi Q, Zhou J, Liang Z (2015) 3D printing fabrication of amorphous thermoelectric materials with ultralow thermal conductivity. Small 11(44):5889–5894. https://doi.org/10.1002/smll.201502153
112. Qiu J, Yan Y, Luo T, Tang K, Yao L, Zhang J, Zhang M, Su X, Tan G, Xie H, Kanatzidis MG, Uher C, Tang X (2019) 3D Printing of highly textured bulk thermoelectric materials: mechanically robust BiSbTe alloys with superior performance. Energy Environ Sci 12(10):3106–3117. https://doi.org/10.1039/c9ee02044f

113. Kee S, Haque MA, Corzo D, Alshareef HN, Baran D (2019) Self-healing and stretchable 3D-printed organic thermoelectrics. Adv Funct Mater 29(51). https://doi.org/10.1002/adfm.201905426
114. Cholleti ER (2018) A review on 3D printing of piezoelectric materials. In: IOP conference series: materials science and engineering, vol 455. https://doi.org/10.1088/1757-899x/455/1/012046
115. Sappati KK, Bhadra S (2018) Piezoelectric polymer and paper substrates: a review. Sensors (Basel) 18(11). https://doi.org/10.3390/s18113605
116. Bodkhe S, Ermanni P (2019) Challenges in 3D printing of piezoelectric materials. Multifunct Mater 2(2). https://doi.org/10.1088/2399-7532/ab0c41
117. Chen C, Wang X, Wang Y, Yang D, Yao F, Zhang W, Wang B, Sewvandi GA, Yang D, Hu D (2020) Additive manufacturing of piezoelectric materials. Adv Funct Mater 30(52). https://doi.org/10.1002/adfm.202005141
118. Lee C, Tarbutton JA (2014) Electric poling-assisted additive manufacturing process for PVDF polymer-based piezoelectric device applications. Smart Mater Struct 23(9). https://doi.org/10.1088/0964-1726/23/9/095044
119. Bodkhe S, Rajesh PSM, Gosselin FP, Therriault D (2018) Simultaneous 3D printing and poling of PVDF and its nanocomposites. ACS Appl Energy Mater 1(6):2474–2482. https://doi.org/10.1021/acsaem.7b00337
120. Kim K, Zhu W, Qu X, Aaronson C, McCall WR, Chen S, Sirbuly DJ (2014) 3D optical printing of piezoelectric nanoparticle - polymer composite materials. ACS Nano 8(10):9799–9806
121. Chen Z, Song X, Lei L, Chen X, Fei C, Chiu CT, Qian X, Ma T, Yang Y, Shung K, Chen Y, Zhou Q (2016) 3D printing of piezoelectric element for energy focusing and ultrasonic sensing. Nano Energy 27:78–86. https://doi.org/10.1016/j.nanoen.2016.06.048
122. Lim S, Son D, Kim J, Lee YB, Song J-K, Choi S, Lee DJ, Kim JH, Lee M, Hyeon T, Kim D-H (2015) Transparent and stretchable interactive human machine interface based on patterned graphene heterostructures. Adv Func Mater 25(3):375–383. https://doi.org/10.1002/adfm.201402987
123. Lim J, Jung H, Baek C, Hwang G-T, Ryu J, Yoon D, Yoo J, Park K-I, Kim JH (2017) All-inkjet-printed flexible piezoelectric generator made of solvent evaporation assisted BaTiO3 hybrid material. Nano Energy 41:337–343. https://doi.org/10.1016/j.nanoen.2017.09.046
124. Chen W, Wang F, Yan K, Zhang Y, Wu D (2019) Micro-stereolithography of KNN-based lead-free piezoceramics. Ceram Int 45(4):4880–4885. https://doi.org/10.1016/j.ceramint.2018.11.185
125. Parra-Cabrera C, Achille C, Kuhn S, Ameloot R (2018) 3D printing in chemical engineering and catalytic technology: structured catalysts, mixers and reactors. Chem Soc Rev 47(1):209–230. https://doi.org/10.1039/c7cs00631d
126. Amin R, Knowlton S, Hart A, Yenilmez B, Ghaderinezhad F, Katebifar S, Messina M, Khademhosseini A, Tasoglu S (2016) 3D-printed microfluidic devices. Biofabrication 8(2):022001. https://doi.org/10.1088/1758-5090/8/2/022001
127. El-Ali J, Sorger PK, Jensen KF (2006) Cells on chips. Nature 442(7101):403–411. https://doi.org/10.1038/nature05063
128. Au AK, Huynh W, Horowitz LF, Folch A (2016) 3D-printed microfluidics. Angew Chem Int Ed Engl 55(12):3862–3881. https://doi.org/10.1002/anie.201504382
129. Dittrich PS, Tachikawa K, Manz A (2006) Micro total analysis systems. Latest advancements and trends. Anal Chem 78(12):3887–3907
130. Nielsen AV, Beauchamp MJ, Nordin GP, Woolley AT (2020) 3D printed microfluidics. Annu Rev Anal Chem (Palo Alto Calif) 13(1):45–65. https://doi.org/10.1146/annurev-anchem-091619-102649
131. Xia Y, Whitesides GM (1998) Soft lithography. Annu Rev Mater Sci 28:153–184
132. Paydar OH, Paredes CN, Hwang Y, Paz J, Shah NB, Candler RN (2014) Characterization of 3D-printed microfluidic chip interconnects with integrated O-rings. Sens Actuat, A 205:199–203. https://doi.org/10.1016/j.sna.2013.11.005

133. Bhattacharjee N, Urrios A, Kang S, Folch A (2016) The upcoming 3D-printing revolution in microfluidics. Lab Chip 16(10):1720–1742. https://doi.org/10.1039/c6lc00163g
134. Sochol RD, Sweet E, Glick CC, Wu S-Y, Yang C, Restaino M, Lin L (2018) 3D printed microfluidics and microelectronics. Microelectron Eng 189:52–68. https://doi.org/10.1016/j.mee.2017.12.010
135. Rogers CI, Qaderi K, Woolley AT, Nordin GP (2015) 3D printed microfluidic devices with integrated valves. Biomicrofluidics 9(1):016501. https://doi.org/10.1063/1.4905840
136. Carrell CS, McCord CP, Wydallis RM, Henry CS (2020) Sealing 3D-printed parts to poly(dimethylsiloxane) for simple fabrication of Microfluidic devices. Anal Chim Acta 1124:78–84. https://doi.org/10.1016/j.aca.2020.05.014
137. Marre S, Jensen KF (2010) Synthesis of micro and nanostructures in microfluidic systems. Chem Soc Rev 39(3):1183–1202. https://doi.org/10.1039/b821324k
138. Hurt C, Brandt M, Priya SS, Bhatelia T, Patel J, Selvakannan PR, Bhargava S (2017) Combining additive manufacturing and catalysis: a review. Catal Sci Technol 7(16):3421–3439. https://doi.org/10.1039/c7cy00615b
139. Lawson S, Li X, Thakkar H, Rownaghi AA, Rezaei F (2021) Recent advances in 3D printing of structured materials for adsorption and catalysis applications. Chem Rev 121(10):6246–6291. https://doi.org/10.1021/acs.chemrev.1c00060
140. Friedmann D, Lee AF, Wilson K, Jalili R, Caruso RA (2019) Printing approaches to inorganic semiconductor photocatalyst fabrication. J Mater Chem A 7(18):10858–10878. https://doi.org/10.1039/c9ta00888h
141. Zhou X, Liu C-j (2017) Three-dimensional printing for catalytic applications: current status and perspectives. Adv Funct Mater 27(30). https://doi.org/10.1002/adfm.201701134
142. Stuecker JN, Miler JE, Ferrizz RE, Mudd JE, Cesarano J (2004) Advanced support structures for enhanced catalytic activity. Ind Eng Chem 43:51–55
143. Lefevere J, Gysen M, Mullens S, Meynen V, Van Noyen J (2013) The benefit of design of support architectures for zeolite coated structured catalysts for methanol-to-olefin conversion. Catal Today 216:18–23. https://doi.org/10.1016/j.cattod.2013.05.020
144. Noyen J, Wilde A, Schroeven M, Mullens S, Luyten J (2012) Ceramic processing techniques for catalyst design: formation, properties, and catalytic example of ZSM-5 on 3-dimensional fiber deposition support structures. Int J Appl Ceram Technol 9(5):902–910. https://doi.org/10.1111/j.1744-7402.2012.02781.x
145. Mehla S, Kukade S, Kumar P, Rao PVC, Sriganesh G, Ravishankar R (2019) Fine tuning H-transfer and β-scission reactions in VGO FCC using metal promoted dual functional ZSM-5. Fuel 242:487–495. https://doi.org/10.1016/j.fuel.2019.01.065
146. Mehla S, Krishnamurthy KR, Viswanathan B, John M, Niwate Y, Kishore Kumar SA, Pai SM, Newalkar BL (2013) n-Hexadecane hydroisomerization over BTMACl/TEABr/MTEABr templated ZSM-12. Microporous Mesoporous Mater 177:120–126. https://doi.org/10.1016/j.micromeso.2013.05.001
147. Mehla S, Krishnamurthy KR, Viswanathan B, John M, Niwate Y, Kumar K, Pai SM, Newalkar BL (2013) n-Hexadecane hydroisomerization over Pt/ZSM-12: role of Si/Al ratio on product distribution. J Porous Mater 20(5):1023–1029. https://doi.org/10.1007/s10934-013-9682-6
148. Mehla S, Krishsna V, Sriganesh G, Ravishankar R (2018) Mesoporous superacid catalysts for valorisation of refinery naphtha stream. RSC Adv 8(59):33702–33709. https://doi.org/10.1039/C8RA07024E
149. Danaci S, Protasova L, Lefevere J, Bedel L, Guilet R, Marty P (2016) Efficient CO 2 methanation over Ni/Al 2 O 3 coated structured catalysts. Catal Today 273:234–243. https://doi.org/10.1016/j.cattod.2016.04.019
150. Elkoro A, Soler L, Llorca J, Casanova I (2019) 3D printed microstructured Au/TiO2 catalyst for hydrogen photoproduction. Appl Mater Today 16:265–272. https://doi.org/10.1016/j.apmt.2019.06.007
151. Arin M, Lommens P, Hopkins SC, Pollefeyt G, Van der Eycken J, Ricart S, Granados X, Glowacki BA, Van Driessche I (2012) Deposition of photocatalytically active TiO2 films by inkjet printing of TiO2 nanoparticle suspensions obtained from microwave-assisted

hydrothermal synthesis. Nanotechnology 23(16):165603. https://doi.org/10.1088/0957-4484/23/16/165603
152. Arin M, Watté J, Pollefeyt G, De Buysser K, Van Driessche I, Lommens P (2013) Low temperature deposition of TiO2 layers from nanoparticle containing suspensions synthesized by microwave hydrothermal treatment. J Sol-Gel Sci Technol 66(1):100–111. https://doi.org/10.1007/s10971-013-2972-2
153. Černá M, Veselý M, Dzik P (2011) Physical and chemical properties of titanium dioxide printed layers. Catal Today 161(1):97–104. https://doi.org/10.1016/j.cattod.2010.11.019
154. Tubío CR, Azuaje J, Escalante L, Coelho A, Guitián F, Sotelo E, Gil A (2016) 3D printing of a heterogeneous copper-based catalyst. J Catal 334:110–115. https://doi.org/10.1016/j.jcat.2015.11.019
155. Zhu C, Qi Z, Beck VA, Luneau M, Lattimer J, Chen W, Worsley MA, Ye J, Duoss EB, Spadaccini CM, Friend CM, Biener J (2018) Towards digitally controlled catalyst architectures: hierarchical nanoporous gold via 3D printing. Sci Adv 4(8):eaas9459
156. Mehla S, Das J, Jampaiah D, Periasamy S, Nafady A, Bhargava SK (2019) Recent advances in preparation methods for catalytic thin films and coatings. Catal Sci Technol 9:3582–3602. https://doi.org/10.1039/C9CY00518H
157. Ambrosi A, Pumera M (2016) 3D-printing technologies for electrochemical applications. Chem Soc Rev 45(10):2740–2755. https://doi.org/10.1039/c5cs00714c
158. Zhakeyev A, Wang P, Zhang L, Shu W, Wang H, Xuan J (2017) Additive manufacturing: unlocking the evolution of energy materials. Adv Sci 4(10):1700187. https://doi.org/10.1002/advs.201700187
159. Deiner LJ, Reitz TL (2017) Inkjet and aerosol jet printing of electrochemical devices for energy conversion and storage. Adv Eng Mater 19(7). https://doi.org/10.1002/adem.201600878
160. Qin LD, Park S, Huang L, Mirkin CA (2005) On-wire lithography. Science 309(5731):113–115. https://doi.org/10.1126/science.1112666
161. Wen LY, Wang ZJ, Mi Y, Xu R, Yu SH, Lei Y (2015) Designing heterogeneous 1D nanostructure arrays based on AAO templates for energy applications. Small 11(28):3408–3428. https://doi.org/10.1002/smll.201500120
162. Pang Y, Cao Y, Chu Y, Liu M, Snyder K, MacKenzie D, Cao C (2019) Additive manufacturing of batteries. Adv Funct Mater 30(1). https://doi.org/10.1002/adfm.201906244
163. Sun K, Wei TS, Ahn BY, Seo JY, Dillon SJ, Lewis JA (2013) 3D printing of interdigitated Li-ion microbattery architectures. Adv Mater 25(33):4539–4543. https://doi.org/10.1002/adma.201301036
164. Wei TS, Ahn BY, Grotto J, Lewis JA (2018) 3D Printing of customized Li-Ion batteries with thick electrodes. Adv Mater 30(16):e1703027. https://doi.org/10.1002/adma.201703027
165. Zhang C, Shen K, Li B, Li S, Yang S (2018) Continuously 3D printed quantum dot-based electrodes for lithium storage with ultrahigh capacities. J Mater Chem A 6(41):19960–19966. https://doi.org/10.1039/c8ta08559e
166. Lyu Z, Lim GJH, Guo R, Kou Z, Wang T, Guan C, Ding J, Chen W, Wang J (2019) 3D-printed MOF-derived hierarchically porous frameworks for practical high-energy density Li-O2 batteries. Adv Funct Mater 29(1). https://doi.org/10.1002/adfm.201806658
167. Wei X, Li D, Jiang W, Gu Z, Wang X, Zhang Z, Sun Z (2015) 3D Printable Graphene Composite. Sci Rep 5:11181. https://doi.org/10.1038/srep11181
168. Maurel A, Courty M, Fleutot B, Tortajada H, Prashantha K, Armand M, Grugeon S, Panier S, Dupont L (2018) Highly loaded graphite-polylactic acid composite-based filaments for lithium-ion battery three-dimensional printing. Chem Mater 30(21):7484–7493. https://doi.org/10.1021/acs.chemmater.8b02062
169. Reyes C, Somogyi R, Niu S, Cruz MA, Yang F, Catenacci MJ, Rhodes CP, Wiley BJ (2018) Three-dimensional printing of a complete lithium ion battery with fused filament fabrication. ACS Appl Energy Mater. https://doi.org/10.1021/acsaem.8b00885
170. Lawes S, Sun Q, Lushington A, Xiao B, Liu Y, Sun X (2017) Inkjet-printed silicon as high performance anodes for Li-ion batteries. Nano Energy 36:313–321. https://doi.org/10.1016/j.nanoen.2017.04.041

171. Brown E, Yan P, Tekik H, Elangovan A, Wang J, Lin D, Li J (2019) 3D printing of hybrid MoS2-graphene aerogels as highly porous electrode materials for sodium ion battery anodes. Mater Design 170. https://doi.org/10.1016/j.matdes.2019.107689
172. Saleh MS, Li J, Park J, Panat R (2018) 3D printed hierarchically-porous microlattice electrode materials for exceptionally high specific capacity and areal capacity lithium ion batteries. Addit Manuf 23:70–78. https://doi.org/10.1016/j.addma.2018.07.006
173. Zhang H, Yu X, Braun PV (2011) Three-dimensional bicontinuous ultrafast-charge and -discharge bulk battery electrodes. Nat Nanotechnol 6(5):277–281. https://doi.org/10.1038/nnano.2011.38
174. Pikul JH, Gang Zhang H, Cho J, Braun PV, King WP (2013) High-power lithium ion microbatteries from interdigitated three-dimensional bicontinuous nanoporous electrodes. Nat Commun 4:1732. https://doi.org/10.1038/ncomms2747
175. Ning H, Pikul JH, Zhang R, Li X, Xu S, Wang J, Rogers JA, King WP, Braun PV (2015) Holographic patterning of high-performance on-chip 3D lithium-ion microbatteries. Proc Natl Acad Sci U S A 112(21):6573–6578. https://doi.org/10.1073/pnas.1423889112
176. Chen Q, Xu R, He Z, Zhao K, Pan L (2017) Printing 3D gel polymer electrolyte in lithium-ion microbattery using stereolithography. J Electrochem Soc 164(9):A1852–A1857. https://doi.org/10.1149/2.0651709jes
177. Cohen E, Menkin S, Lifshits M, Kamir Y, Gladkich A, Kosa G, Golodnitsky D (2018) Novel rechargeable 3D-microbatteries on 3D-printed-polymer substrates: feasibility study. Electrochim Acta 265:690–701. https://doi.org/10.1016/j.electacta.2018.01.197
178. Zhu C, Liu T, Qian F, Chen W, Chandrasekaran S, Yao B, Song Y, Duoss EB, Kuntz JD, Spadaccini CM, Worsley MA, Li Y (2017) 3D printed functional nanomaterials for electrochemical energy storage. Nano Today 15:107–120. https://doi.org/10.1016/j.nantod.2017.06.007
179. Nathan-Walleser T, Lazar I-M, Fabritius M, Tölle FJ, Xia Q, Bruchmann B, Venkataraman SS, Schwab MG, Mülhaupt R (2014) 3D micro-extrusion of graphene-based active electrodes: towards high-rate AC line filtering performance electrochemical capacitors. Adv Func Mater 24(29):4706–4716. https://doi.org/10.1002/adfm.201304151
180. Sun G, An J, Chua CK, Pang H, Zhang J, Chen P (2015) Layer-by-layer printing of laminated graphene-based interdigitated microelectrodes for flexible planar micro-supercapacitors. Electrochem Commun 51:33–36. https://doi.org/10.1016/j.elecom.2014.11.023
181. Liu T, Zhu C, Kou T, Worsley MA, Qian F, Condes C, Duoss EB, Spadaccini CM, Li Y (2016) Ion intercalation induced capacitance improvement for graphene-based supercapacitor electrodes. Chem NanoMat 2(7):635–641. https://doi.org/10.1002/cnma.201600107
182. Zhao C, Wang C, Gorkin R, Beirne S, Shu K, Wallace GG (2014) Three dimensional (3D) printed electrodes for interdigitated supercapacitors. Electrochem Commun 41:20–23. https://doi.org/10.1016/j.elecom.2014.01.013
183. Yang Y, Chen Z, Song X, Zhu B, Hsiai T, Wu P-I, Xiong R, Shi J, Chen Y, Zhou Q, Shung KK (2016) Three dimensional printing of high dielectric capacitor using projection based stereolithography method. Nano Energy 22:414–421. https://doi.org/10.1016/j.nanoen.2016.02.045
184. Zhu C, Liu T, Qian F, Han TY, Duoss EB, Kuntz JD, Spadaccini CM, Worsley MA, Li Y (2016) Supercapacitors based on three-dimensional hierarchical graphene aerogels with periodic macropores. Nano Lett 16(6):3448–3456. https://doi.org/10.1021/acs.nanolett.5b04965
185. Yao B, Chandrasekaran S, Zhang J, Xiao W, Qian F, Zhu C, Duoss EB, Spadaccini CM, Worsley MA, Li Y (2019) Efficient 3D printed pseudocapacitive electrodes with ultrahigh MnO2 loading. Joule 3(2):459–470. https://doi.org/10.1016/j.joule.2018.09.020
186. Wang Y, Yao M, Ma R, Yuan Q, Yang D, Cui B, Ma C, Liu M, Hu D (2020) Design strategy of barium titanate/polyvinylidene fluoride-based nanocomposite films for high energy storage. J Mater Chem A 8(3):884–917. https://doi.org/10.1039/c9ta11527g
187. Kalsoom U, Nesterenko PN, Paull B (2018) Current and future impact of 3D printing on the separation sciences. TrAC, Trends Anal Chem 105:492–502. https://doi.org/10.1016/j.trac.2018.06.006

188. Koning S, Janssen H-G, Brinkman UAT (2009) Modern methods of sample preparation for GC analysis. Chromatographia 69(S1):33–78. https://doi.org/10.1365/s10337-008-0937-3
189. Wang H, Cocovi-Solberg DJ, Hu B, Miro M (2017) 3D-printed microflow injection analysis platform for online magnetic nanoparticle sorptive extraction of antimicrobials in biological specimens as a front end to liquid chromatographic assays. Anal Chem 89(22):12541–12549. https://doi.org/10.1021/acs.analchem.7b03767
190. Martinez-Jarquin S, Moreno-Pedraza A, Guillen-Alonso H, Winkler R (2016) Template for 3D printing a low-temperature plasma probe. Anal Chem 88(14):6976–6980. https://doi.org/10.1021/acs.analchem.6b01019
191. Su CK, Peng PJ, Sun YC (2015) Fully 3D-printed preconcentrator for selective extraction of trace elements in seawater. Anal Chem 87(13):6945–6950. https://doi.org/10.1021/acs.analchem.5b01599
192. Martínez-Jarquín S, Moreno-Pedraza A, Cázarez-García D, Winkler R (2017) Automated chemical fingerprinting of Mexican spirits derived from Agave (tequila and mezcal) using direct-injection electrospray ionisation (DIESI) and low-temperature plasma (LTP) mass spectrometry. Anal Methods 9(34):5023–5028. https://doi.org/10.1039/c7ay00793k
193. Calderilla C, Maya F, Cerda V, Leal LO (2017) 3D printed device including disk-based solid-phase extraction for the automated speciation of iron using the multisyringe flow injection analysis technique. Talanta 175:463–469. https://doi.org/10.1016/j.talanta.2017.07.028
194. Calderilla C, Maya F, Cerda V, Leal LO (2018) 3D printed device for the automated preconcentration and determination of chromium (VI). Talanta 184:15–22. https://doi.org/10.1016/j.talanta.2018.02.065
195. Wang Z, Wang J, Li M, Sun K, Liu CJ (2014) Three-dimensional printed acrylonitrile butadiene styrene framework coated with Cu-BTC metal-organic frameworks for the removal of methylene blue. Sci Rep 4:5939. https://doi.org/10.1038/srep05939
196. Thakkar H, Eastman S, Al-Naddaf Q, Rownaghi AA, Rezaei F (2017) 3D-printed metal-organic framework monoliths for gas adsorption processes. ACS Appl Mater Interfaces 9(41):35908–35916. https://doi.org/10.1021/acsami.7b11626
197. Couck S, Lefevere J, Mullens S, Protasova L, Meynen V, Desmet G, Baron GV, Denayer JFM (2017) CO2, CH4 and N2 separation with a 3DFD-printed ZSM-5 monolith. Chem Eng J 308:719–726. https://doi.org/10.1016/j.cej.2016.09.046
198. Salazar-Aguilar AD, Quintanilla A, Vega-Díaz SM, Casas JA, Miranzo P, Osendi MI, Belmonte M (2021) Iron-based metal-organic frameworks integrated into 3D printed ceramic architectures. Open Ceram 5. https://doi.org/10.1016/j.oceram.2020.100047
199. Mardani S, Ojala LS, Uusi-Kyyny P, Alopaeus V (2016) Development of a unique modular distillation column using 3D printing. Chem Eng Process - Process Intensif 109:136–148. https://doi.org/10.1016/j.cep.2016.09.001
200. Worawit C, Cocovi-Solberg DJ, Varanusupakul P, Miro M (2018) In-line carbon nanofiber reinforced hollow fiber-mediated liquid phase microextraction using a 3D printed extraction platform as a front end to liquid chromatography for automatic sample preparation and analysis: a proof of concept study. Talanta 185:611–619. https://doi.org/10.1016/j.talanta.2018.04.007
201. Lahtinen E, Hänninen MM, Kinnunen K, Tuononen HM, Väisänen A, Rissanen K, Haukka M (2018) Porous 3D printed scavenger filters for selective recovery of precious metals from electronic waste. Adv Sustain Syst 2(10). https://doi.org/10.1002/adsu.201800048
202. Belka M, Ulenberg S, Baczek T (2017) Fused deposition modeling enables the low-cost fabrication of porous, customized-shape sorbents for small-molecule extraction. Anal Chem 89(8):4373–4376. https://doi.org/10.1021/acs.analchem.6b04390
203. Konieczna L, Belka M, Okonska M, Pyszka M, Baczek T (2018) New 3D-printed sorbent for extraction of steroids from human plasma preceding LC-MS analysis. J Chromatogr A 1545:1–11. https://doi.org/10.1016/j.chroma.2018.02.040
204. De Middeleer G, Dubruel P, De Saeger S (2017) Molecularly imprinted polymers immobilized on 3D printed scaffolds as novel solid phase extraction sorbent for metergoline. Anal Chim Acta 986:57–70. https://doi.org/10.1016/j.aca.2017.07.059

205. Woodhouse M, Herman GS, Parkinson BA (2005) Combinatorial approach to identification of catalysts for the photoelectrolysis of water. Chem Mater 17:4318–4324
206. Woodhouse M, Parkinson BA (2008) Combinatorial discovery and optimisation of a complex oxide with water photoelectrolysis activity. Chem Mater 20:2495–2502
207. Katz JE, Gingrich TR, Santori EA, Lewis NS (2009) Combinatorial synthesis and high-throughput photopotential and photocurrent screening of mixed-metal oxides for photoelectrochemical water splitting. Energy Environ Sci 2(1):103–112. https://doi.org/10.1039/b812177j
208. He J, Parkinson BA (2011) Combinatorial investigation of the effects of the incorporation of Ti, Si, and Al on the performance of alpha-Fe2O3 photoanodes. ACS Comb Sci 13(4):399–404. https://doi.org/10.1021/co200024p
209. Liu X, Shen Y, Yang R, Zou S, Ji X, Shi L, Zhang Y, Liu D, Xiao L, Zheng X, Li S, Fan J, Stucky GD (2012) Inkjet printing assisted synthesis of multicomponent mesoporous metal oxides for ultrafast catalyst exploration. Nano Lett 12(11):5733–5739. https://doi.org/10.1021/nl302992q
210. Gregoire JM, Xiang C, Liu X, Marcin M, Jin J (2013) Scanning droplet cell for high throughput electrochemical and photoelectrochemical measurements. Rev Sci Instrum 84(2):024102. https://doi.org/10.1063/1.4790419
211. Gregoire JM, Xiang C, Mitrovic S, Liu X, Marcin M, Cornell EW, Fan J, Jin J (2013) Combined catalysis and optical screening for high throughput discovery of solar fuels catalysts. J Electrochem Soc 160(4):F337–F342. https://doi.org/10.1149/2.035304jes
212. Seley D, Ayers K, Parkinson BA (2013) Combinatorial search for improved metal oxide oxygen evolution electrocatalysts in acidic electrolytes. ACS Comb Sci 15(2):82–89. https://doi.org/10.1021/co300086h
213. Xiang C, Suram SK, Haber JA, Guevarra DW, Soedarmadji E, Jin J, Gregoire JM (2014) High-throughput bubble screening method for combinatorial discovery of electrocatalysts for water splitting. ACS Comb Sci 16(2):47–52. https://doi.org/10.1021/co400151h
214. Tully JJ, Meloni GN (2020) A scientist's guide to buying a 3D printer: how to choose the right printer for your laboratory. Anal Chem 92(22):14853–14860. https://doi.org/10.1021/acs.analchem.0c03299
215. Mehla S, Kandjani AE, Babarao R, Lee AF, Periasamy S, Wilson K, Ramakrishna S, Bhargava SK (2021) Porous crystalline frameworks for thermocatalytic CO_2 reduction: an emerging paradigm. Energy Environ Sci 14(1):320–352. https://doi.org/10.1039/d0ee01882a
216. Mahyuddin MH, Shiota Y, Yoshizawa K (2019) Methane selective oxidation to methanol by metal-exchanged zeolites: a review of active sites and their reactivity. Catal Sci Technol 9(8):1744–1768. https://doi.org/10.1039/c8cy02414f
217. Acar C, Dincer I (2019) Review and evaluation of hydrogen production options for better environment. J Clean Prod 218:835–849. https://doi.org/10.1016/j.jclepro.2019.02.046
218. Venvik HJ, Yang J (2017) Catalysis in microstructured reactors: short review on small-scale syngas production and further conversion into methanol, DME and Fischer-Tropsch products. Catal Today 285:135–146. https://doi.org/10.1016/j.cattod.2017.02.014
219. Pan X, Jiao F, Miao D, Bao X (2021) Oxide-Zeolite-based composite catalyst concept that enables syngas chemistry beyond fischer-tropsch synthesis. Chem Rev 121(11):6588–6609. https://doi.org/10.1021/acs.chemrev.0c01012

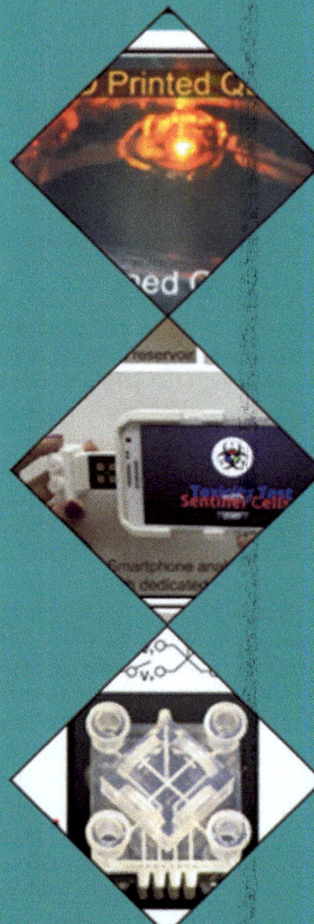

Emerging Technological Applications of Additive Manufacturing

Functional Surfaces, Sensing, Optical Devices, Electronics Devices, Microfluidics and Flow Devices, Catalysis, Batteries, Capacitors, Separators, High-throughput Experimentation

Sunil Mehla, PR. Selvakannan, Maciej Mazur, Suresh K. Bhargava

7

Chapter 8
Additive Manufacturing as the Future of Green Chemical Engineering

Sunil Mehla, Ravindra D. Gudi, D. D. Mandaliya, Takashi Hisatomi, Kazunari Domen, and Suresh K. Bhargava

Contents

S. Mehla · S. K. Bhargava (✉)
Centre for Advanced Materials and Industrial Chemsitry, RMIT University, Melbourne, VIC, Australia
e-mail: suresh.bhargava@rmit.edu.au

R. D. Gudi
Department of Chemical Engineering, Indian Institute of Technology Bombay, Mumbai, Maharastra, India
e-mail: ravigudi@iitb.ac.in

D. D. Mandaliya
L. D. College of Engineering, Gujarat, India

T. Hisatomi · K. Domen
Shinsu University, Matsumoto, Japan

K. Domen
University of Tokyo, Tokyo, Japan

S. K. Bhargava et al. (eds.), *Additive Manufacturing for Chemical Sciences and Engineering*, https://doi.org/10.1007/978-981-19-2293-0_8

Abstract Chemical sciences and engineering have made outstanding contributions in fulfillment of the needs of our technological society and upliftment of the quality of human life by enabling critical technologies for processing and manufacturing of high-volume essential commodities such as detergents, creams, sanitizers, pharmaceuticals, textiles, plastics, rubbers and so on. The global models of chemical manufacturing have been continuously evolving since the colonial rule to adapt to the socio-economic and political climate. In the face of the twenty-first century, chemical manufacturing is experiencing global challenges of different kinds ranging from depletion of natural resources, climate change and growing population to sustainable living, environmental compliance, and personalized healthcare. Thus, a great need has been felt for reinvigoration of the chemical processing and manufacturing industry in response to which the philosophies of industry 4.0 and smart manufacturing have seen global acceptance and recognition. However, this transcendence from the existing state of chemical manufacturing to smart manufacturing heavily relies on the availability and development of advanced manufacturing and sophisticated equipment, process intensification and efficient micro reaction technologies. Study, design, and engineering of equipment to manipulate chemical and physical phenomenon is at the heart of chemical engineering since equipment and reactor design directly governs process and energy efficiencies, performance, and process economics. Conventional subtractive and casting based manufacturing technologies are proving inadequate to meet the needs of modern chemical engineering demanding capabilities to fabricate sophisticated equipment with complex internal geometries, rapid prototyping, and performance testing. These manufacturing related issues can be effective phased out with the adoption and integration of additive manufacturing with chemical processing and manufacturing industries. While the automotive, aerospace, engineering and biomedical industries have hastily adopted AM technologies for associated benefits, the chemical industries have been lagging behind in the uptake due to lack of agility and adaptability in industrial chemical processes. Integration of AM with chemical industry is anticipated to bring widespread process automation, improved feedstock and product inventories, digital process control and process intensification. Future chemical reactors will be complex tailor-made devices with design optimized for fluid and particle flow, heat and mass transport and reaction thermodynamics. When reactions are catalytic in nature, the performance of the reactor becomes critically dependent on several of the catalyst properties, both physical and chemical. While the intrinsic chemical properties in terms of selectivity and activity are dependent on the choice of catalyst, other important considerations related to reactor efficiencies and design, are related to the structural properties of the catalyst, i.e., on how the active phase of the catalyst is dispersed in the support, that influences the available surface area, porosity, heat, and mass transfer characteristics as well as the hydrodynamic pressure dissipations. This aspect of structured catalytic reactor design, that is based on a functional integration of the catalytic as well as reactor design steps, is also receiving increasing focus from an advanced manufacturing viewpoint towards achieving higher efficiencies and lower energy consumption. In this chapter the role of additive manufacturing in chemical reactor and equipment design ranging from simulation and reactor modelling to demonstrated examples of additively manufactured micro mixers, micro heat exchanges and microreactors are elaborated which distinctly project additive manufacturing as the future of green chemical engineering.

Keywords Industry 4.0 · Smart manufacturing · Process intensification · Micro reaction technology · Additive manufacturing · Green hydrogen

Abbreviations

3DFD	Three-dimensional Fibre deposition
3D	Three-dimensional
ABS	Acrylonitrile Butadiene Styrene
AM	Additive Manufacturing
CAD	Computer Aided Design
CFD	Computational Fluid Dynamics
CLIP	Continuous Liquid Interface Production
CMA	Complex Metallic Alloys
CNC	Computer-numerically Controlled Machining
DIW	Direct Ink Writing
DLP-SLA	Dynamic Laser Projection Stereolithography
DMF	Dimethylformamide
DNS	Direct Numerical Simulations
EDM	Electrode Discharge Machining
FCC	Fluidized Catalytic Cracker
FDM	Fused Deposition Modelling
FEM	Finite Element Method
FTS	Fischer-Tropsch Synthesis
FVM	Finite Volume Method
HIPS	High-impact Polystyrene
LBM	Lattice Boltzmann model
LVG	Longitudinal Vortex Generation
MJM	Multijet Modelling
MOF	Metal Organic Framework
MTO	Methanol to Olefin reaction
NCM	Nanochannel Microreactor
PA	Polyamide
PC	Polycarbonate
PDMS	Polydimethylsiloxane
PEEK	Polyether ether ketone
PEFC	Polymer electrolyte fuel cells
PET	Polyethylene terephthalate
PETG	Polyethylene terephthalate glycol

PETT	Polyethylene co-trimethyleneterephthalate
PIV	Particle Image Velocimetry
PLA	Polylactic Acid
PLIF	Planar Laser Induced Fluorescence
PPSF	Polyphenylsulfone
PP	Polypropylene
PVC	Polyvinylchloride
QC	Quasicrystals
Re	Reynolds Number
RTD	Residence Time Distributions
RWP	Rectangular Winglet Pair
SLA	Stereolithography
SLM	Selective Laser Melting
SLS	Selective Laser Sintering
SMS	Selective Mask Sintering
UAR	Unsteady Asymmetric Regime
USR	Unsteady Symmetric Regime
UV	Ultraviolet
We_G	Weber number

8.1 Need for Integration of Additive Manufacturing with Chemical Engineering

A chemical is a substance with a characteristic composition and distinct property. Chemicals are the building blocks of our modern technological society and it would be impossible to sustain our current lifestyle without them. Chemical engineering is a field of study which uses the fundamental knowledge from chemical sciences and deals with the design, manufacturing, safety, operation and maintenance of chemical processing instruments, equipment, and plants. Thus, chemical engineering operates at the interface of chemical sciences, mathematics, economics, mechanical and civil engineering. Chemical engineers have long been the leaders of process-based manufacturing of chemicals, fuels, foods, and consumer products, as opposed to manufacturing of parts, components and machines by machining and assembly. The improved availability of chemical products at affordable prices and the rapid progress of humanity since the second world war is a grand testimonial to the wide success of chemical engineering industry.

However, the needs and trends of our modern society are enforcing a reassessment and reinvigoration of existing chemical processes and unit operations. For instance, the rise of new feedstocks such as biomass and light molecules that must be processed into high demand fuels, chemicals and products with established infrastructure is demanding a higher degree of agility and adaptability from existing processes. The diversity in the type and availability of feedstock is anticipated to increase even further in the future with the increased production of waste materials such as plastic, electronic and industrial waste. A second driver for systemic technological innovations in chemical processes is the rise of expectations around energy efficiency, process efficiency, sustainability, and environmental footprint. Finally, dynamic shifts in customer demands are the third biggest driver demanding rapid and effective reinvigoration of the existing chemical processing industry. For example, demand for paper and printed material is decreasing whereas demand for electronic devices is increasing at the same time. In response to the need of the society, latest developments in chemical engineering are going beyond the traditional approaches of innovations in chemical processes and operations. Many universities and industrial research centers are working on novel equipment and technologies that could potentially transform the very concept of chemical plants and lead to compact, safe, energy-efficient, sustainable, and environment-friendly processes. These developments share the common theme of 'process intensification'—an approach that has been hailed in the past few years as the holy grail of chemical engineering. Similarly, a need for more agile and adaptable processes has been widely acknowledged. Such requirements of the chemical industry can only be met through systemic process, material, and technological innovations.

Additive manufacturing has recently emerged as a powerful manufacturing technology and being globally hailed as the harbinger of the fourth industrial revolution. The enormous potential of this manufacturing philosophy has seen its diffusion into almost all major sectors of science and engineering such as aerospace, automobile, biomedical, construction, electronics, consumer goods, arts and so on. Additive manufacturing not only enables near total design freedom and capability to manufacture sophisticated geometries but also facilitates some unique benefits such as flexibility to manufacture with a wide range of materials, manufacturing using functionally graded materials, mass customization, digital storage of object design and on-site, on-demand manufacturing. Thus, additive manufacturing presents tremendous opportunities for the entire chemical processing industry which is powerful enough to usher the much-needed materials, processes, and technological innovation in the chemical processing sector. In this chapter, we take a closer look at the role of additive manufacturing in reinvigoration of chemical engineering and the chemical processing industry. The challenges and enormous potential of AM processes have been elaborated in view of latest developments in chemical engineering.

8.1.1 The Rise of Digitally Controlled Manufacturing

Manufacturing, trade, and logistics are some of the oldest industries which have seen continual change in their modes of operation with socio-economic disposition. For instance, the old colonial model was to grow and extract resources from backward nations with free labor, export raw material to developed industrialized nations and refine, process, and sell finished products in markets and outlets of high demand. Even after the cessation of the colonial rule in the mid-twentieth century, some remnants of this old manufacturing model can still be seen today. However, most modern industries have now adopted new models of operation such as establishment of manufacturing facilities as close to the resources as possible, utilization of cheap labor and economy of scale. The emergence of digitally controlled chemical manufacturing and processes is related to the need for process automation and remote control, the roots of which were already laid in the rapidly surging chemical industry post the second world war. Continuous reaction and separation processes which are widely used for large scale production and manufacturing are well suited to be controlled automatically and remotely. The early chemical engineering industries achieved this automation using mechanical and pneumatic controllers that were operated manually. By 1970s several industries were getting increasingly inclined towards adopting digital automation for process control. By 1980s the increase availability of solid-state microelectronics and computing technology enabled the systemic transformation of industrial production and analytical equipment by allowing varying degrees of automation and digital control. Today, digitally controlled manufacturing has become the norm for the chemical processing industry wherein all process parameters are digitally recorded, analysed, and displayed for monitoring in centralized control rooms with facilities to observe past trends and extrapolate the performance of the plant into the future. This allows operation, maintenance, and technical teams to constructively work in tandem for continuous manufacturing operations round the clock with minimal down time and production losses. Digitally controlled manufacturing has laid a strong foundation for the chemical processing industry upon which modern generation movements of smart manufacturing and process intensification are well established.

8.1.2 Industry 4.0 and Smart Manufacturing

Industry 4.0 or the fourth industrial revolution (4IR) refers to the reorganization, modernization, and digitization of the entire industrial value chain. Industry 4.0 is now getting global recognition and widely anticipated to alleviate the prevailing sustainability, socio-political, economic, environmental, and developmental challenges faced by our modern technological society. A schematic depicting the four industrial revolutions is shown as Fig. 8.1. The first industrial revolution was brought by the rise of widespread mechanization, steam engines, manufacturing plants, mining and metallurgy, iron processing and utilization, textile industry and machine tools. The second industrial revolution was characterized by the rise of electricity and electrical tools, mass production, globalization, telegraph, production lines, turbines, fuel, gas, and water supply. The third industrial revolution was empowered by the computing and networking advancements and characterized by global connectivity, computer networks (WAN, LAN, MAN, Internet), digital control and manufacturing, robotics, and programmable logic controllers (PLC), widespread digitization and automation. The fourth industrial revolution takes a step further into the future and aims to go beyond connecting people into connecting devices, sensors, machines,

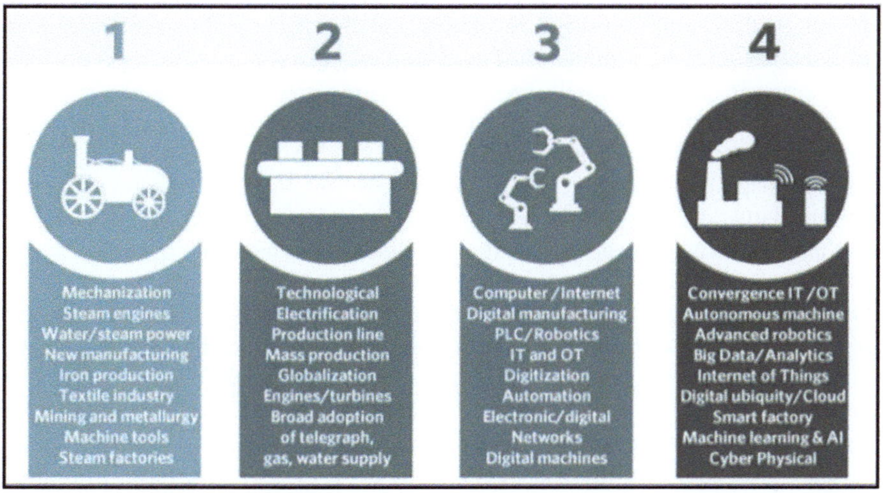

Fig. 8.1 The four industrial revolutions as characterized by major developments and main movements. Reproduced from https://www.i-scoop.eu/industry-4-0/

servers, robots, and interfaces to bring a unification of the digital and physical environments. Thus, Industry 4.0 embodies principles and elements from at least nine industrial digital technologies such as advanced robotics, additive manufacturing, augmented reality, simulation, horizontal and vertical integration, industrial internet, cloud systems, cyber security, big data, and data analytics as shown in Fig. 8.2 [1].

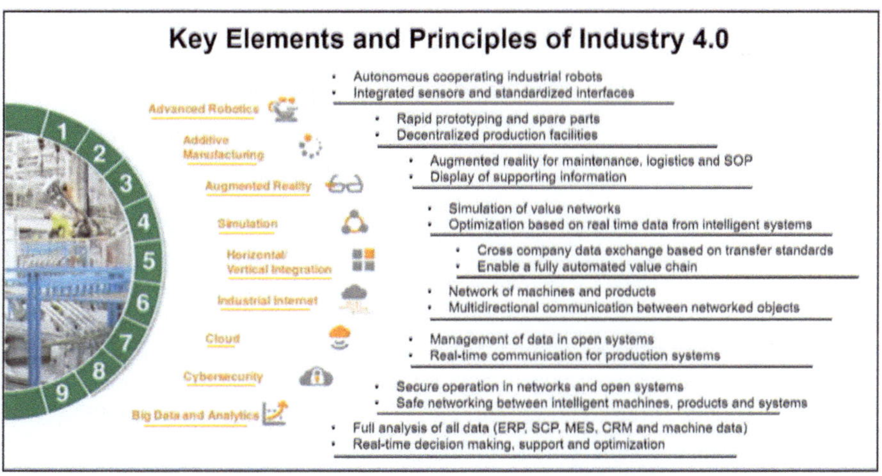

Fig. 8.2 A schematic depicting the embodiment of principles and elements from nine industrial digital technologies in Industry 4.0 or the fourth industrial revolution. Reproduced with permission from Ref. [1]

Enabled by the fourth industrial revolution, modern technology accelerator platforms such as internet of things (IOT), cyber physical systems (CPC), deeply integrated systems, artificial intelligence, and machine learning are facilitating wide scale automation and process optimization in entirely novel ways leading to ample opportunities for innovation and development.

Taking inspiration from the fourth industrial revolution, new ideologies stemming from innovative combination of interdisciplinary aspects and technologies such as smart manufacturing have already started to shape the chemical engineering industry. The developing field of 'smart manufacturing' refers to data-driven manufacturing which treats data as the key asset and integrates information and operational technology (OT). Conventional manufacturing relies on analytics and effective management based on product, processes and their life cycle structures. In contrast, smart manufacturing relies on analytics and effective management based on data structures and their life cycles. The data collected from industry and sector wide physical and digital systems contains all encapsulating information about products, processes, safety, logistics, inventory, timelines, manpower, efficiencies and so on. A mere uptake of 5–10% in smart manufacturing practices by the US manufacturing industry is anticipated to result in energy and economic savings worth 200 billion dollars over ten years. In the coming years, emerging technologies for process data acquisition, analytics and control in the chemical processing industry are anticipated to transform the fine-chemical and pharmaceutical industry into a data-rich gold mine with far reaching implications on the quality and quantity of human life. The adoption of simulation and system wide integration practices by chemical manufacturers for generation of large datasets has incredible potential in facilitating rapid improvements in product quality, process safety, economic viability, and industry sustainability. Fully embracing these technological advancements ensures significant improvements in the speed and reliability of chemical process development, maintenance, and management. Large challenges still exist which need to be circumvented to realize the true potential of Industry 4.0 and smart manufacturing. For instance, smart manufacturing requires right data in the right form at the right time, thus imposing a huge constraint on the data industry to manage and communicate gigantic, incredibly complex, and interconnected data.

As one of the industrial digital technology enabling the ethos of the fourth industrial revolution, additive manufacturing is anticipated to steer rapid progress in the chemical industries by facilitating rapid prototyping, localized and decentralized production of spare parts and components, improvement in efficiencies and life cycles of reactor, processing equipment and devices, efficient inventory management, supply-chain optimization, and system wide process intensification. Intensification of processes is anticipated to be another major disruptor in the upcoming future and can be accomplished in many ways such as hybridization of unit operations, effective mass, and energy transfer, shrinking processes and economy of scale.

8.1.3 Process Intensification

Process intensification refers to increase in intensity or severity of process parameters with the goal to improve process efficiency. Gerven and Stankiewicz defined process intensification as a drastic improvement in equipment and efficiency of a chemical process by one or more of the four fundamental approaches i.e. spatial, thermodynamic, functional, and temporal [2]. These approaches have also been laid down as four fundamental principles of process intensification as maximization of the effectiveness of intra- and inter molecular events, homogeneous processing environment for each molecule, optimization of driving force at each scale and maximization of the specific surface area to which these forces apply and maximization of the synergetic effects from partial processes [2]. In a nutshell, process intensification is the application of the best possible technological solutions to classic chemical engineering problems in transport phenomena, reaction engineering, and separation science. Some other popular and well accepted definitions of process intensification are given below:

1. Improvement in the cost-effectiveness, safety, and sustainability of a chemical engineering process by reduction in equipment size, energy consumption, or waste formulation.
2. Application of modern and radically innovative principles in process and equipment design which can benefit (often with more than a factor two) process and chain efficiency, capital and operating expenses, quality, wastes, process safety and more.
3. A paradigm shift in chemistry, process science, and equipment design which facilitates reduction in operation and capital expenses and waste generation while leading to improvements in process efficiency and safety.

However, there are several hurdles to process intensification and it may not always lead to process improvements and expected benefits. For instance, basic chemical engineering teaches us that reaction rates double approximately every 10 °C. Importantly, the rate of all competing reactions is increased at elevated temperatures and increase in reaction temperature may result in loss of desired product selectivity. Similarly, process intensification can also lead to more co-products, by-products, and over-reacted material, which might be not desirable. An intensification change might also lead to catastrophic consequences in circumstances where normal control systems prove insufficient. In the development of new chemical processes or new routes to existing chemical processes, side reactions and byproducts may not reveal themselves until an emergency is at hand. Waste generation is another persistent issue that can be enhanced by process intensification and lead to increased treatment and disposal costs. Consider the simple example of change in size of shipping containers and corresponding shift of risk from one population to another. Use of smaller shipping containers for just-in-time manufacturing has strong appeal for the minimization of hazardous materials inventory. However, smaller containers require more frequent connects and disconnects, and potentially more time on the road. All these factors simply result in shifting of risk from one impacted population to another such as from storage and inventory manager to the materials handling, packaging and delivery manager. The process innovation may still prove worthwhile on the whole with fine tuning and improvisation; however, careful observations must be made to ensure that tradeoffs are carefully and adequately examined to reduce hazards and avoid any shifting of hazards from one geographical location or population to another. Such issues are manageable and form an important aspect for successful realization of process intensification. For full realization of the promise of process intensification it is essential to ensure that reactive chemistry (intentional and unintentional) issues, waste generation and disposal, potential for uncontrolled release to the environment, storage, shipping, and transport options are adequately identified, assessed, and mitigated. Multifunctional microreactors [3], reactive distillation [4], and reactive separation [5] are some well-known examples of process intensification that have been enthusiastically pursued by the chemical engineering research and industrial community some examples of which are shown in Fig. 8.3.

Additive manufacturing refers to the wide variety of manufacturing technologies that can directly produce complex three-dimensional parts with near-complete design freedom in an additive manner. Additive manufacturing is a powerful tool for rapid prototyping and has tremendous potential in realization of process intensification principles by means of materials, equipment, and process innovations. On the materials front, additive manufacturing allows fabrication of sophisticated structures and geometries in a wide variety of materials such as polymers, metals, ceramics, glasses, and respective composite materials. Such flexibility and capability to manufacture parts and components in different materials is essential for large scale and far-reaching implementation of additive manufacturing in chemical processes. Material compatibility under reaction conditions takes the center stage in the design of chemical processing structural parts and components such as a chemical reactor, battery, fuel cell, or a separator. The material compatibility in turn depends on various factors such as strength under reactive environment, conductivity, temperature, corrosion resistance and surface properties.

Figure 8.4 depicts the variation of mechanical strength with maximum operating temperatures for different groups of structural materials classified as metals, polymers, ceramics, and composites that can be utilized by AM technologies for manufacturing under different conditions such as ambient, mild, and extreme. On the equipment front additive manufacturing allows fabrication of parts and components with complex and customizable geometries, including those difficult to fabricate by conventional manufacturing technologies. The design freedom offered by AM allows production of bespoke reactors with enhanced control over reactant residence times manifested through custom flow paths and reaction volume. Complex internal geometries necessary for turbulent and static mixing are also well within the scope of AM technologies. AM also offers considerable benefits in terms of fabrication cost and production time. AM manufactured parts can often be taken from concept to realization within hours, with huge time-savings both during the design and manufacturing stages.

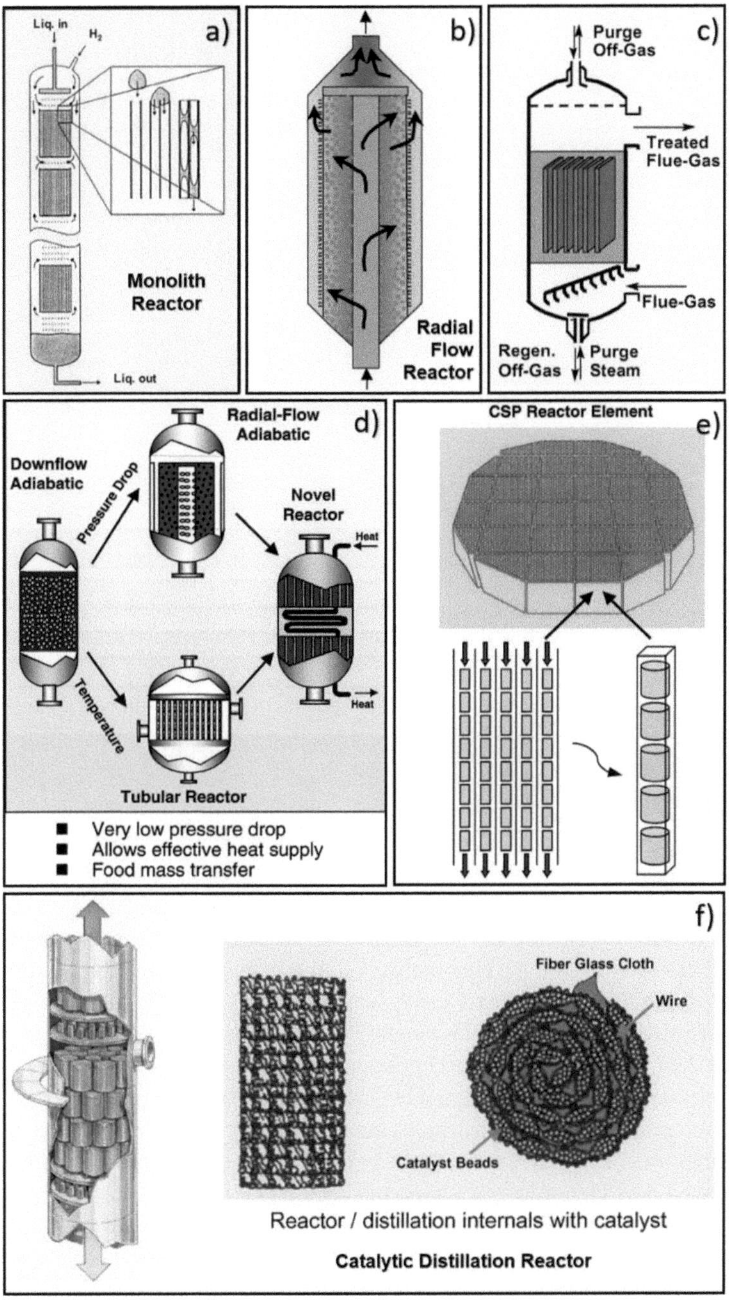

Fig. 8.3 Examples of process intensification using multifunctional reactors; **a** Monolith reactor; **b** Radial flow reactor; **c** Parallel passage reactor; **d** Improvisation from a downflow adiabatic reactor to radial-flow adiabatic, tubular, and novel multifunctional reactors to reduce pressure drop, enhance heat, and mass transfer; **e** Depiction of a reactor packing element in composite structured packing reactor; and **f** Catalytic distillation unit. Reproduced from Ref. [3] with permission from Elsevier

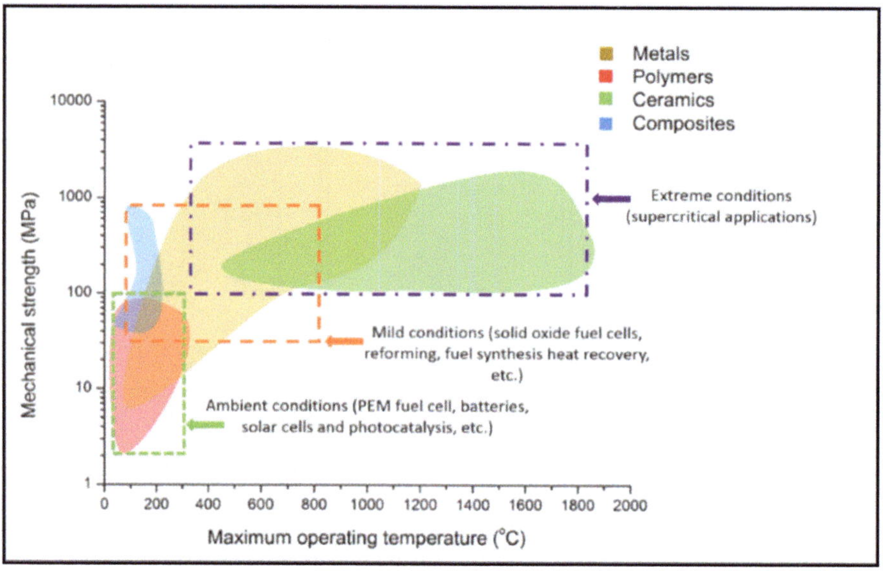

Fig. 8.4 Comparison of different structured materials that can be employed for additive manufacturing in terms of mechanical strength and maximum operating temperature. Reproduced from Ref. [6] with permission from Wiley

The gradual inception and integration of additive manufacturing into research laboratories has fueled significant research and development efforts for innovative manufacturing of reactor systems. Commercial manufacturing technologies are currently already available for both milliliter-scale and microliter-scale flow reactors. Milliliter-scale reactors often employ simplified shapes of polymer and metal tubing or etched glass chips. Although such tubing is commercially available at low cost, it does not allow the same level of design complexity as possible for microfluidics with intricate mixing pathways. In contrast to milliscale reactors, micro reactors are commonly designed using inert polymers (e.g., PDMS) or glass with two-dimensional pattern of flow channels. Design of such micro reactors using well-developed AM processes allows fabrication of sophisticated designs with intricate mixing pathways, online analysis, and bespoke temperature-controlled zones. Instead of design and manufacturing constraints, this technology is currently only constrained by cost and time of manufacturing which varies based on the degree of sophistication and employed AM technology. Finally, material and equipment innovations brough about by additive manufacturing can be judiciously employed to pioneer process innovations that uplift economic viability, process sustainability, process efficiency, product selectivity, regulatory compliance, and environment compatibility. In the following sections of this chapter main attention has been given to global progress and developmental efforts in the design and fabrication of microreactors using additive manufacturing.

8.2 Micro Reaction Technology

Micro process engineering refers to the science and engineering of processes and unit operations in small (micro) volumes characterized by typical vessel dimensions of size smaller than 1 μm. Such processes are generally designed to be continuous in nature and conducted in parallel to achieve an overall high process throughout. In chemical processes, the fundamental physical phenomenon such as mass transfer, heat transfer, maldistribution, temperature gradients and process fluid composition homogeneity begin to play an increasingly important role in controlling the chemical processes and quality of end products as the scale of operation is increased [7]. Thus, most existing commercial chemical manufacturing and processing industries which operate at large macroscopic scales are fundamentally operating under sub-optimal regimes. Conversely, the same governing phenomenon are relatively easier to control and manipulate at smaller scales such as in millimeter and micrometer scales. Thus, global research and development efforts in process engineering are unanimously focused on the advancement of micro processing technology. Efficient completion of chemical processes at the micro scale further depends on availability of adequate processing equipment and devices. Broadly, such equipment for microscale processes can be classified as micromixers, micro heat exchangers, micro reactors, and microfluidics. A micromixer is a device which is designed for adequate mixing of input fluids. Micromixers form an import component of most micro processing equipment where more than one fluid is involved. Micro heat exchangers are small scale devices designed for adequate transfer of heat between the process fluids or between the process fluid and the device structural components. Micro heat exchangers become important for heating and heat recovery operations during fluid handling in micro processing equipment. A micro reactor or a micro channel reactor or a microstructured reactor is a microscale device specifically designed for conducting reactions. Design of microreactors is a sophisticated process since both physical (mass transfer, heat transfer, maldistribution, temperature gradients and reactant composition homogeneity) and chemical (reaction kinetics, thermodynamics, phase, catalyst, product selectivity) phenomenon need to be optimized for high microreactor performance. Thus, microreactors are universally acknowledged as the great enablers and harbingers of process intensification. Microfluidics are microscale devices specifically designed for handling extremely small volumes of fluids. Such devices are often employed in analytical technologies and equipment for related benefits in reduction of analysis time, analyte volume, cost and so on. Microfluidics take advantage of active and passive microfluidic components for mixing, gradient generation, automated actuation, and on-demand fluid supply.

Additive manufacturing has the unique capability to directly transform functional materials into useful geometries, shapes, configurations, and devices based on a digital design. Thus, it is no wonder that such a flexible and adaptable technology has stimulated wide interest in micro processing technology to provide tailored solutions. As a manufacturing technology, additive manufacturing outperforms other manufacturing techniques due to reduced wastage of materials, decentralized and customized production, capability to manufacture complex shapes and patterns and availability of wide range of materials that can be printed. Thus, AM is a powerful tool for micro process engineering and finding great acceptance in rapid prototyping, lab to pilot scale experimentation, and small-to-medium scale production of reactors, microreactors, microfluidics, mixers, and heat exchangers. In this chapter major attention has been given to the use of additive manufacturing for design and fabrication of micro reactors. Since micro reactor design also involves design elements from micro mixing, microfluidics and micro heat exchanging technology, the presented discussion provides useful knowledge about all elements of micro processing technology.

8.2.1 Microreactors

Microreactors are micro scale chemical reactors where the reactor dimensions are smaller than 1 mm. Microreactors demonstrate several advantages over conventional continuous reactors such as improved heat and mass transfer [8, 9], safe operation under severe process conditions, safer handling of hazardous chemicals [10, 11], efficient mixing [12], enhanced catalyst-reactant interactions [13], improved catalyst lifetimes [14, 15] and improved activity and selectivity [16, 17] leading to a reduced E-factor [18]. High mass transfer originates in microreactors from laminar flow [19, 20] which reduces gradient formation in concentration, temperature, pressure and volume leading to a higher efficiency in chemical reactions [21]. At the interface of microchannel wall with the process fluid, viscous and inertial forces exert a dragging force on the flowing fluid. This drag reduces considerably as the distance from the microchannel wall increases. When the microchannel dimensions are small enough, the dimension of the interfacial film at the fluid-microchannel wall interface becomes comparable to microchannel dimensions. As a result, the drag arising from fluid viscosity and inertial imposes a laminar flow conditions in microchannels of sufficiently small dimensions. This laminar flow results in several aforementioned advantages of microchannel reactors such as good temperature control ensuring high product selectivity. Heat transfer in microreactors is highly efficient due to high surface area to volume ratios [22] and near isothermal conditions in the reactor channels [23]. In contrast large axial and radial gradients persist in conventional reactors which lead to sub-optimal operation with relatively lower process efficiencies. In addition to better spatial distribution of pressure, temperature, flow rate, flow mixing, and reactant residence times, microscale dimensions also manifest smaller forces or lower pressure on reactor walls [24]. For instance, reactions can be carried out safely even at pressures above 400 bar in capillary microreactors [21]. Similarly, detonation of mixtures of highly reactive gases such as a stoichiometric mixture of hydrogen and oxygen gases does not propagate in microchannels of sufficiently small diameter [25–28]. This characteristic safety feature in micro reactors is sometimes compared to the principle of intrinsic safety in electrical circuits. Even highly exothermic reactions can be conducted almost isothermally and safely by employing adequate cooling strategies such as using the adjacent microchannels for flow of a cold fluid with sufficiently high heat capacity to allow desired temperature control.

The possibility of sophisticated and intricate microreactor internal geometries in AM fabricated microreactors also present tremendous opportunities for intentional insertion of non-isothermal characteristics into the reaction media and other similar experiments which cannot be performed with geometrically simplified microreactors fabricated using conventional technologies. Recent literature has seen a tremendous surge in research in the field of hierarchical materials because significant benefits have been realized by improving the diffusion of reactants and products from catalytically active sites [31–34]. Microreactor technology not only provides advantages in process parameters discussed above but also provides a means to relieve diffusion constraints in catalytic reactor systems. Therefore, micro reaction technology can be indisputably regarded as a superior reaction technology compared to conventional continuous flow reaction technology as depicted in an infographic in Fig. 8.5 [29, 30].

Despite the several aforementioned advantages of microreactors, several issues and challenges still persist which have prevented the rapid uptake of microreactors in industrial and commercial production. One inherent limitation is the issue of limited throughput because of the small volumes involved. In principle, utilization of parallel microchannel reactors can resolve this limitation. However, practical considerations such as periodic maintenance, space availability, shape and size constraints and geometrical uniformity in parallel microreactor systems may impose a limit on the maximum number of parallel systems that can be employed. At the very least the current state of micro reaction technology is developed enough to replace some parts of existing industrial processes as demonstrated in the case of the

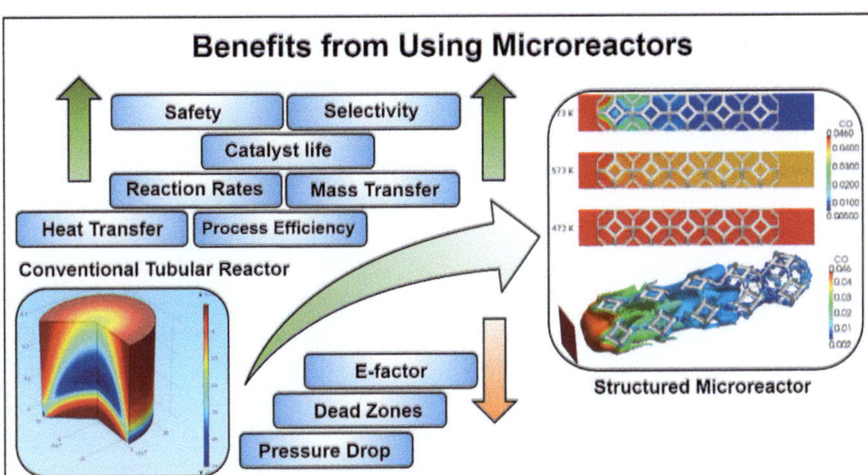

Fig. 8.5 An infographic depicting the benefits associated with the use of micro reactors in comparison to conventional reactor technologies. Reproduced from Refs. [29, 30] with permission from Elsevier

industrial production of writing ink [35]. In addition, heat transfer, maldistribution and catalyst issues can also impose additional constraints on the performance of microreactors. For instance, the reduction of hydraulic diameter of the microchannel beyond a limit is understood to deteriorate heat transfer characteristic and reported to negatively impact reaction yields [36]. Similarly, fluid maldistribution in micro reactors arising from its geometric features, wall roughness or roughness arising from material or catalyst packing can impact reaction kinetics and reported to negatively impact product selectivity [37]. Deviation in flow characteristics between parallel micro channels from above factors and manufacturing defects may prove to be an additional factor which can impact product selectivity. Finally, functionalization of micro reactor channels with catalyst and adequate integration of catalyst into the reactor micro channels can have a large impact on the performance of micro reactors. Many different methods are currently available that can be used for functionalization of microreactor channels with catalytic thin films and coatings [38]. The stability and durability of such catalytic films over the reaction life cycle are of critical importance since catalyst leaching can cause severe plugging, blocking, back pressure and safety issues in microreactors in addition to drop in product yield and selectivity. The morphology, porosity and diffusional characteristics of catalyst films and coatings are understood to have a significant impact on the performance of the microreactor [39–42]. Subsequent sections throw further light on the design and fabrication of microreactors using additive manufacturing starting with the role of modelling and simulation.

8.2.2 Modelling and Simulation in Microreactor Design

Design of chemical reactors whether for large scale testing and production or lab to pilot scale testing is a sophisticated design problem involving multistage theoretical and experimental studies. There is a tremendous diversity in the shape, size, and efficiencies of chemical reactors due to varying reaction kinetics and energetics. For optimal accommodation of the variety of chemical reactions, it becomes essential for reactors to manifest in different shapes, sizes, and hydrodynamics. Within the control volume of the reactor, design efforts are always oriented towards facilitation of various reaction objectives such as yields, purities, selectivities and energetics, while easing up constraints imposed upon by thermodynamics. Traditionally the task of overall reactor design, especially for the case of catalytic reactions, has been performed in two directions of rather limited focus on their own, i.e., catalyst design and reactor design. However, current research efforts are directed towards integrating these tasks in ways that simultaneously promote the objectives related to reaction efficiencies and also those related to achieving energy efficiencies by reducing pressure drops and improving heat transfer characteristics. The use of mathematical models to generalize the overall task of reactor design including both of the above objectives has been an active area of research [43–46]. These mathematical models when used in conjunction with advanced optimization approaches for realizing the above objectives, facilitate the development of interesting reactor configurations as well as catalyst distributions within them. For instance, an additively manufactured self-catalytic reactor concept as shown in Fig. 8.6 i.e., a reactor with specific composition which acts both as the reactor and the catalyst was demonstrated for Fischer–Tropsch synthesis of value-added chemicals (C_2–C_{5+} range) in good yields

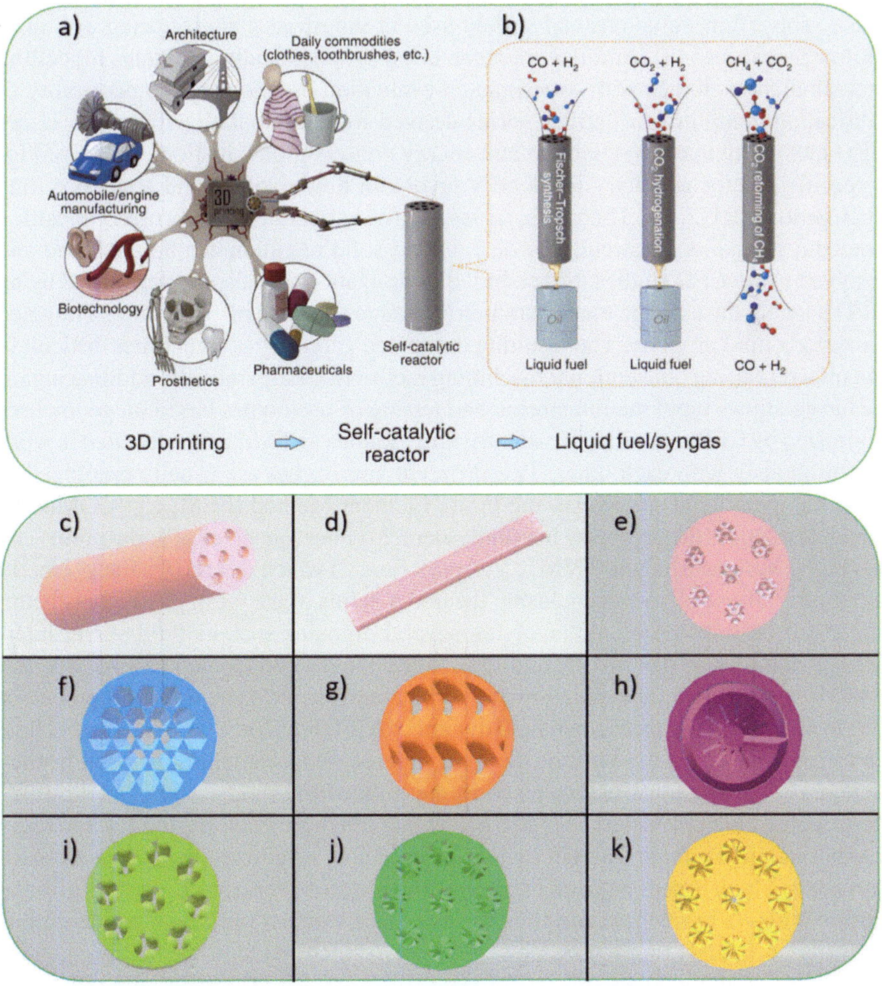

Fig. 8.6 **a** Illustration of the concept of self-catalytic reactors i.e. reactors in which catalytic active channels and walls for applications in different sectors; **b** Self-catalytic reactors for Fischer–Tropsch synthesis, CO_2 hydrogenation and dry reforming of methane; **c** Construction of a self-catalytic reactor composed of cobalt metal (Co-SCR-1); **d** Longitudinal cross-section of the Co-SCR-1 reactor; horizontal cross-section of the **e** Co-SCR-1 reactor; **f** Co-SCR-2 reactor; **g** Co-SCR-3 reactor; **h** Co-SCR-4 reactor; **i** Co-SCR reactor; **j** Co-SCR-5 reactor; **k** Co-SCR-6 reactor. Reproduced from Ref. [47] with permission from Springer Nature

starting from light gas feedstocks such as methane, carbon monoxide and carbon dioxide with hydrogen [47]. Thus, additive manufacturing of metallic microreactors such as with selective laser sintering and selective laser melting can greatly aid in the integration of catalyst and reactor design research and scientific communities. Computational fluid dynamics (CFD) is one of the most important simulation tools for performing theoretical studies on the efficiency and performance of reactors and flow component designs based on fluid flow behavior. CFD uses numerical analysis to solve flow equations and widely used in industrial manufacturing and automotive industries [48]. Additive manufacturing synergistically integrates modelling and simulation studies with experimental evaluation by allowing the possibility of fabricating complex geometrical forms derived from computational studies. Using CFD modelling mass, momentum and energy conservation equations are solved for a specific reactor geometry to identify favorable flow regimes and residence time distributions (RTD) [7]. Therefore, issues such as temperature gradient, composition gradient, stagnation, recirculation of fluid and solid retention can be modelled and resolved using CFD studies. Shape based optimization can also be performed using CFD wherein the user can make parametric changes in geometry to achieve a targeted value for a fluid property. The resulting optimized geometry is sometimes difficult to manufacture using conventional machining methods. Comparatively, additive manufacturing allows rapid manufacturing and testing of prototypes based on geometries optimized by CFD modelling. Evidently, CFD studies and AM are well suited to work synchronously with each other. Two different approaches are usually employed in these computational studies i.e. the Finite Element Method (FEM, e.g. COMSOL) which discretizes the geometry into finite elements interconnected at nodal points and the Finite Volume Method (FVM, e.g. Ansys Fluent) which discretizes the geometry into finite volume elements enclosing the nodal points of the mesh [7]. FEM is better suited for bespoke problems involving several domains such as fluid stirrer interactions in a fluid under stirring and fluid motion under ultrasonic pressure whereas FVM is better suited for evaluation of hydrodynamic studies and transport processes [7]. In addition to computational fluid dynamics (CFD) studies which can predict heat and mass transfer in the bulk of the fluid, microkinetic modelling is often needed to address the reactions taking place in the bulk of the fluid when catalytic reactions are involved [49]. The scenario becomes even more complicated for heterogeneous catalytic systems wherein multiscale modeling is required to capture different surface and bulk processes such as adsorption, reaction, desorption, and diffusion at different length (pore size) and time scales (contact times) [49]. As a result, empirical kinetic models employing power law or surface reaction centric Langmuir–Hinshelwood-Hougen-Watson model need to be employed. Finally, the molecular parameters extracted from such molecular and microkinetic models are passed to macroscopic CFD models to develop a complete picture of the undergoing physical and chemical processes and the reactor shape and dimensions are finalized based on the outcomes [50]. The shape of a chemical reactors involving design parameters such as reactor dimensions, shape of channels, contact angle of internal walls, mixer structure, packing material, porosity and shape is governed by several physico-chemical factors such as reaction kinetics, reaction conditions, fluid phase, safety, and most critically feasibility and economics of manufacturing the reactor [7]. While the reaction kinetics remain same as the reactor is scaled-up, physical processes such as heat and mass transfer are largely scale dependent [51]. Thus, beyond a certain optimum size, reactors often need to be scaled-out (i.e., increased in numbers) to achieve the target production. Motivated by the benefits of parallel processing a modern approach is to use an array of parallel microreactors. The research on microreactors both for experimentation and production is rapidly increasing because of the advantages of efficient heat and mast transfer, efficient mixing, high control over reaction parameters, better stability of intermediates and improved process safety [52, 53]. This microreactor revolution has been robustly empowered by the 3D printing revolution which enables capacity to print microreactors as well as 3D microstructured catalysts in complex shapes and geometries.

Going beyond the traditional modelling with the finite element and the finite volume methods, tremendous opportunities exist for substantial improvement in microreactor efficiencies by means of integrated modelling studies that consider the impact of all physical and chemical phenomenon such as reaction kinetics, adsorption and desorption kinetics, interfacial interactions, hydrodynamics, mass transfer and heat transfer. Together with CAD design and manufacturing capabilities enabled by additive manufacturing it is anticipated that new innovations in microreactor configurations with superior performance and process sensitivities are going to break the barriers imposed by conventional chemical reactor technology. For the integrated catalyst and reactor design approach, species, momentum, and energy conservation equations must be integrated with microreactor shape and dimensions to describe reaction behavior in three-dimensional space. In the next subsection, total continuity, species continuity, momentum and energy conservation equations for a multi-species reactive mixture flowing in an arbitrary Eulerian volume in fixed space are described initially, followed by a brief description of the modelling methodology. As examples of the generalized all-intensive modelling, a varying area plug flow reactor and annular reactor are also described in brief.

8.2.3 Equation Based Generalized Reactor Modelling

The dynamic equations shown in Table 8.1 are the starting point for analyzing reactor species concentration and temperature profiles. Other decisions such as heat removal rate, optimum operating conditions, feed temperature etc. can be optimized by placing these modelling equations as constraints in an appropriate single-criteria or multiple criteria-based objective function of an optimization formulation.

In the above generalized equations, ρ is mixture density, and $\boldsymbol{v}$ is the mass average-velocity. The total density of the system can be related to total concentration in the system, using $\rho = C_T M$, where C_T is total molar concentration, and M is the average molecular weight. C_j is the concentration of species j, $\boldsymbol{N}_j$ is molar flux due to convection and diffusion, $\boldsymbol{N}_j = C_j \mathbf{v}_j^* = C_j \mathbf{v}^* + \mathbf{J}_j^*$, where $\mathbf{v}_j^*$ is the molar average

Table 8.1 Dynamic generic conservation equations for flow in a reactor

Total continuity	$\frac{\partial \rho}{\partial t} = -(\nabla \cdot \rho \boldsymbol{v})$
Species continuity	$\frac{\partial C_j}{\partial t} = -\nabla \cdot \left(C_j \boldsymbol{\upsilon}^* + \mathbf{J}_j^*\right) + R_j \quad j = 1, 2, 3 \ldots, n_s$
Momentum balance	$\frac{\partial \rho \boldsymbol{v}}{\partial t} = -[\nabla \cdot (p\boldsymbol{\delta} + \boldsymbol{\tau} + \rho \boldsymbol{v})] + \rho \boldsymbol{g}$
Energy balance	$\rho \hat{C}_p \left(\frac{\partial T}{\partial t} + (\upsilon \cdot \nabla \mathrm{T})\right) = -$ $(\nabla \cdot \boldsymbol{q}) - (\tau : \nabla \upsilon) + \left(\frac{\partial \ln \hat{V}}{\partial \ln T}\right)_{p,x_j} \left(\frac{\partial p}{\partial t} + (\upsilon \cdot \nabla p)\right) +$ $\sum_{j=1}^{n_s} \bar{H}_j [(\nabla \cdot \mathbf{J}_j) - R_j]$

velocity of mixture, and $\mathbf{J}_j^* = C_j(\boldsymbol{\upsilon}_j - \boldsymbol{\upsilon}^*)$ is the molar diffusional flux. $\boldsymbol{\upsilon}_j$ is the velocity of species j. If species j moves at the same velocity as the molar average velocity, i.e., $\boldsymbol{\upsilon}^* = \boldsymbol{\upsilon}_j$, then, $\mathbf{J}_j^* = 0$. Further, for zero diffusional flux, $\boldsymbol{\upsilon}^* = \boldsymbol{\upsilon}$, i.e., molar average velocity is equal to mass average velocity. The term R_j represents the molar rate of production of species j from all the reactions. p is thermodynamic pressure of the mixture, $\boldsymbol{\delta}$ is Kronecker delta operator, $\boldsymbol{\tau}$ is viscous stress tensor, and $\boldsymbol{g}$ is gravitational vector. $\widehat{C}_p$ is the mass averaged mixture heat capacity, T is the temperature, $\boldsymbol{q}$ is heat flux vector, $\widehat{V}$ is the specific volume of the system, for ideal gas system $\left(\frac{\partial \ln \widehat{V}}{\partial \ln T}\right)_{p,x_j} = 1$, $\overline{H}_j$ is the molar enthalpy of species j. $\mathbf{J}_j$ is the mass diffusional flux of species j. The effect of reaction on the enthalpy of the system is captured by the last term in the above equation. The expression for viscous stress tensor is $\boldsymbol{\tau} = -\mu(\nabla \boldsymbol{\upsilon} + (\nabla \boldsymbol{\upsilon})^T)$, where μ is the viscosity of the mixture. The term $(\boldsymbol{\tau} : \nabla \boldsymbol{\upsilon})$ describes the degradation of mechanical energy into thermal energy. For chemical reactors, with lower velocity gradients, and lower viscosity of species, the contribution of these terms can be neglected as they are very small.

The conservation equations described above need to be integrated for the reactor shape as well as catalyst loadings/distribution under consideration. These decisions on the reactor shape and the catalyst distribution would then become decision variables and forced into an outer loop of an optimization formulation. The shape description inherently includes the description of the boundaries. The boundaries can be made to allow or restrict transport of heat and mass across them. Further the rate of production of species R_j participating in the reactions needs to be calculated from the rate expression of the reactions. The rate expressions themselves could be based on a catalyst activity calculation that is either aggregated in terms of an effective factor or could actually be set by the outer loop optimizer to determine the catalyst distribution. The conservation equations are triple volume integrated around the actual shape of the reactor, then the volume integral can be converted to surface integral or line integral based on reactor modelling consideration, i.e., 3-D, 2-D, or 1-D descriptions of the reactor by using the Gauss-Ostrogradskii Divergence Theorem.

Mandaliya proposed the instantiation of the above generalization equation for 5 specific instances of reactors [45] from which two instances are briefed below for illustration of the generalized reactor modelling concept.

8.2.3.1 Generalized Reactor Modelling for a Varying Area Reactor

Varying area reactor refers to a chemical reactor whose cross-sectional shape and hence the cross-sectional area varies with the depth of the reactor as depicted in Fig. 8.7. For instance, the fluidized catalytic cracker (FCC) riser reactor which is used commercially for cracking of viscous vacuum gas oils and crude oil feedstocks into liquified petroleum gas, cracked naphtha, diesel and other heavy hydrocarbon streams is an example of a varying area reactor. The varying area reactor is relatively uncommon but a useful reactor type for reactions where there is a net change in the moles of the reacting species. For instance, the decomposition of Hydrogen Iodide is a reversible gas phase reaction where net mole decrease during the reaction [54]. From Table 8.1, the conservation equations have been area averaged to arrive at the one-dimensional modelling equations for varying cross section area plug flow reactor. The area averaging routine has been implemented in the generalized approach. The total continuity equation is,

$$(\nabla \cdot \rho \boldsymbol{v}) = 0$$

which reduces to,

$$\frac{d}{dz}(\rho v_z S) = 0$$

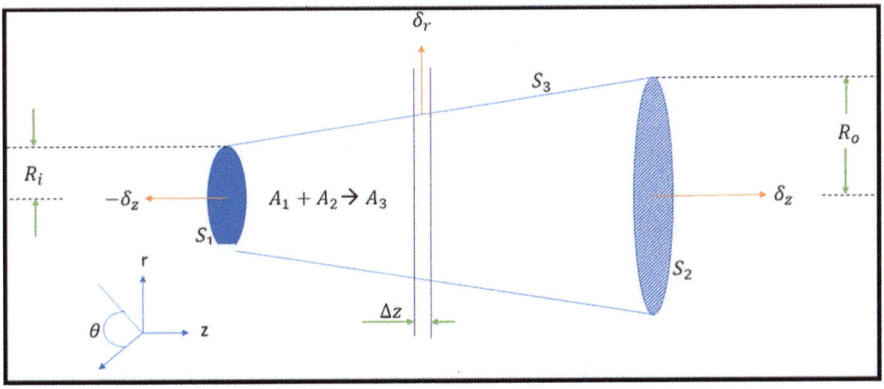

Fig. 8.7 Schematic description of a varying area plug flow reactor

S is the cross-sectional area at length z. Above is total continuity equation for varying area plug flow reactor.

To arrive at the one-dimensional species conservation equation for varying area reactor, the volume average of species continuity equation is taken as,

$$\iiint_V (\nabla \cdot (C_j \boldsymbol{\upsilon}^* + \mathbf{J}_j^*))dV = \iiint_V R_j dV \quad j = 1, 2, 3$$

Substituting $C_j = \frac{\rho_j}{M_j}$ into the above equation and defining mass rate of production of species as, $\widehat{R}_j = R_j M_j$,

$$\frac{1}{S}\frac{\partial(\rho_j v_z S)}{\partial z} = \hat{R}_j$$

The above equation is the species balance equation for a varying area reactor.

The orientation of reactor is horizontal, i.e., potential energy is same at entry and exit of reactor. Omitting the viscous stress contributions $\boldsymbol{\tau}$, the one-dimensional momentum balance equation can be arrived as,

$$\rho v_z \frac{dv_z}{dz} = -\frac{dp}{dz}$$

The above is the momentum balance for steady state varying area reactor.

For the steady state energy conservation, we shall input the thermal conductivity of wall $k = 0$, and neglecting diffusional fluxes $\nabla \cdot \mathbf{J}_j$, for ideal gases, $\left(\frac{\partial \ln \widehat{V}}{\partial \ln T}\right)_{p,x_j} = 1$, and neglecting irreversible conversion of kinetic to internal energy $\boldsymbol{\tau} : \nabla\boldsymbol{\upsilon}$, the energy balance equation shall reduce to,

$$\nabla \cdot \left(\rho \hat{C}_p T \boldsymbol{\upsilon}\right) = \boldsymbol{\upsilon} \cdot \nabla p - \sum_{j=1}^{n_s} \overline{H}_j R_j$$

Applying the above Equation over control volume of varying area reactor,

$$\iiint_V \left(\nabla \cdot \left(\rho \hat{C}_p T \boldsymbol{\upsilon}\right) - \boldsymbol{\upsilon} \cdot \nabla p + \sum_{j=1}^{n_s} \overline{H}_j R_j\right) dV = 0$$

Substituting the relationship $\nabla p \cdot \boldsymbol{\upsilon} = \nabla \cdot (p\boldsymbol{\upsilon}) - p(\nabla \cdot \boldsymbol{\upsilon})$ and using The Gauss-Ostrogradskii Divergence Theorem on the left-hand side of above equation,

$$\iint_S \left(\rho \hat{C}_p T \boldsymbol{\upsilon} \cdot \hat{\boldsymbol{n}}\right) dS - \iint_S (p\boldsymbol{\upsilon} \cdot \hat{\boldsymbol{n}}) dS + p \iint_S (\boldsymbol{\upsilon} \cdot \hat{\boldsymbol{n}}) dS + \iiint_V \sum_{j=1}^{n_s} \overline{H}_j R_j dV = 0$$

The final form of energy conservation equation can be written as,

$$\rho \upsilon_z \hat{C}_p \frac{dT}{dz} - v_z \frac{dp}{dz} + (-\Delta \overline{H}_1) r_1 = 0$$

The above equation is the integral energy balance equation for steady state varying area reactor. Table 8.2 compiles the equations required to reduce the degree of freedom to zero to obtain a unique solution of the system of differential equations. Even though total mass concentration first decreases and then increases to attain a stable value, the total molar concentration of the mixture continues to decrease from the inlet to outlet of the reactor. Figure 8.8 depicts the simulation results obtained for variation of different reaction parameters across the length of the varying area plug flow reactor [45].

8.2.3.2 Generalized Reactor Modelling for an Annular Flow Reactor

An annular flow reactor refers to a chemical reactor which is composed of a concentric tube (core) in tube (annulus) construct for flow of different process fluids such as reactants and products as depicted in Fig. 8.9. Annular flow reactors are often preferred for process intensification objectives such as in the case of an exothermic reaction where reaction heat can be simultaneously removed and used for pre-heating of the feedstock. In this section application of the generalized reactor modelling framework to an annular flow reactor is briefed. The reacting mixture enters into the core, and reverses direction at the end of reactor, and flows into the annulus. For exothermic reactions, this reactor configuration enables efficient heat utilization, i.e., high temperature products in the annulus shall heat the reactants entering in the core. The generic conservation equation incorporates the heat exchange taking place between core and annulus. Using the generalized conservation described in Table 8.1, overall equations can be generated and summarized in Table 8.3 [45].

The system of equations is complete with kinetic equations and shape statements. The kinetic equation is, $r_i = kC_j$, where k is rate constant. The reactor was simulated using the parameters listed in Table 8.4. The simulation of reaction heat, heat removed, and total heat generated in the reaction as well as variation in the temperature across the length of reactor for the core and annulus section of the reactor are depicted in Fig. 8.10.

8.2.4 *Critical Elements in Microreactor Design*

A wide variety of microreactors in different shapes, sizes, and internal geometries have been reported in the literature for a diverse range of chemical reactions ranging

Table 8.2 Conservation equations for varying area reactor

Total continuity	$\frac{1}{\rho}\frac{d\rho}{dz} + \frac{1}{\upsilon}\frac{d\upsilon}{dz} = -\frac{1}{S}\frac{dS}{dz}$
Species A_1 continuity	$\frac{1}{\rho_{A_1}}\frac{d\rho_{A_1}}{dz} + \frac{1}{\upsilon}\frac{d\upsilon}{dz} = \frac{R_{A_1}}{\rho_{A_1}\upsilon} - \frac{1}{S}\frac{dS}{dz}$
Species A_2 continuity	$\frac{1}{\rho_{A_2}}\frac{d\rho_{A_2}}{dz} + \frac{1}{\upsilon}\frac{d\upsilon}{dz} = \frac{R_{A_2}}{\rho_{A_2}\upsilon} - \frac{1}{S}\frac{dS}{dz}$
Species A_3 continuity	$\frac{1}{\rho_{A_3}}\frac{d\rho_{A_3}}{dz} + \frac{1}{\upsilon}\frac{d\upsilon}{dz} = \frac{R_{A_3}}{\rho_{A_3}\upsilon} - \frac{1}{S}\frac{dS}{dz}$
Average molecular weight	$-\frac{dM_{avg}}{dz} = M_{avg}^2\left[-\left(\frac{\rho_{A_1}}{M_{A_1}} + \frac{\rho_{A_2}}{M_{A_2}} + \frac{\rho_{A_3}}{M_{A_3}}\right)\frac{1}{\rho^2}\frac{d\rho}{dz} + \frac{1}{\rho M_{A_1}}\frac{d\rho_{A_1}}{dz} + \frac{1}{\rho M_{A_2}}\frac{d\rho_{A_2}}{dz} + \frac{1}{\rho M_{A_3}}\frac{d\rho_{A_3}}{dz}\right]$
Ideal gas law	$\frac{dp}{dz} = \frac{\rho R}{M_{avg}}\frac{dT}{dz} + \frac{RT}{M_{avg}}\frac{d\rho}{dz} - \frac{\rho RT}{M_{avg}^2}\frac{dM_{avg}}{dz}$
Momentum balance	$\rho\upsilon\frac{\partial\upsilon}{\partial z} = -\frac{\partial p}{\partial z}$
Energy balance	$\rho\hat{C}_p\upsilon\frac{\partial T}{\partial z} = \upsilon\frac{\partial p}{\partial z} - \Delta\overline{H}_1 r_1$
Change in cross sectional area	$\frac{1}{S}\frac{dS}{dz} = \frac{\frac{2}{L}\left(\frac{R_0}{R_i}-1\right)}{\left[1+\left(\frac{R_0}{R_i}-1\right)\frac{z}{L}\right]}$

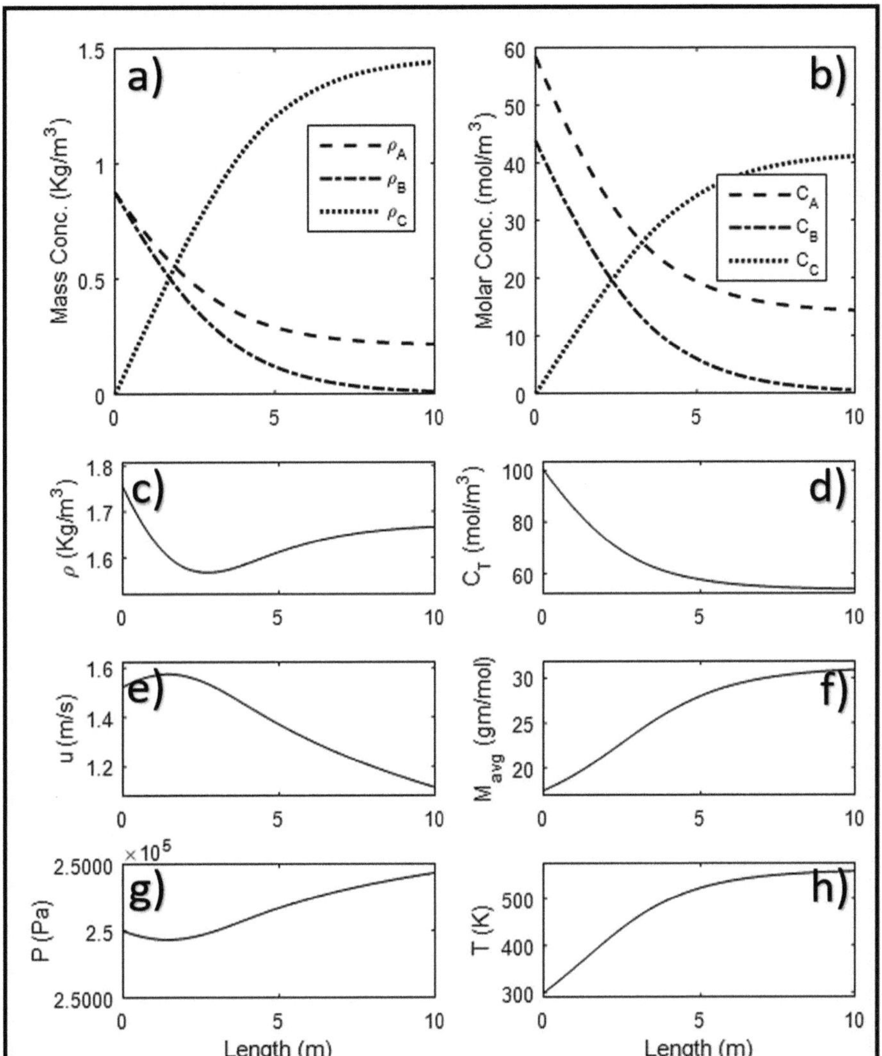

Fig. 8.8 Simulation of variation in reaction parameters across the reactor length for the base case in a varying area plug flow reactor; **a** Species mass concentration; **b** Species molar concentration; **c** Total mass concentration; **d** Total molar concentration; **e** Velocity; **f** Average molecular weight; **g** Total pressure; **h** Temperature along the length of reactor. Adapted with permission from Ref. [45]

from gas phase reactions to gas–liquid-solid slurry phase reactions. In general, efficient transport of mass and heat between reactants, products and microreactor components such as catalyst packing or coating, channels, baffles, and walls is essential for the microreactor to exhibit high performance. However, exact behavior of reaction medium and specific microreactor must be evaluated on a case to case basis due to the

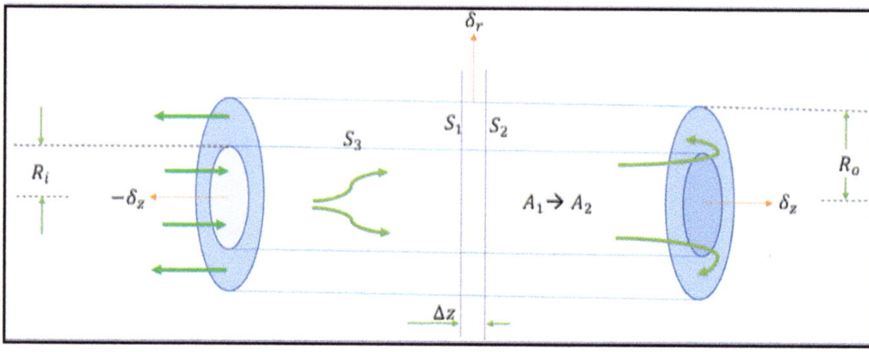

Fig. 8.9 Schematic description of an annular flow reactor

Table 8.3 Modelling equations for an annular reactor

Core section	
Total continuity	$\frac{d(C_T^{in} \upsilon^{in})}{dz} = 0$
Species A_1 continuity	$\frac{d\left(C_j^{in} \upsilon^{in}\right)}{dz} = R_j\left(C_j^{in}, T^{in}\right)$
Momentum balance	$C_T^{in} M^{in} \upsilon^{in} \frac{d\upsilon^{in}}{dz} = -\frac{dp^{in}}{dz}$
Energy balance	$C_T^{in} C_p^{in} \upsilon^{in} \frac{dT^{in}}{dz} = (-\Delta \overline{H}_1) r_1 + U\left(\frac{4}{D_i}\right)\left(T^{out} - T^{in}\right)$
Annular section	
Total continuity	$\frac{d(C_T^{out} \upsilon^{out})}{dz} = 0$
Species continuity	$\frac{d\left(C_j^{out} \upsilon^{out}\right)}{dz} = R_j\left(C_j^{out}, T^{out}\right)$
Momentum balance	$C_T^{out} M^{out} \upsilon^{out} \frac{d\upsilon^{out}}{dz} = -\frac{dp^{out}}{dz}$
Energy balance	$C_T^{out} C_p^{out} \upsilon^{out} \frac{dT^{out}}{dz} = (-\Delta \overline{H}_1) r_1 - U\left(\frac{4}{D_i}\right)\left(T^{out} - T^{in}\right)$

sophisticated nature and wide range of micro to macroscale interactions that govern the performance. In this section important design elements for microreactors which are critical for their high performance and efficient operation are described based on demonstrated examples.

Table 8.4 Parameters used for simulation of an annular reactor

Variable	Value	Unit
M_{A_1}	15	gmmol^{-1}
M_{A_2}	15	gmmol^{-1}
α_{11}	−1	
α_{12}	1	
T(inlet)	500	K
P(inlet)	200,000	Pa
$A_1 : A_2$ inlet mass ratio	1:0	
k_0	3.06599e6	$\text{kmol}^{-1}\text{m}^{-3}\text{s}^{-1}$
Universal gas constant, R	8.314	$\text{kJkgmol}^{-1}\text{K}^{-1}$
Activation energy, E/R	8000	K
Inlet velocity, υ^{in}	5.10	ms^{-1}
Core section diameter of reactor, D_i	0.3	m
Annular section diameter of reactor, D_o	0.4	m
Length of reactor, L	2	m
Overall heat transfer coefficient, U	0.070	$\text{kWm}^{-2}\text{K}^{-1}$
C_p	28.77	$\text{kJkgmol}^{-1}\text{K}^{-1}$
$-\Delta\overline{H}_1$	5000	kJkgmol^{-1}

8.2.4.1 Impact of Microreactor Design on Mixing Efficiency

Microreactors typically compose of either an array of microchannels or a continuous network of microchannels. Microchannel path variation refers to the variation in the path of the microchannels or more generally a variation in the patterns of geometrical arrangements in microchannel networks. Microchannel geometry, path variations and flow obstruction geometry significantly impact the performance of a microreactor because they have a direct impact on mixing efficiency, back pressure, number, and distribution of dead zones and even reactant conversion. Fluid mixing also impacts mass and heat transfer which govern the reaction kinetics and thermodynamics. The dependence of microreactor performance on microchannel geometry and path variations arises from different physico-chemical factors and governing forces such as size effect, rarefaction effect, viscous effect, hydrodynamic and electrostatic forces [55]. The behavior of fluid and especially multi-component fluids in channels is fairly sophisticated and as a result a large proportion of research has been dedicated to development of optimal microchannel geometries and path variations for mixing since it directly impacts mass transfer, heat transfer and back pressure. Researchers have often relied on the characterization of flow regimes in microchannels which allows development of useful correlations to predict the behavior of fluid with reasonable degree of accuracy. For instance, Reynolds number (Re) and Weber

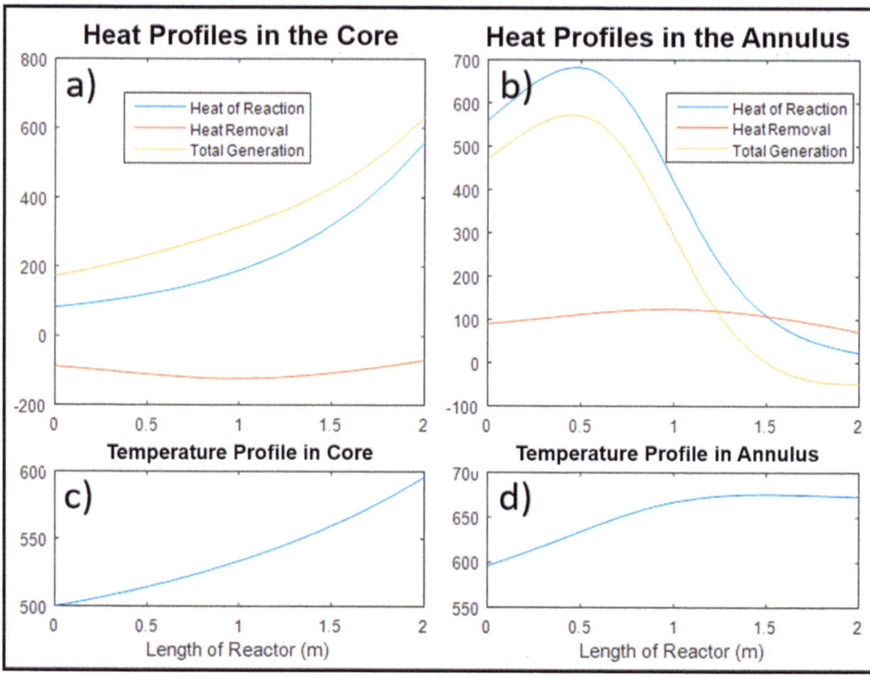

Fig. 8.10 Simulation of variation in heat profiles and temperatures in the core and anulus section of the annular flow reactor with reactor length; heat of reaction, heat removal and total heat generation in **a** Core section of the reactor; **b** Annulus section of the reactor; and temperature profile in **c** Core section of the reactor; and **d** Annulus section of the reactor. Adapted with permission from Ref. [45]

number (We_G) are widely used to characterize flow regimes. Liu et al. identified that serpentine microchannels perform best for $6 < Re < 70$ and in this range, a 3D serpentine microchannel provided much better mixing than a square wave and straight microchannels [56]. For low Reynolds numbers molecular diffusion was identified to be the dominant mechanism responsible for mixing while viscous effects acted to inhibit mixing [57]. At higher Reynolds numbers in the range of $15 < Re < 276$, transverse flow was observed which significantly enhanced mixing by chaotic advection [57]. The Weber number (We_G) is a dimensionless number that describes the ratio of vapor inertia to surface tension forces and can be used to predict the flow regime. For instance, pure intermittent flow was observed for $We_G < 6$ and pure annular flow was observed for $We_G > 35$ [58].

$$We_G = \frac{\rho_G \times U_G^2 \times D_H}{\sigma}$$

In an interesting study, An et al. used CFD simulations to examine the overall performance of microreactors with a square cross-sectional channel in eight different configurations namely parallel, pin hole, wavy, oblique fin, serpentine, coiled, coiled with serpentine, and coiled with double serpentine channel geometries [59]. In addition to mixing efficiencies and back pressure, reaction rates manifested as reactant conversion were also observed to vary with microchannel geometries as depicted in Fig. 8.11. For all geometries the performance decreased with increase in Reynold number. The pinhole channel geometry was identified to result in highest overall performance resulting from high reaction rates and minimal pressure drops and the effect was more pronounced at higher Reynold numbers. All coiled geometries exhibited high reaction rates but suffered from significant pressure drops [59]. Thus, it is evident from the above discussion that microchannel geometry and path variations are important design elements which must be optimized based on theoretical and experimental simulations for designing high performance microreactors.

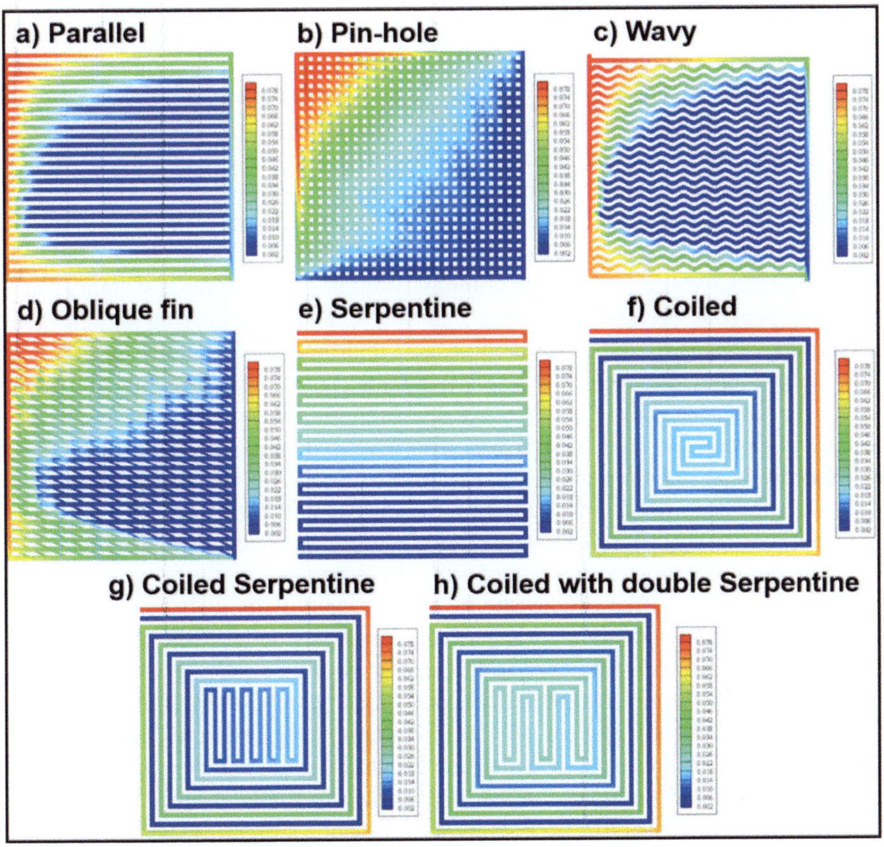

Fig. 8.11 Effect of microchannel cross-section on reaction rates manifested as mole fraction of methane at $z = 5 \times 10^{-4}$ m for different types of investigated microchannel cross-sections such as **a** Parallel; **b** Pin-hole; **c** Wavy; **d** Oblique fin; **e** Serpentine; **f** Coiled; **g** Coiled serpentine; and, **h** Coiled with double serpentine. Reproduced from Ref. [59] with permission from Elsevier

Generation of micro droplets inside microfluidics channels is another area of research which greatly benefits from the knowledge accumulated on the behavior of fluid mixing in microchannels. Micro droplets are often leveraged as microscale compartmentalized reactors inside which important reagents can rapidly combine to generate miniscule quantities of desired product such as metal nanoparticles, semiconductor nanoparticles and even polymeric and biological macromolecules of importance. Zhao et al. developed a three-dimensional Lattice Boltzmann model (LBM) for multiphase/multicomponent fluid mixtures and utilized it for studying droplet generation and complex hydrodynamics inside microdroplets generated from T-shaped and modified T-shaped junctions [60]. Breaking of bilateral symmetric fluid distribution and fluid recirculation inside the microdroplets were identified as critical factors that improved mixing efficiencies.

Zhang et al. presented a thorough account of the state of art in design and scale-up of microchemical systems and design of micromixers for efficient mass and heat transfer [61]. A micromixer relies on its geometrical shape to efficiently contact multiple streams and rapidly generate a single compositionally homogenous mixed stream [62, 63]. Mixers can be classified as active mixers which rely on external energy input or as passive mixers which rely on its geometrical shape for mixing. Micromixers can be evaluated based on their performance for mixing efficiency, flow hydrodynamics, mass, and heat transfer. For passive mixers, a coil design is the most common for milliscale and macroscale reactors because of ease of fabrication whereas microchannels are common for microreactors and lab on chip devices. Packed bed passive micromixers are commonly employed in catalytic systems where the heterogeneous powder or shaped catalysts enables the mixing of reactants. Geometrical shapes of some common passive micromixers are shown in Fig. 8.12 [61]. Simple contacting structures like T-junctions, Y-junctions, cross junctions, and co-flow junctions as shown in Fig. 8.12a are routinely employed in reactor systems because of ease of fabrication. However, flow hydrodynamics and mixing efficiency vary largely with differences in stream contact orientation and poorly segmented or annual flow can result when multiphase and viscous fluids are involved [64]. Multilaminate micromixing structures like the triangular and slit interdigital mixer based on shear thinning of fluid into films result in superior mixing even at low flow rates but can prove costly to fabricate by conventional methods because of the 3D structure [65]. Split-recombine micromixing structures such as the AccoMix and Caterpillar mixers repeatedly cut, stretch, and stack fluid streams producing excellent mixing even for multiphase and viscous fluids but lead to high pressure drops and back pressures [65]. Helical and curved micromixing structures like the Lonza [66] and the tangential mixer [67] are easy to fabricate and enhance mixing performance by chaotic advection in curved geometries especially at high flow rates but with poor performance at low flow rates. Periodic static structures take advantage of both chaotic advection and shearing effect to generate efficient mixing of multiphase fluids but also generate a high pressure drop [68]. In addition to passive mixers, active mixers based on the input of external energy in different forms have also been widely employed for mixing applications such as ultrasonic energy [69], acoustic variations [70], periodic oscillations [71], electric fields [72], and magnetic fields [73]. Sochol et al. demonstrated the use of multijet modelling (MJM) for fabrication of geometrically complex but functionally advantageous integrated microfluidic components with both passive and active components such as fluidic capacitors, fluidic diodes and fluidic transistors [74]. Complete on-chip automation of designed fluidic component by actuation with pressure was demonstrated. Theoretical and experimental results indicated that 3D fluidic capacitors and transistors show improved component performance compared to their planar analogues [74]. Gong et al. also harnessed the power of rapid prototyping using DLP-SLA of a customized resin and used it for fabrication of valves, pumps and multiplexers based on series and parallel arrangements of a membrane based valve [75]. Pump flow rates as high as 40 μL/min could be obtained and the valve showed good performance even after 1 million actuations [75]. With the capability to manufacture complex geometries at low cost, additive manufacturing has enormous potential for rapid fabrication and performance testing of efficient passive or active micro mixing systems and associated microreactors whether planar or three-dimensional in design based on CFD studies or purely experimental based evaluation.

8.2.4.2 Impact of Microreactor Geometry on Heat Transfer

The importance of microreactor geometry in governing the performance of microreactor was clearly evident from the detailed comparative study by An et al. However, many researchers regard the microchannel cross-section as the most important geometrical factor because of its strong influence on heat transfer in addition to mixing efficiency and back pressure. This strong dependence arises because cross-section geometry directly controls the surface area of the fluid that comes in contact with the channel walls. In-fact many generalized correlations have been developed based on thorough studies to correlate hydraulic diameter (D_h) with flow regimes and efficiency of heat transfer. In addition to micro reaction technology, the phenomenon of heat transfer in microchannels also takes center stage in other chemical systems such as phase change in microchannels, power generation, electronics cooling, mobile and stationary heating, cooling, refrigeration, and chemical processing. Coleman and Garimella conducted thorough studies on characterization of flow regimes and their transition behavior in biphasic air–liquid fluid flow in round and rectangular microchannels [76]. It is well known that pressure drop in biphasic flow is much higher than single phase flow because of several microscopic and interfacial physico chemical forces. They identified that at macroscopic scales ($D_h > 10$ mm) pipe diameter and surface tension play a negligible role whereas their contribution in controlling flow patterns and transition becomes increasingly important as the channel diameter is decreased. Different flow regimes in biphasic fluid flow as shown in Fig. 8.13 were characterized as stratified flow regime consisting of stratified smooth and stratified wavy flow, intermittent flow regime consisting of the elongated bubble or plug flow and slug flow, annular flow regime consisting of the annular and wavy annular flow and dispersed flow regime consisting of the bubble and dispersed flows [76]. It was identified that flow regime transitions occur

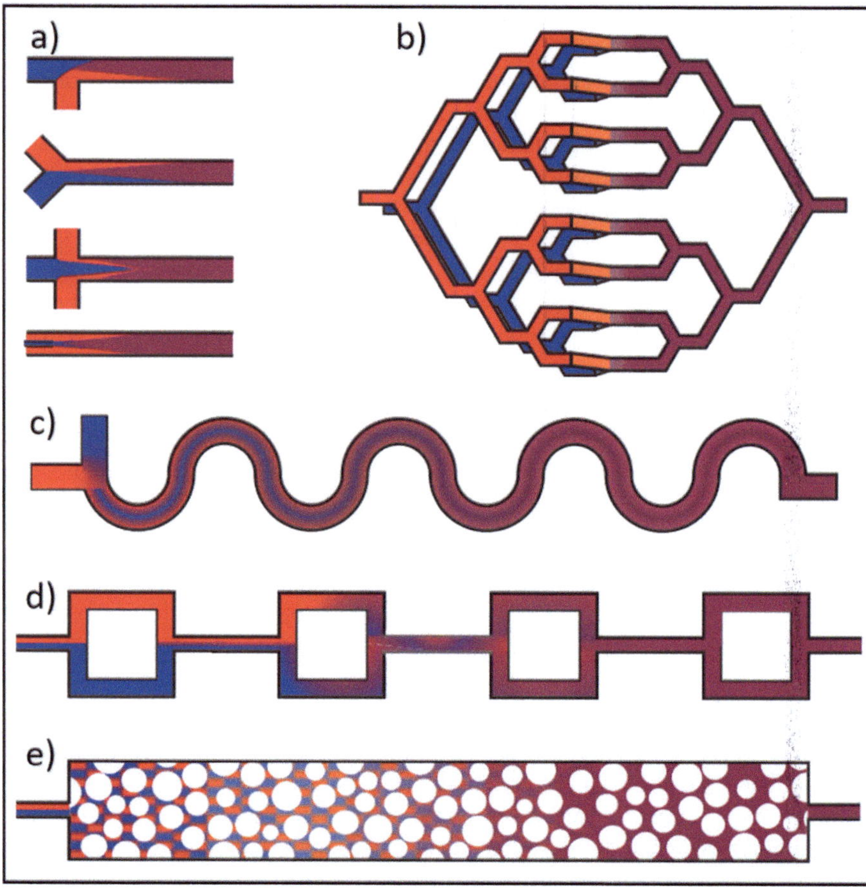

Fig. 8.12 Geometrical shapes of some common passive micromixers; **a** Elementary contacting structures like T-junction, Y-junction, cross junction, and co-flow junction; **b** Multilayer mixing structures; **c** Curved or serpentine micromixer; **d** Square split-recombine micromixer; and **e** Randomly sized periodic static mixers. Adapted from Ref. [61] with permission from Annual Reviews

at different combinations of superficial liquid and gas velocities. Combined effect of surface tension and tube diameter enforced flow regime transition to the dispersed flow regime on decrease of tube diameters to higher values of superficial liquid velocities whereas transition to annular flow regime occurred at a nearly constant value of superficial gas velocity and reached a limit value with decrease of tube diameter. Same forces suppressed the stratified flow regime in small diameter tubes and increased the size of intermittent flow regimes. The effects were compounded even more for rectangular shaped tubes in comparison to round tubes [76]. Garimella and Fronk further provided a detailed account of the single- and multi-constituent condensing fluid and mixtures in different types of microchannels [77]. Phase change

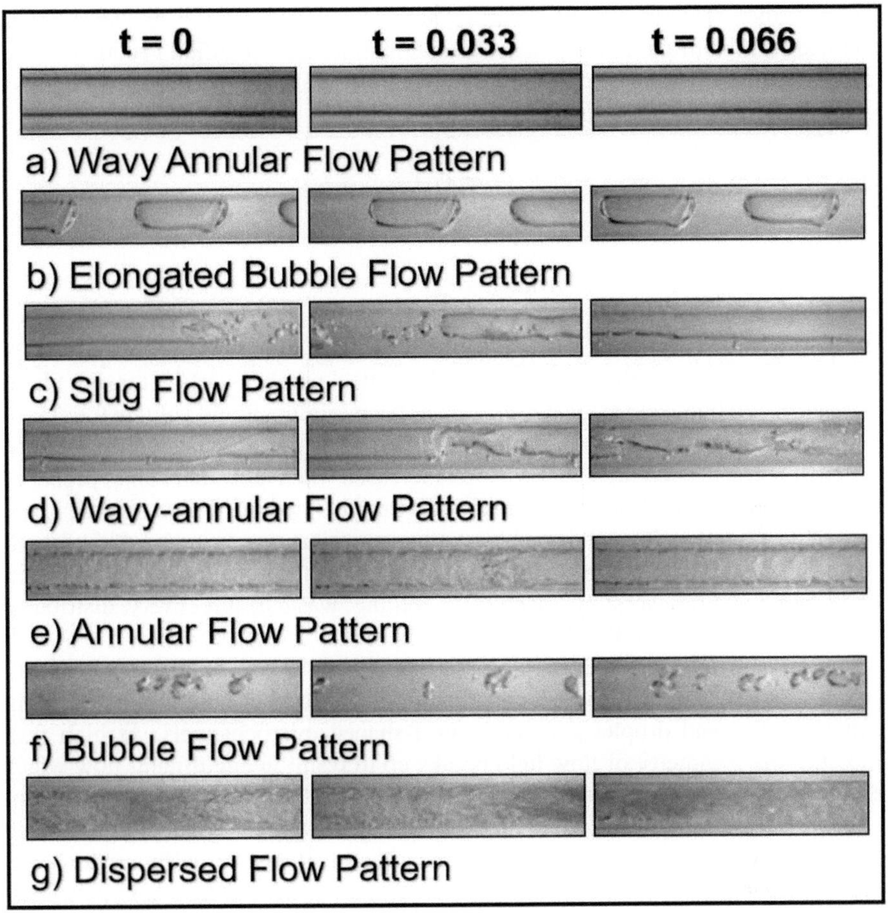

Fig. 8.13 Time-lapse photographs of flow patterns in a round microchannel from start of flow at intervals of 0.033 s depicting; **a** Wavy annular; **b** Elongated bubble; **c** Slug; **d** Wavy-annular; **e** Annular; **f** Bubble flow; and **g** Dispersed flow patterns. Reproduced from Ref. [76] with permission from Elsevier

behavior such as condensation is largely scale dependent because of the dominant role played by surface tension at the microscale. It was concluded that conventional models describing condensing fluids were more suited for shear-dominated annular flow and gravity dominated stratified flow whereas intermittent flow regime becomes more dominant as the microchannel diameter decreased [77].

Comparatively, pressure drop could still be estimated using conventional models without large deviations. Considering the importance of flow regime and transitions in governing convective heat transfer, the group further developed dimensionless transition criteria to identify condensation flow regimes and their transitions for various fluids and impact of operating conditions and channel sizes based on hydraulic diameter (D_h), quality (χ), mass flux (G), saturated liquid and vapor density (ρ_L,

ρ_G), dynamic viscosity (μ_L, μ_G), surface tension (σ), vapor phase Weber number, modified Martinelli parameter, critical Bond number and modified Froude number [58]. The Bonds number is a dimensionless number that compares the magnitude of gravitational and surface tension forces and defined as

$$Bo = \frac{(\rho_L - \rho_G) \times g \times D_H^2}{\sigma}$$

$$Bo_{critical} = \frac{1}{\left(\frac{\rho_L}{\rho_L - \rho_G} - \frac{\pi}{4}\right)}$$

When the Bonds number was less than the critical number, surface tension forces dominated, and liquid slugs formed as soon as the minimum liquid volume fraction to form them was reached. For Bond numbers greater than the critical number the liquid volume fraction needed for slug formation increased when tube diameter approached 50%. The developed flow regime maps allowed estimation of flow regimes with reasonable accuracy for condensable synthetic refrigerants in tubes of hydraulic diameters smaller than 5 mm which are highly relevant for heating, ventilation, air-conditioning and refrigeration, electronics cooling, waste heat recovery, distributed generation, and other similar industries [58].

Engulfment flow refers to the enclosure and flow of one fluid over another in a microchannel geometry and becomes important to address fluid mixing behavior without baffles and droplet generation in T-shaped microchannels. At high flow velocities the symmetry of flow field breaks up to result in engulfment flow which results in improved mixing performance. Soleymani et al. employed CFD simulations to identify the critical value of a dimensionless quantity, the identification number (k), beyond which flow regimes transition from vortex flow to engulfment flow in T-shaped micromixers [78]. It was identified that formation of vortices in microchannels and engulfment was not only dependent on the Reynolds number but also dependent on volumetric flow rates and geometric parameters of the mixer such as the aspect ratio of both inlet and mixing channels. The proposed identification number (k) took both the volumetric flow rates and geometrical parameters into consideration and engulfment of flow occurred beyond the critical value of 100 for the identification number [78]. The difference between path lines and streamlines as simulated for vortex and engulfment flows are depicted in Fig. 8.14 and the proposed dimensionless identification number was defined as

$$k = \mathrm{Re}^{0.82}(\mathrm{B/C})^{-0.79}\left(\mathrm{D_h(inlets)}/\mathrm{D_h(mixing)}\right)^{-1.5}(\mathrm{A/C})^{0.15}$$

where A = Width of the mixing channel

B = Width of the inlet channels

C = Depth of all channels

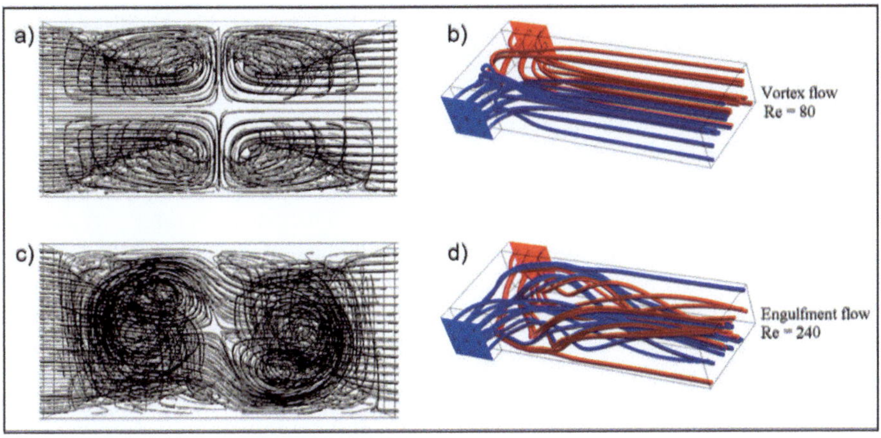

Fig. 8.14 Differences in development of vortices for vortex and engulfment flows in the laminar flow range in terms of path lines and streamlines at the entrance of the mixing channel with increasing Reynolds number; **a** Path lines and **b** Streamlines for vortex flow and Reynolds number of 80; **c** Path lines and **d** Streamlines for engulfment flow and Reynolds number of 240. Reproduced from Ref. [78] with permission from Elsevier

Dh (mixing) = Hydraulic diameter of the mixing channel

Dh (inlets) = Hydraulic diameter of the inlet channels.

Cherlo and Pushpavanam observed that the proposed identification number accurately predicted onset of engulfment only when the depth of the channels (C) was lower than the widths of the mixing (A) and inlet (B) channels (A/C and B/C > 1). When the above aspect ratios were both less than unity, they observed a delay and even absence of engulfment even after further increase of Reynolds number [79]. Fani et al. further explored the use of direct numerical simulations (DNS), sensitivity and linear stability analysis for identification of unsteady flow regimes in T-shaped micromixers at high Reynolds number because of high mixing efficiencies exhibited in this unsteady regime [80]. Two distinct time periodic regimes were identified as a function of the Reynold number. Unsteady asymmetric regime (UAR) was identified for 230 < Re < 400 where flow was asymmetric in space and periodic in time with a frequency monotonically increasing with Reynolds number. Further increase of Reynolds number ($440 \leq Re \leq 480$) led to a different time-periodic regime, termed the unsteady symmetric regime (USR) which was characterized by a continuous and periodic switch from a symmetric configuration to a slightly asymmetric one that resembled the engulfment regime [80]. Bello-Ochende et al. used the finite volume method to optimize a three-dimensional microchannel heat sink for the minimization of wall peak temperature to coolant fluid and identified the optimal aspect ratio and hydraulic diameter (channel shape) [81]. Degrees of freedom were observed to have a strong effect on peak wall temperature and maximum thermal conductance. It was concluded that global thermal conductance could be maximized by optimization of the aspect ratio (H/G) and the channel hydraulic diameter (D_h) for laminar forced-conjugate heat transfer. Moreover, the optimal microchannel shape and maximized thermal conductance were identified to be a function of the applied pressure difference and solid volume fraction and under the studied conditions optimal aspect ratio increased with increase of pressure difference and optimal range for solid volume fraction was found to be between 0.4 and 0.5 [81]. It was argued that for hydraulic diameters less than the optimal values loss of conductance occurred because of fluid thinning and for hydraulic diameters greater than the optimal value reduction of specific surface area was assigned responsible for the reduction in thermal conductance. Similarly, Salimpour et al. used the finite volume method for geometric optimization of an array of circular, square, and isosceles right triangle shaped microchannels to predict the optimal hydraulic diameter and dimensionless heat transfer per unit volume [82]. It was observed that optimal channel size decreased with increase in pressure difference and the channel with square cross-section exhibited the highest heat transfer per unit volume [82]. Kim and Mudawar developed heat diffusion models based on power series solution to tackle the singularity in solution of one-dimensional heat diffusion equation for prediction of temperature distribution in an array of circular microchannels commonly found in electronic heat sinks [83]. Comparison of microchannels with respect to diameter, spacing and Biot number indicated superior cooling performance when microchannel array spacing to diameter ratio was decreased [83]. Chen et al. observed that the microchannel surface topography and roughness also played an important role in governing convective heat transfer and more frequent variations in surface irregularities resulted in a favorable angle between the velocity and temperature gradients leading to a better synergy of heat and fluid flow [84]. They introduced the fractal Weierstrass–Mandelbrot function to characterize the microchannel roughness and observed that flow heat transfer increased with increasing fractal dimension of the surface till a value of three [84]. It is evident from the above examples that heat transfer also depends to a large extent on microreactor geometry, process conditions and resultant flow regimes. Thus, design flexibility and rapid prototyping enabled by additive manufacturing has incredible potential in microreactor design.

8.2.4.3 Baffles and Flow Obstructions in Microreactors

Baffles are obstructions that are used to restrain or regulate flow in the channels of microreactors and micro flow devices. Fins or ribs on the other hand are surfaces that extend from a solid body and used to improve heat transfer from object to the surrounding. Baffles, ribs, and fins are often employed in microreactor design because of several benefits that arise from their implementation such as increase in surface area aiding heat transfer and higher catalyst loadings, enhanced reaction rates and enhanced mixing of fluids. Leung et al. conducted theoretical and experimental studies on crossflow and mixing behavior in rotating radial channels with different obstructions and/or width-constriction geometries in a centrifugal microfluidic platform and identified microchannel length, length of obstruction, width constriction, rotation speed, Coriolis and local centrifugal acceleration, viscous friction effect and jet–jet impingement as critical design elements [85]. Microchannels with repeated patterns of obstruction followed by width-constriction exhibited the best mixing results. The combination of local centrifugal acceleration, Coriolis acceleration, flow splitting at stagnation point of an obstruction and flow recombination due to jet–jet impingement led to excellent mixing and high crossflow [85]. Naher et al. conducted a comparative study between a straight microchannel, microchannel with rectangular baffles and microchannel with fins inclined at an angle of 150° from channel wall and identified microchannel internal geometry as a critical factor that governed mixing efficiency even at constant fluid inlet velocity, viscosity and pressure of fluids based on experimental and theoretical CFD studies [86]. Creation of swirling vortexes within the microchannels increased diffusional gradients and greatly aided diffusive mixing. However, creation of swirling vortexes within dead zones led to large pressure drops which enforced lateral flow that decreased mixing efficiency [86]. Vortex generation such as longitudinal vortex generation (LVG) is often employed in macroscopic heat exchangers to enhance heat transfer. Hsiao et al. extended the concept to micro heat exchangers and investigated the impact of LVG in microchannels using a pair of divergent rectangular winglet (RWP) baffles located at the bottom of the microchannel and at an angle to fluid flow [87]. Both theoretical and experimental results obtained for a wide range of Reynolds number indicated improved mixing efficiency for both convection-dominated, and diffusion-dominated cases in comparison to microchannel without any baffles. Formation of LVGs was confirmed for large enough Reynolds number and static Taguchi analysis allowed ranking of relative influence of design parameters as asymmetry index (d) > angle of attack (h) > winglet height (h) > winglet spacing (D) [87]. Kim et al. used experimental and theoretical studies based on an advanced response method for optimization of a micromixer with rectangular ribs incorporated to enhance mass diffusion and advective flow at low Reynolds numbers [88]. The parametric optimization study identified optimal values of critical design elements rib angle, rib height, rib width, and rib spacing. CFD modelling and optimization of geometrical parameters in a grooved staggered herringbone micromixer by Aubin et al. underpins the importance of baffle geometry in governing mixing quality and dead zones [89]. Spatial data statistics, maximum striation thickness and residence time analyses were employed to quantify mixing quality. Large groove depth and width allowed rapid reduction in maximum striation thickness without significant increase in pressure drops. Wide grooves on the other hand led to increase in dead zones in the microchannel whereas deep grooves improved spatial mixing quality [89].

Tsai and Wu proposed the use of radial baffles in a curved C-shaped planar microchannel to improve mixing efficiency by multidirectional vortex generation [90]. Interestingly, the location of radial baffles on the external or internal walls greatly impacted mixing quality and resulted into formation of an extra vortex based on the magnitude of Reynolds number as shown in Fig. 8.15. Theoretical modelling and experimental results indicated that Dean vortices in the curved channel, the expansion vortices after the baffles, and the converging–diverging flow through the passages contribute towards enhancement of fluid mixing. The optimal micromixer

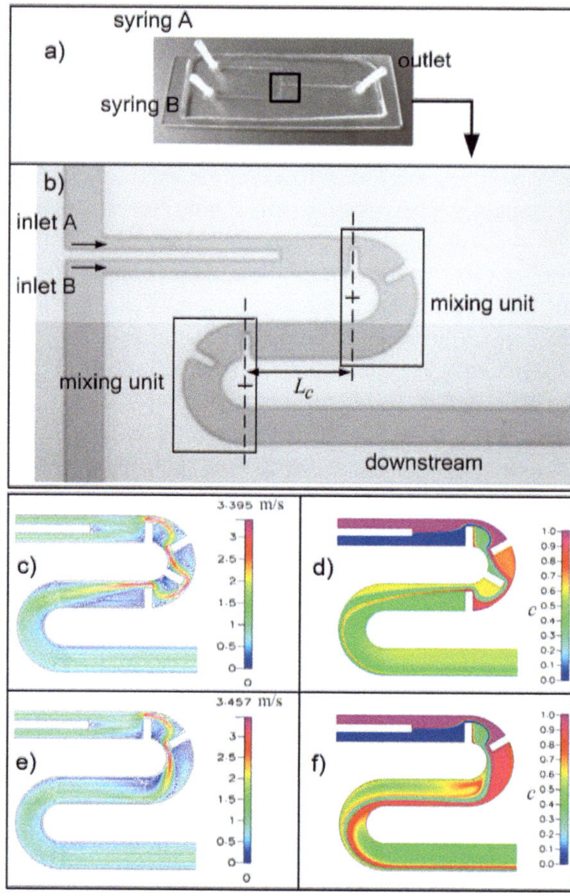

Fig. 8.15 Fluid mixing in a serpentine microchannel composed of C-shaped micromixer channels with radial baffles, radius (R_c) of 160 μm and microchannel length of 3605 μm; **a** Image of the fabricated microfluidic devices used for the study; **b** Optical microscope image of the microchannel with two C-shaped micromixers one of which is inverted; **c** Flow pattern and **d** Concentration distribution in the C-shaped micromixer with four equally spaced radial baffles at a Reynolds number of 81; **e** Flow pattern and **f** Concentration distribution in the C-shaped micromixer with two equally spaced radial baffles at a Reynolds number of 81. Reproduced from Ref. [90] with permission from AIP publishing

design allowed exit flow to be fully mixed (>90%) within an overall channel length of only 4255 μm including the curved-straight-curved regions [90]. Based on CFD simulations and experimental studies using colored gold nanoparticles and copper sulphate solutions in a microfluidic channel, Jeon et al. observed that microchannel geometry and obstructions can be successfully employed to enhance mixing at low pressure drops for lab-on-chip and on-chip total analysis applications [91]. Compared to a straight channel, zigzag geometry resulted in most homogenous mixture followed by contraction-enlargement geometry, but a reverse trend was observed for pressure drops. Interestingly, zigzag geometry achieved efficient mixing based on rapid velocity shifts at 90° bends but the mechanism resulted in higher pressure drops. Comparatively, contraction-enlargement geometry achieved effective mixing without significant increase in pressure drops through velocity and pressure variations arising from the changing cross-section [91]. Such studies throw light on the importance of tree-like fractal structures with periodic and regular bifurcation for uniform flow distribution and low pressure drops that are abundantly observed in nature such as human lungs or tree roots. Chaun et al. proposed the use of porous fins in place of solid fins for reduction of pressure drop in microchannel heat sinks and found a strong dependence of drag reduction efficiency on porous fin parameters based on the Forchheimer–Brinkman–Darcy model [92]. The presence of porous fins led to non-zero velocity of the coolant at the channel-fin interface, enforcing a coolant slipping condition the mechanism of which was different from slipping arising on ultra hydrophobic surfaces but had the similar drag reducing effect. Porous fin with higher permeability, smaller quadratic drag factor, or a larger width led to large slip lengths and corresponding large reduction in pressure drop. Compared to the conventional solid fin design a drop in pressure in the range of 43–48% was observed with only a 5% increase in thermal resistance [92]. Karakaya and Avci conducted a theoretical (CFD) study and estimated hydrogen yields from catalytic oxidation of methanol, ethanol and propane in a custom microchannel reactor consisting of wall separated parallel exothermic oxidation and endothermic reforming microchannels of square shape integrated with respective oxidation and reforming catalysts [36]. Interestingly, microchannel dimensions, distance between channels and use of micro-baffles affected heat transfer to the reforming zone and hydrogen yields. Increase in hydrogen yields with decrease in channel size was assigned to improvement in mixing driven heat transfer coefficients arising from increase in superficial fluid velocities. However, beyond a critical dimension further reduction in channel size led to sharp drops in residence times and hydrogen yields.

Under the same conditions use of micro-baffles was observed to significant improvements in reforming reaction rates and hydrogen yields [36]. The high exothermicity of the Fischer–Tropsch Synthesis (FTS) reaction poses serious reactor design challenges because of the strong dependence of FTS product distribution on reaction temperature. Thus, many researchers have enthusiastically pursued the use of microchannel reactors because of associated benefits. Gumuslu and Avci modelled the FTS reaction in a heat-exchange integrated microchannel network composed of horizontal groups of square-shaped cooling and wall-coated, catalytic reaction microchannels [93]. Thicker walls with high thermal conductivities and presence of micro-baffles in the catalytic microchannels were observed to facilitate near-isothermal conditions. The microchannel reactor was also identified to respond strongly to coolant flow when side lengths of cooling channels were large enough [93]. Thus, it is evident from the above discussion that flow obstruction geometries such as baffles, fins and rib play an important role in improving mixing efficiency and heat transfer by generation of different types of vortices, flow churning and chaotic advective flow in microchannels in addition to provide addition surface area for catalyst integration and fluid contacting.

8.3 Additive Manufacturing of Micro Reactors

It is clear from the previous sections that the performance of microreactors has a strong dependence on its structure and geometry. Additive manufacturing refers to the group of manufacturing technologies that can directly produce complex three-dimensional solids with near-complete design freedom in an additive manner. Thus, Additive manufacturing is anticipated to revolutionize microreactor design and dramatically change the technological landscape in micro reaction technology. Humanity is currently undergoing through a difficult phase faced by arduous global challenges such as climate change, environmental degradation, rapid extinction of flora and fauna, non-uniform distribution of resources across the planet and exhaustion of critical resources. Thus, the onus is on additive manufacturing empowered chemical reaction sciences and engineering to steer global developmental efforts towards a green, environment friendly, sustainable, inclusive, equitable and prosperous world. In the next section, the current state of research and developmental efforts in additive manufacturing for micro reaction technology is critically analysed and individually addressed for major additive technologies such as fused deposition modelling, direct ink writing or robocasting, stereolithography and selective laser sintering and selective laser melting.

8.3.1 Fused Deposition Modelling for Microreactors

The main principle of extrusion based additive manufacturing processes is extrusion of a semi-solid through a nozzle onto a stage where the extruded semi-solid can settle into a solid structure. The movement of the nozzle or the stage in XYZ direction can be used to create the 3D object as per CAD design. Based on the material used for extrusion, this AM process can be classified into fused deposition molding specific to thermoplastics or direct ink writing or robocasting for direct AM of the other gel-like semi-solid functional materials. Fused deposition modeling (FDM) developed by Scott Crump of Stratasys, is based on the extrusion of thermoplastic polymers heated above their glass transition temperatures through a nozzle and onto a cooled stage. Resistive heaters are typically used to heat the thermoplastic polymer filament near the nozzle which starts to flow once its glass transition temperature is reached. The 3D object design is achieved by either movement of the nozzle or the stage as per the CAD design. Each consecutive layer is fused to the previously deposited layer by the thermosetting action of the polymer at lower temperature. FDM manufactured objects are less dense and more prone to failure under mechanical stress compared to their injection molding counterparts because of inadequate fusion between different layers in FDM. Fused deposition modelling is one of the most common and popular AM technology used by hobbyists and enthusiasts. A wide range of materials such as polylactic acid (PLA), acrylonitrile butadiene styrene (ABS), polyphenylsulfone (PPSF), polyamide (PA), polyethylene terephthalate (PET), polyethylene terephthalate glycol (PETG), polyethylene co-trimethyleneterephthalate (PETT) polycarbonate (PC), high- impact polystyrene (HIPS), polypropylene (PP), polyetherimides such as ULTEM and nylon can be used for printing. As a result, a variety of different colors, mechanical and electronic properties can be achieved from derived AM constructs.

Kitson et al. demonstrated a unique combination of additively manufactured modular reaction ware and integrated catalyst material by developing a trifunctional pump-free catalytic reactor including purification components for organic synthesis [94, 95]. The key idea behind the use of this unique reactor design was to use gravity for enabling flow of liquid between feed, catalytic and purification chambers. The whole setup including the catalyst bed was additively manufactured using fused deposition modelling of polypropylene for non-reactive chambers and an acetoxy-silicone polymer composite with integrated catalysts (Pd/C and montmorillonite) for printing of the catalytic chamber as shown in Fig. 8.16. Polypropylene was chosen because of its compatibility with reactants, products, and solvents of the reaction whereas the choice of silicone polymer (Loctite 5366) was based on the feasibility of integration of catalytic materials. By rotation of the reactor setup in steps of 90 degrees, the liquid could be moved from the feed chamber to the catalysis chamber and finally to the purification chamber to complete the reaction process. The same group also demonstrated the fabrication of an array of hydrothermal batch reactors using fused

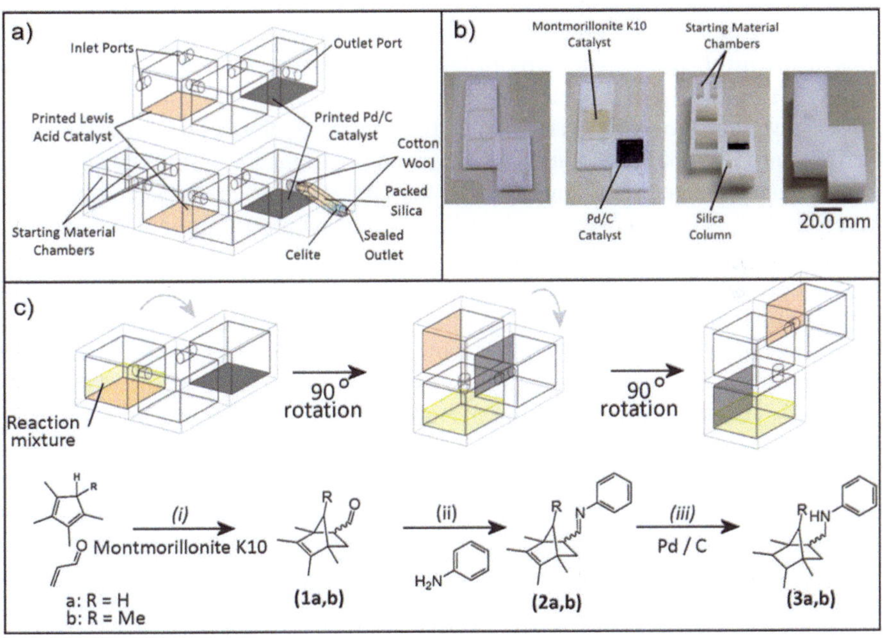

Fig. 8.16 **a** Design of a custom sequential reactor powered by manual rotation additively manufactured by fused deposition modelling; **b** Depicting of individual chambers for reactants and catalysts; **c** Schematic diagram of the multistep reaction sequence. Adapted from Ref. [94] with permission from Royal Society of Chemistry

deposition modelling of propylene and used it to discover two new coordination polymers and optimised the scaled-up synthesis of one of these materials using the same 3D printer [96]. The authors also demonstrated single use application of the fabricated hydrothermal reactors wherein the reactants were loaded in the vessels during printing and the reactor sealed by completing the 3D fabrication as per design. The low cost of fabrication of such polypropylene reactors makes such disposable reactors a niche application. Kitson et al. also demonstrated two and three inlets continuous microreactors fabricated using fused deposition modelling of propylene for reductive amination and alkylation reactions and synthesis of polyoxometalate and gold nanoparticles [97]. These reactors could also be used in disposable manner by loading of reagents during printing and permanent sealing. Rossi et al. investigated the use of continuous flow microreactors fabricated by fused deposition modelling of polylactic acid, high impact polystyrene and nylon for the synthesis of biologically active chiral 1,2 amino alcohols such as norephedrine, metaraminol and methoxamine by a two-step nitroaldol and hydrogenation reaction sequence [98]. Optimised reactors and process allowed high yield of products with moderate diastereoselectivity and enantiomeric excess up to 90% [98]. The main advantages of FDM for manufacturing are its low cost, low waste, high speed of printing, possibility for recycle of printed or unused material, no post processing steps and user friendliness. However, the available range of materials is limited and especially not suitable for use in chemical reactions involving solvents, hydrocarbons, and other reactive chemicals. FDM printers are mechanically robust, reliable and repeatability of finished product is reasonably good. Some degree of control is available for improving product quality parameters such as density and build resolution by optimizing layer thickness, raster patterns and nozzle diameter. Capel et al. investigated the fabrication of milli-scale (millimeter scale) reactor using stereolithography (SLA), multi-jet modelling (MJM), selective laser melting (SLM), selective laser sintering (SLS) and fused deposition modelling (FDM) and compared their performances in a range of organic and inorganic reactions [99]. The fundamental article usefully highlighted the advantages and shortcomings of each manufacturing technology for application in reaction technology. SLA and MJM fabricated constructs generally composed of UV curable resins with epoxy and acrylate backbones showed poor chemical stability and significant swelling in strong solvents as well as poor thermal stability due to their low glass transition temperatures. SLM fabricated constructs generally composed of aluminium, titanium and steel showed excellent thermal stabilities and chemical stabilities due to native oxides on their surface. The only shortcoming of such constructs was the need for post fabrication processing such as polishing and grinding, however, resolution up to +/−0.1 mm was observed to be possible. SLS fabricated constructs generally composed of NYLON-11, NYLON-12, polypropylene, PEEK, and ceramics were observed to have mediocre thermal and chemical stabilities. With build resolutions up to +/−0.2 mm, these constructs were thus observed suitable for low temperature reactions. FDM fabricated constructs generally composed of ABS and polycarbonate also exhibited mediocre thermal and chemical stabilities due to ester linkages being susceptible to degradation [99].

8.3.2 Direct Ink Writing or Robocasting for Microreactors

Direct ink writing (DIW) or Robocasting is a material extrusion additive manufacturing technology like FDM which employs extrusion of a functional gel or material paste through a nozzle. The extruded material, usually called ink, is composed of a solvent and solid component which is deposited onto a solid surface post extrusion. In DIW, the rheological properties of the ink and composition are critical to successful printing and optimization of ink composition (binders, flow agents, solvent, solid content) is generally needed based on targeted application. A vast variety of approaches have been reported in the literature depending on the composition of material used such as the use of heating for printing of Gelatin based suspensions, use of crosslinked hydrogels as inks or novel mixtures of inorganic particles. The greatest advantage of DIW as an AM technology is the possibility of ink customization to incorporate additional functionalities as per need which is the most important objective for applications in chemical science. A variety of functional entities such as proteins, metal oxides, metal nanoparticles, semiconductor nanoparticles, graphene, carbon nanotubes, quantum dots, zeolites and MOFs have been additively manufactured using DIW. Symes et al. demonstrated manufacturing of customized multifunctional batch reactionware using robocasting for organic synthesis, inorganic synthesis and in situ reaction monitoring using spectroelectrochemistry as shown in Fig. 8.17 [100]. The printed bespoke reactionware had good control over mixing of reagents, flow rates and purification of reaction products and the desirable functions built directly into the reactionware design.

Moreover, they were able to demonstrate higher selectivity of one reaction product over another by simple modification of reactor design but keeping all other reaction parameters constant [100]. Such reconfigurable, low-cost, and highly accessible reactionware have revolutionized research by bringing mechanical engineering close to chemical and reaction engineering. Research on robocasting is also being enthusiastically pursued by the catalysis and chemical manufacturing community. Chemical processing and manufacturing industry typically employ catalysts in the form of extrudates or a packed bed of catalyst pellets to retain catalytic activity but reduce pressure drop. Catalyst powders which have higher catalytic activities then pellets and extrudates cannot be directly used since they can lead to complete blockage or enormous back pressures along with a high possibility of blockage of downstream equipment due to catalyst leaching and carryover in process fluid. Recent research has therefore explored the use of structured catalytic supports like monoliths, foams and meshes. Such structures greatly aid pressure drop issues but exhibit inferior catalytic activity. Coating of these structures with catalytically active coatings and thin films is a possible solution to the above problems [38]. However, deposition of uniform catalytic coatings, its stability, life-time, and low volumetric catalyst loadings are serious challenges that have presented a major hurdle in the uptake of this technology by the industrial production sector [101, 102]. An alternative to the use of catalyst coated structured materials is the use of robocasted catalyst structures which

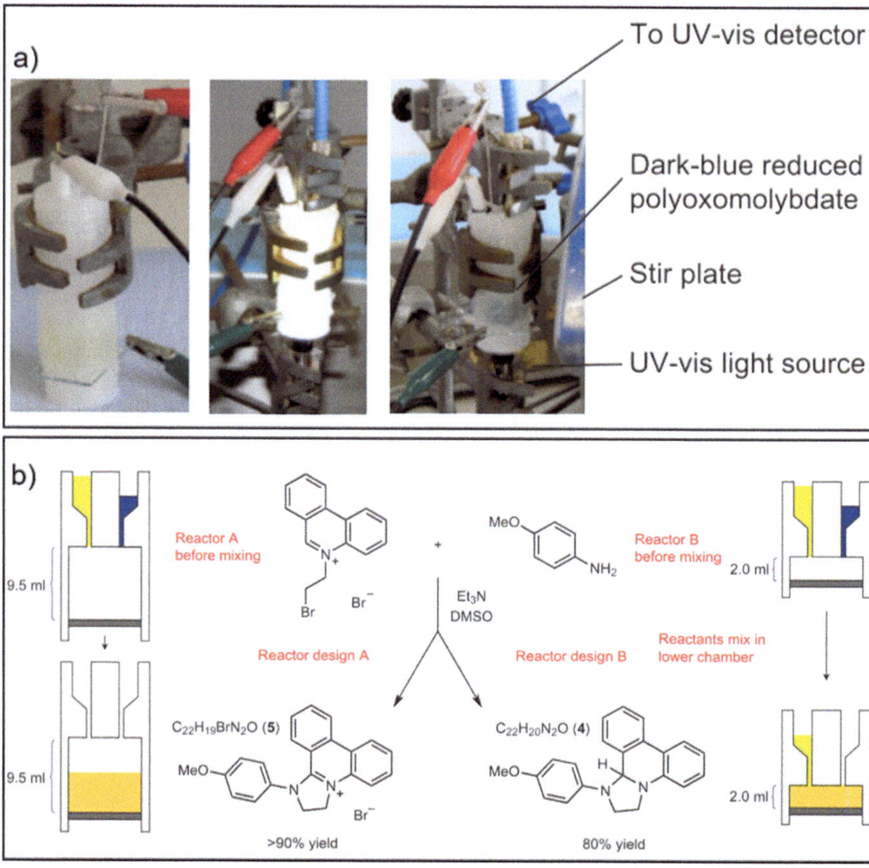

Fig. 8.17 **a** Images of custom designed reactionware additively manufactured using direct ink writing for electrochemistry and in situ spectroscopy studies; **b** Illustration of change in product selectivity with change in reactor size due to differences in reaction kinetics. Reproduced from Ref. [100] with permission from Springer Nature

can potentially resolve both major challenges of high catalyst loading and back pressures. For instance, Lefevere et al. investigated the use of robocasted zeolite ZSM-5/binder composites as self-supported catalysts for the methanol-to-olefin (MTO) reaction as shown in Fig. 8.18 [103]. The composition of the catalyst composite and the architecture of the printed macroporous catalyst structure such as filament size, porosity, and shape of the channels in the direction of the flow were identified to play an important role in governing the activity of the catalyst for the MTO reaction [103]. Konarova et al. used direct ink writing fabricated structured catalysts by incorporating nickel and molybdenum ions into water soluble polyvinylalcohol with starch and subsequent pyrolysis of organic compounds to obtain a highly porous carbon scaffold with catalyst loadings up to 25 wt% [104]. The advantages of highly porous and interconnected flow channels allowed higher CO conversion by a factor

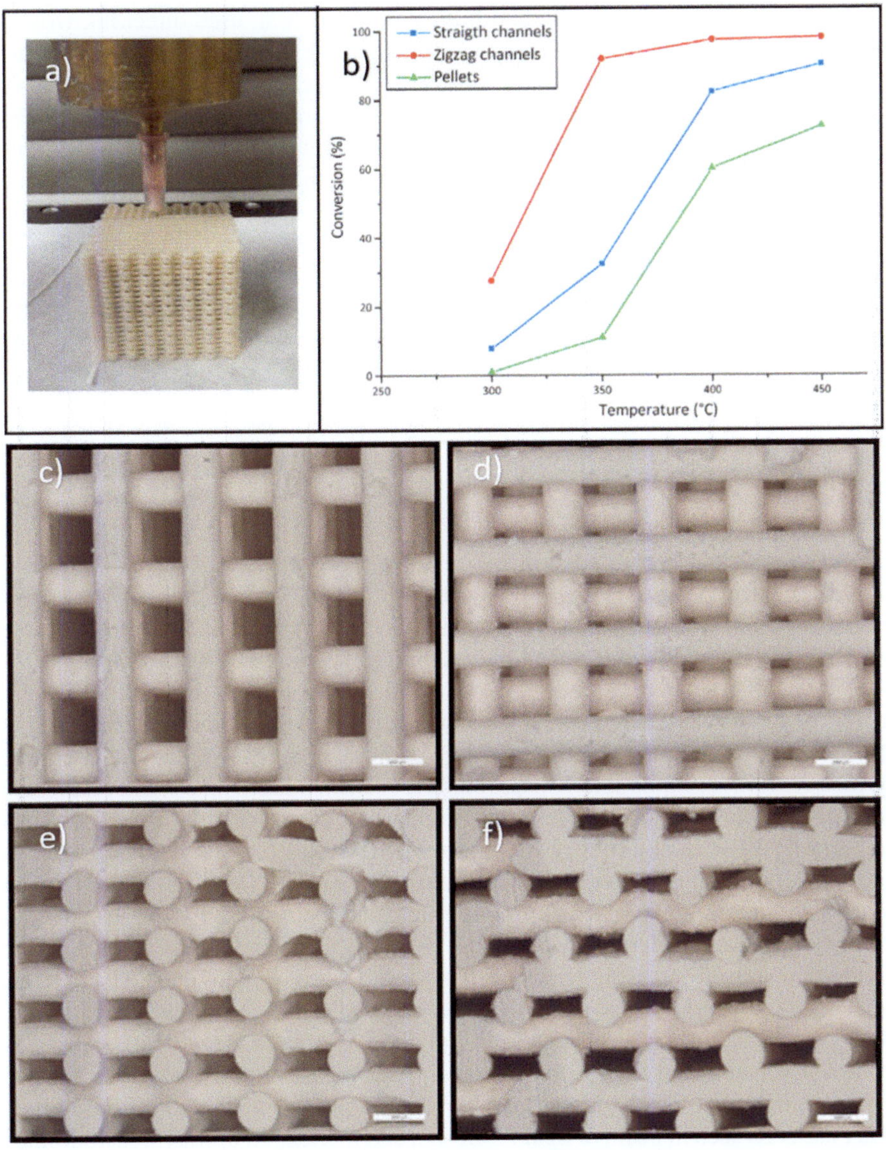

Fig. 8.18 **a** Depicting of robocasting of zeolite ZSM-5 as a self-supported catalyst in a woodpile structure; **b** Comparison of methanol conversion in the methanol to olefin reaction for self-supported ZSM-5 catalyst in woodpile configuration with straight channels, zigzag channels and pellets; **c** Top view of woodpile configuration with straight channels; **d** Top view of woodpile configuration with zigzag channels; **e** Side view of woodpile configuration with straight channels; and **f** Side view of woodpile configuration with zig-zag channels. Reproduced from Ref. [103] with permission from Elsevier

of 2.2 for syn gas synthesis reaction. Similarly, Zhou et al. used direct ink writing of a custom ink made by incorporating monodisperse silica spheres in starch and gelatin as a carbon source to obtain catalytic 3D carbon structures with controlled micro and macro porosity [105]. Silica was used as a hard template to induce porosity and removed post printing using sodium hydroxide. Catalytic activity of monolithic carbon structures obtained by pyrolysis under nitrogen was demonstrated for metal free oxidation of benzyl alcohol to benzaldehyde [105]. AM monoliths and structured catalyst bodies as packing in reactors have enormous potential in solving modern day challenges in gas processing such as carbon dioxide utilization, methane oxidation and hydrogen production where high throughout and catalytic activity can significantly impact the process economics [106]. More information on similar robocasted catalytic systems is available in the literature [107, 108].

8.3.3 Stereolithography for Microreactors

Stereolithography (SLA) is one of the first commercialized AM technologies and was invented by Chuck Hull of 3D systems in 1987. Since most resins used for SLA have poor mechanical properties SLA is most suited for rapid prototyping, visualisation of solids and architectures, planning, and optimisation studies. In SLA, a UV beam directly exposes the resin in the desired pattern and cures it up to a depth governed by the energy source. New resin is applied to the surface and the stage is moved upwards or downwards for exposure of the next layer. A recent improvement, termed continuous liquid interface production (CLIP), utilizes an oxygen permeable transparent window which creates a dead zone below the exposed layer so that it does not attach to the stage [109]. This improvement makes the layered configuration of SLA a highly precise and efficient printing technology. Vertical resolutions up to 25 μm and building speeds up to 1.5 cm/h can be achieved today using SLA by some desktop printers. However, one of the greatest limitations of SLA is the limited number of available resins (either acrylic, epoxy or urethane based), their high cost and possibility of using only one resin at a time. Most resins are brittle, demonstrate significant shrinkage after polymerization and significant swelling in common solvents and chemicals. However, in comparison to FDM, SLA is more accurate, reliable, repeatable and allows fabrication with higher resolution. Thus, SLA has been seen in positive light for design and fabrication of microreactors since the associated benefits of high accuracy and resolution allow fabrication of small microchannels in sophisticated internal geometries and optical clarity allows visual inspection of fabrication and reaction processes [110]. Recent research on the use of SLA in microreactor design has focused on the development of resins with better chemical and mechanical properties like commercially available polymers which are known to be chemically stable such as ABS, PC and PEEK. Ceramic filled resins are another alternative which can result in good mechanical properties and chemical resistance. Capel et al. described the use of commercially available '3D Systems Accura 60' resin for fabrication of a microreactor consisting of a convoluted tube of length 3300 mm, diameter 3 mm and a total volume of 23 mL [99]. Accura 60 was stable up to 150 °C with a heat deflection temperature of 50 °C and resistant to swelling in a variety of organic solvents up to 24 h except dichloromethane and tetrahydrofuran. These issues enforced the limited use of developed reactors up to 30 bar pressure with some preferred reactants and products such as oxidation of aldehydes in methanol with bleach and tetrabutyl ammonium bromide. Interestingly, authors observed yields and selectivities comparable and superior to a conventional flow microreactor of similar size. Similarly, two-step synthesis of the red colored dye Sudan-1 from aniline and hydrochloric acid in dimethylformamide and sodium nitrite in dimethylformamide was demonstrated in a modified version of the above reactor where the second inlet was incorporated half-way through the reactor [99]. Such a design allowed in-situ quenching of the unstable diazonium intermediate where 2-naphtol in DMF was introduced in the next step. The modified reactor demonstrated comparable yields and selectivities to conventional reactor without any stability and deterioration issues. Likewise, a split and recombine mixer-reactor was fabricated as shown in Fig. 8.19a which was used for studying elemental iodine production from potassium iodide and potassium persulfate in the presence of parallel competitive reaction of Iodine reduction back to iodide by sodium thiosulphate [99]. The split and recombine design reduced the reaction time by up to 17% arising from improved mixing efficiency.

Recently, Chaparro-Garnica et al. disclosed the use of an attractive photosensitive liquid resin (VisiJet FTX Green) for fabrication of an advanced nanochannel microreactor (NCM) which was stable in oxidising and reducing environments up to 300 °C.

The fabricated reactor with complex internal geometries demonstrated superior performance in comparison to conventional monolith reactors for preferential oxidation of carbon monoxide in syn gas mixtures [111]. The conventional monoliths used in the study were designed with a honeycomb pattern with a cell density of 45 cpsi and a void fraction of 0.43 with cylindrical grooves along the wall for improved catalyst loadings. Advanced NCM monoliths were designed to consist of a square array of cylindrical discs which were connected on alternate sides with slits for gas transport and exhibited a void fraction of 0.53. A comparison of the hydrodynamic performance of both monolith reactors is shown in Fig. 8.19. Maldistribution and plug formation were evident for conventional monoliths design whereas highly turbulent flow improved mass and heat transfer to enable reduction of temperature at active sites and enhancement of reaction rate for NCM monoliths [111]. It is evident from this outstanding research that once SLA has circumvented the challenge of chemically stable resins, this technology has the potential to become a widely adopted and impactful additive manufacturing technology for micro reactor design. Knitter and Liauw demonstrated an interesting indirect rapid prototyping process which was reliant on initial preparation of molds by stereolithography, transfer of structures to silicone rubber molds and final reactor fabrication by low pressure injection molding of ceramic loaded thermo-plastic binder to obtain modular ceramic microchannel reactors with chemical and thermal stability up to 1000 °C [112]. Oxidative coupling

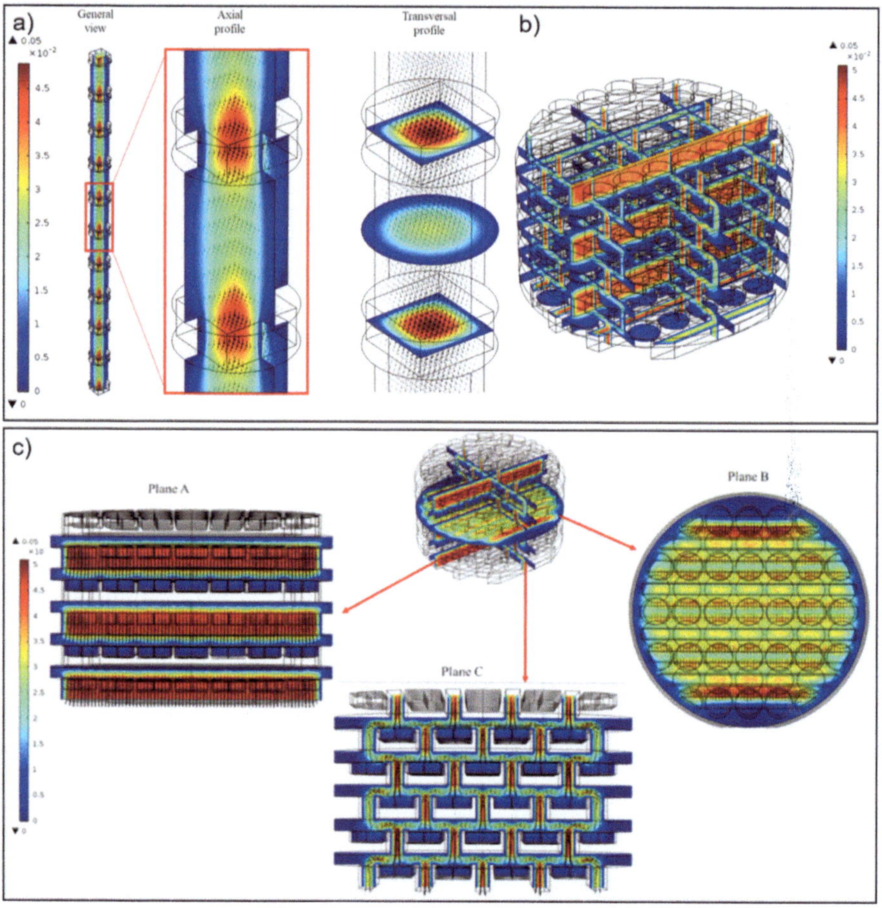

Fig. 8.19 **a** Depiction of the overall view, axial and transversal profile for the velocity vector for one channel of a conventional monolithic reactor; **b** Magnitude of gas velocity; and **c** Velocity profiles for the advanced nanochannel microreactor. Volumetric flow is 200 mL/min in all cases. Reproduced from Ref. [111] with permission from American Chemical Society

of methane catalysed by lithium aluminate and selective oxidation of isoprene catalysed by titania-supported vanadia were demonstrated at pressures up to 50 bar without any observation of wear and corrosion in the developed reactors.

8.3.4 Selective Laser Sintering for Microreactors

Developed by Carl Deckard and Joseph Beaman in mid 1980s, selective laser sintering (SLS) is based on the selective binding of powdered material in the desired pattern using an IR laser to induce localized sintering of particles. Sintering is a process where different particles of material start binding with each other to form bigger particles under high temperature or pressure. The laser is rastered over the surface tracing the pattern governed by the CAD file to generate one layer. Subsequently, more powder is applied by a blade or a roller and the next layer is sintered onto the previous layer as per the CAD design. The 3D object is manufactured in a layer-by-layer fashion and the powdered material which is not sintered after the fabrication process is removed. For optimum processing of materials by SLS, materials should have a large difference between the melting temperature and the crystallization temperature, high melting enthalpy, low heat conductivity, high surface tension but low surface energy [113]. High melting enthalpy and low heat conductivity enable localized heating of the material needed for localized sintering. The interconnection of semi-molten material is enabled by high surface tension while the low surface energy enables sharpness of object edges. SLS can be used for a variety of materials such as polyamide (PA), polycarbonate (PC), polyvinylchloride (PVC), nylon, polyester, acrylonitrile butadiene styrene (ABS), metal and ceramic powders. A variation of SLS is the selective mask sintering (SMS) where the energy source for sintering is changed to an IR lamp and a mask is used for simultaneous exposure of pattern in the entire layer as per CAD file. The mask is produced at each stage by application of toner ink to a glass slide in a pattern which inverse of the desired pattern to be printed for each layer. The sintering of material layer takes place by exposure of the desired pattern through the mask. After sintering the stage is moved downwards in a range of 50–120 μm depending on the material and next layer is exposed through the mask as per the CAD design. SMS greatly enhances the speed of manufacturing objects because of the simultaneous exposure of layers. The power density consumption is also reduced since a wide spectrum light source (visible to microwave) is used in place of a laser. Usually, some IR absorbing materials such as carbon black are used as additives up to a content of 2 wt% to enhance the absorption of IR radiation.

Hutter et al. investigated the performance of a chemical reactor with shaped internals directly fabricated by SLS and compared it to reactors packed with structured metal foams of different sizes with respect to axial dispersion, residence time distribution [114]. An image of the fabricated reactor is shown as Fig. 8.20. In all cases narrow residence time distributions were obtained indicating a global plug flow behavior and high fluid velocities enabled by high porosity of packed material. They observed superior performance of the SLS fabricated reactor for catalytic reactions with comparable surface area to per unit volume of catalyst, but pressure drops over 10 times smaller. Moreover, additive manufacturing of reactor allowed flexibility in design of reactor internals, wide choice of materials, efficient heat transfer and minimal fluid channeling due to good contact with reactor walls [114]. The performance evaluation of foam packed microreactors fabricated either by SLS or three-dimensional fibre deposition (3DFD) two phase flows by Potdar et al. also resulted in similar performance enhancement for mass transfer and pressure drops [115].

They relied on the use of optical measurements using particle image velocimetry (PIV) and planar laser induced fluorescence (PLIF) as well as chemical methods for measuring interfacial mass transfer and hydrodynamic properties of dual-phase

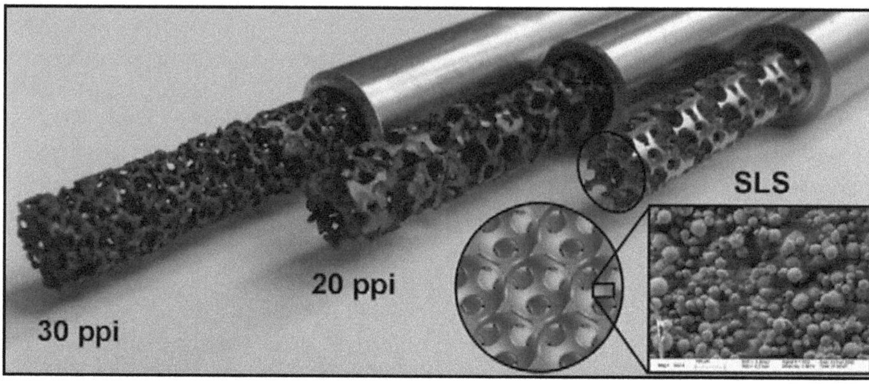

Fig. 8.20 Depiction of a fully sintered structured reactor manufactured using selective laser sintering along with reactors packed with commercially available metal foams for comparison. Reproduced from Ref. [114] with permission from Elsevier

liquid flows. Porous foam packed reactors were observed to induce slug break-up resulting in lower effective slug lengths and higher inter-phase penetration especially at high flow rates compared to hollow tubular reactors. The shape of the packing greatly affected the velocity profile in the slugs and porous packing with no structural variations in the streamline direction were observed to be susceptible to channeling. Thus, rotational structures parallel and normal to mean flow direction allowed efficient mass transfer which was much inferior in their absence. It was concluded that porous foam packed structural reactors achieve same order of interfacial mass transfer as packed bed reactors but an order of magnitude lower energy dissipation [115]. Ideally, the microreactor research community experimenting with additive manufacturing should base their micromixer and microreactor components design on available CFD studies such as the NETmix reactor [116] demonstrated by Fonte et al. using a network of 3D mixing elements which are easy to fabricate using additive manufacturing. However, despite the wide variety of materials compatible with SLS, it is most commonly used for fabrication using polymeric materials such as Nylon [99]. When SLS is used for fabrication polymer-composite parts it is referred to as indirect selective laser sintering which uses metallic and ceramic powders along with polymer powders. Taking advantage of this powerful capability to fuse powdered particles, new technologies have now been commercialized which allow additive manufacturing with complex metallic alloys (CMA) and quasicrystals (QC) [117]. CMAs and QC have been viewed as advanced functional materials with superior mechanical and physical properties such as low coefficients of friction, corrosion resistance, low wettability, and good wear resistance. However, their brittleness limits their use in production of functional solid parts and components. Incorporation of CMA and QC powders allows AM of parts and components with useful lightweight alloys such as a metallic solid consisting of Al–Cu–Fe–B. Kenzari et al. demonstrated AM of light weight matrix composites with CMA and QC whose mechanical properties were similar to steel-brass composites used in the industry today but with a volumetric density half of these composites [117]. Similarly, Amorim et al. demonstrated indirect SLS with molybdenum and copper-nickel alloy (Mo-NiCu) powders as the metal-matrix composite material for fabrication of electrode discharge machining (EDM) electrodes [118]. After optimization of operational parameters, SLS could be successfully employed for AM with the novel Mo-CuNi alloys and fabricated electrodes performed better than SLS fabricated Cu electrodes but inferior to solid copper electrodes [118].

Compared to other additive manufacturing technologies, SLS presents some benefits such as improved range of compatible materials and no requirement for support structures relieving all geometrical restrictions since the structures are inherently supported by the surrounding powder. SLS allows layer thicknesses between 0.1–0.2 mm but the wall thickness which is governed by the laser spot size is typically of the size around 0.6 mm which is inferior to other AM technologies. The produced parts are mechanically stable and tough but subject to distortion over time. Another issue in SLS is the uncontrolled appearance of porosity which may cause leakage under elevated pressures. This porosity arises from the high viscosity of thermoplastic melts and incomplete particle fusion within the build [99]. Another major issue which is a big hurdle in the wide adoption of SLS for microreactor fabrication is the removal of un-sintered powdered particles and dense semi-solids formed from thermoplastic melts in small sized microreactor channels which are typically removed from macroscopic builds using agitation, ultrasonication or compressed air. Despite, the above mentioned shortcoming SLS is widely used for rapid prototyping and fabrication of macroscopic parts and components. The alleviation of above mentioned issues might lead to a greater uptake by the chemical science and engineering community.

8.3.5 Selective Laser Melting for Microreactors

Selective laser melting (SLM) or layered additive manufacturing (LAM) is a powder-based AM technique similar to SLS but uses a higher power laser to completely melt powder particles into a single solid body [119]. SLM is commonly preferred for metal powders and presents some significant benefits compared to SLS. Compared to thermoplastic melts, melted metals have much lower viscosity allowing formation of dense parts with isotropic mechanical properties. Re-melting of the underlayer when the upper layer is exposed to laser also allows alleviation of interlayer-binding issues and formation of a dense solid with minimal porosity using SLM with properties comparable to bulk metals [120]. The greatest advantage of SLM compared to other AM technologies is the possibility of microreactor fabrication in metals such as stainless steel [121], titanium [122] and aluminium [123] which are well known to be chemically inert and mechanically stable. SLM fabricated metallic constructs also exhibit minimal shrinkage and deforming issues. SLM allows printing of layers as thin as 0.02–0.1 mm with dimensional accuracy as low as 0.05 mm. However, similar to SLS, the laser spot size in SLM ranges between 0.3–0.7 mm which imposes a minimum thickness limit on component walls. SLM printed parts also tend to be rougher owing to unwanted adhesion of metal powdered particles and the stair stepping effect which might make them unfavorable in some applications. Capel et al. investigated fabrication of a metallic microreactor with the Renishaw AM250 consisting of a channel diameter of 3 mm and concluded that the undesired solidification of liquid metal into metallic chunks and removal of unwanted metal powdered particles from microreactor channels still remains a big issue for microreactor fabrication using SLM [99]. The use of a more advanced Realizer SLM 50 by the same group resolved these blockage issues and allowed successful development of a steel microreactor as shown in Fig. 8.21 with channel diameters ranging from 1–2 mm [99]. AM of complex geometrical shapes with good mechanical properties and capability to manufacture these in a wide range of material choices is also attractive for the energy storage and conversion industry where these is a large demand for high performance metal electrodes for batteries, supercapacitors, and fuel cells. Scotti et al. disclosed the use of SLM for fabrication of stainless steel (SS316L) micro fuel cells with complex flowfield plate designs containing enclosed cavities and channels as shown in Fig. 8.22 [124]. The use of stainless steel allowed fabrication of extremely durable, robust, chemically, and thermally stable micro fuel cells and the use of SLM for fabrication allowed rapid prototyping and performance testing of fuel cells with the capability to fabricate encapsulated micro channels which is not possible with conventional technologies. The best performing micro fuel cell exhibited a current density of 1.515 A cm^{-2} and maximum power density of 363 mW cm^{-2} [124]. The enhanced performance was assigned to efficient contact between the patterned flowfield plate and gas diffusion layer. Dawson et al. explored the use of SLM for fabrication of metallic bipolar plates for polymer electrolyte fuel cells (PEFCs) and compared its performance with plates manufactured from CNC machining [125]. A bipolar plate is used in PEFCs to provide stable and low

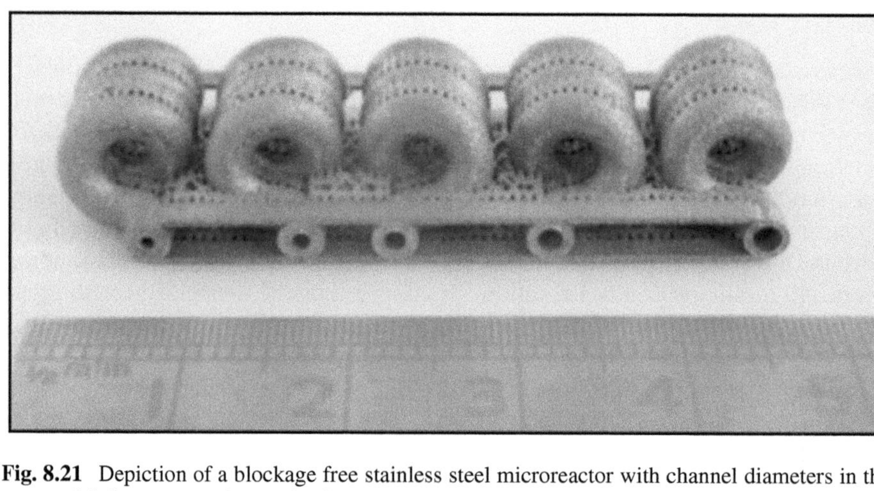

Fig. 8.21 Depiction of a blockage free stainless steel microreactor with channel diameters in the range of 1–2 mm manufactured using selective laser melting. Reproduced from Ref. [99] with permission from Royal Society of Chemistry

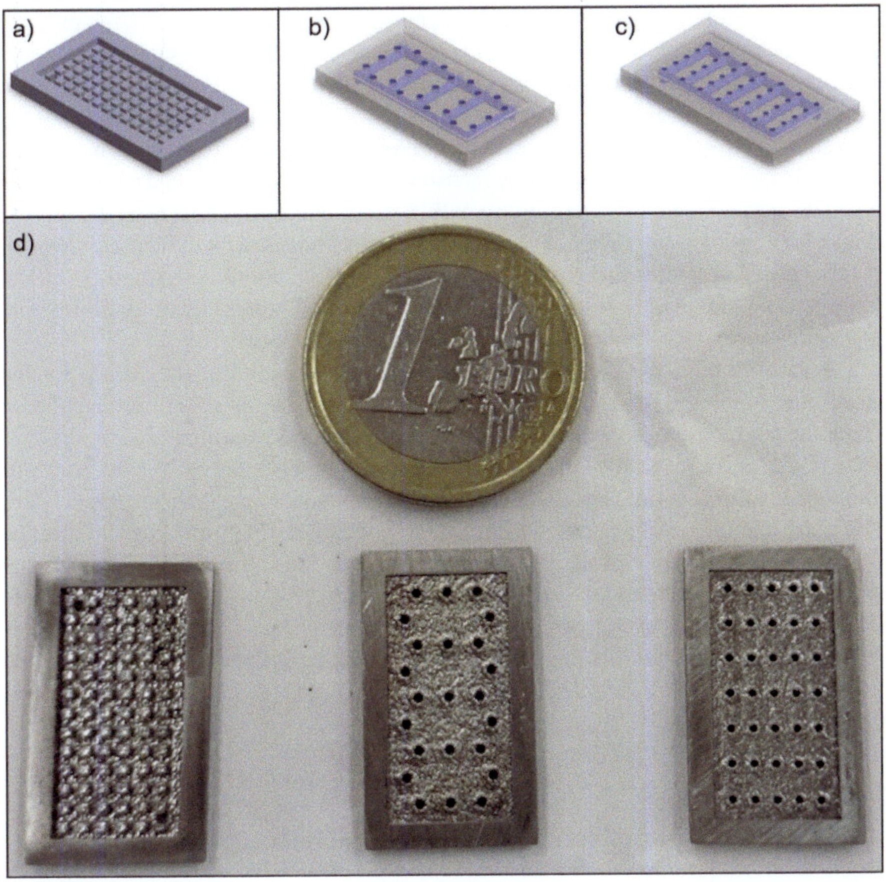

Fig. 8.22 Design of three stainless steel flowfield plates additively manufactured using selective laser melting for a micro fuel cell; **a** Type G; **b** Type E1; **c** Type E2; and **d** Images of the finished flowfield plates and comparison with a 1 Euro coin for reference. Reproduced from Ref. [124] with permission from Elsevier

resistance electrical connection between cells, to facilitate efficient and safe transfer of reactants and products to and from membrane electrode assemblies and to aid thermal management in stacks of fuel cells. Traditionally, bipolar plates are manufactured in carbon composites. However, recent research has seen the exploration of metallic bipolar plates for benefits related to cost, electrical performance, and size. The authors identified the performance of SLM printed bipolar plates to be comparable and even higher than the plates manufactured using CNC machining [125]. However, the additive manufacturing philosophy of SLM allowed other benefits such as rapid prototyping and performance testing along with large reduction in tooling costs during manufacturing. Calignano et al. extended the use of SLM towards fabrication of a low density porous anode network composed of AlSi10Mg alloy for microbial fuel cells that exhibited energy recovery as high as 3 kWh m^{-3} per day, sufficient enough to power sensors and lower power appliances in remote and harsh environments such as deep sea exploration, manned space missions, on-board boats, exploration of remote rivers and lakes and wastewater treatment with energy harvesting [126]. Microbial fuel cell technology utilizes microorganisms such as bacteria to digest organic or inorganic feedstocks such as pollutants and contaminants in wastewater, sludge, soil, or biomass while simultaneously generating electricity. The technology heavily relies on the development of adequate anodic materials that are also efficient bacteria carriers. For promotion of bacterial activity, it is critical to develop anodic material with high surface roughness, good biocompatibility and efficient electron transfer between bacteria and electrode surface.

The authors fabricated a cellular diamond lattice structure of AlSi10Mg for the anode material using the EOSINT M270Xt direct metal laser sintering machine on which the bacterial consortium was deposited using spray coating [126].

From the above examples it is evident that in its current state SLM has enormous potential for fabrication of milliscale microreactors and energy conversion devices with good chemical resistance, mechanical and heat transfer properties and thermal stability which makes it attractive for rapid uptake in the commercial and industrial scale processing and manufacturing of chemicals. Nevertheless, build resolution with SLM is still limited due to laser spot size and in its current state SLM cannot compete with conventional metal forming technologies like computer-numerically controlled machining (CNC) or electro-discharge machining (EDM) [99]. Future developments are anticipated to resolve these resolution issues.

8.4 Additive Manufacturing for Green Hydrogen

The impact of anthropogenic activities on our climate and environment is now well recognised and global efforts are being rallied in pursuit of new technologies that can facilitate responsible production and consumption. Rapid technological advancements are anticipated in the near future in the transportation and energy sectors which still remain as one of the major contributors towards carbon dioxide generation. In this regard global research efforts are ardently focused on development of technologies in two critical areas i.e., utilization of carbon dioxide as a feedstock for a circular carbon economy and large scale production of green hydrogen as an alternative fuel and reactant for consumption in transportation, chemicals, and energy sectors. The utilization of carbon dioxide as a feedstock also depends largely on availability of green hydrogen since most CO_2 utilization schemes employing CO_2 conversion rely on the use of hydrogen for reduction of CO_2 to useful fuels and chemicals [106]. In the current scenario, the cost of so-called green methanol produced from green hydrogen is 1.3–2.6 times higher than fossil-derived methanol, with 73% of the green methanol costs associated with obtaining the renewable H_2 [127]. With this state of economics for green hydrogen large uptake by relevant industrial sectors is not sustainable. Nevertheless, hydrogen is without any doubt the fuel of the future due to its high energy density and environmentally benign water as the main product of its combustion. Undergoing research efforts are anticipated to break this techno-economic barrier within the next 10–15 years [128]. Thus, production of green hydrogen and related research and development efforts are anticipated to rapidly expand within the decade. This surge will not only drive research efforts in innovation of materials and catalysts for hydrogen production but also fuel research into development of novel reactors and fuel cells for green hydrogen production, hydrogen detection and control instruments, process instrumentation, automation, hydrogen storage and transportation sectors. Thus, additive manufacturing with its ability to manufacture with a wide choice of materials and immense capacity for customization is bound to play an important role in the development and rise of the green hydrogen economy through innovations in efficient high flow microreactors, catalysts, sensors, membranes, valves, storage, and transportation hardware.

Conventionally hydrogen (grey hydrogen) is produced through steam reforming of hydrocarbons or coal gasification. Both these processes are not sustainable in the long run owing to their large carbon footprint. Comparatively, production of green hydrogen mainly relies on the utilization of water as a feedstock through one of the available routes such as electrolysis of water, photocatalysis, photoelectrocatalysis, photobiological or photothermal hydrogen generation [129]. In addition, large scale green hydrogen production also needs to address other critical aspects such as cost-competitiveness with grey hydrogen, carbon footprint and process sustainability. These factors are supported by the use of sunlight as source of input energy in photocatalysis based routes (called solar hydrogen) or renewable electricity in the case of electrolysis (electrocatalysis) based routes. The use of sea water or wastewater for production of hydrogen through photocatalysis routes has also been investigated but usually leads to drop in production rates due to negative interference from unwanted ions.

Solar conversion efficiencies are usually compared in terms of solar-to-hydrogen (STH) or quantum efficiencies (QE). Solar driven water electrolysis (photovoltaic electrolysis PV-E) currently exhibits the highest STH efficiencies but are associated with high costs due to the sophistication and high cost of electrolyzers. Similarly, photoelectrochemical cells show unfavourable economics compared to grey hydrogen. As a result the relatively simple and cost-effective photocatalytic routes for green hydrogen production through solar water splitting have been abundantly studied since the discovery of the Honda-Fujishima effect [130, 131]. To put this technology to practical use, it is however important to not only develop photocatalytic materials that efficiently split water under sunlight, but also develop novel and efficient microreactors that can be spread over a large area without sacrificing the performance of the water-splitting photocatalysts. Recently, safe operation of a pilot scale 100 m^2 array of solar water splitting panel reactors was demonstrated which exhibited STH efficiencies around 0.5% with autonomous hydrogen collection using commercial polyimide membranes [132]. The panel reactor as shown in Fig. 8.23 employed an aluminum doped strontium titanate catalyst and is the largest reported solar hydrogen production unit with demonstrated continuous operation over several

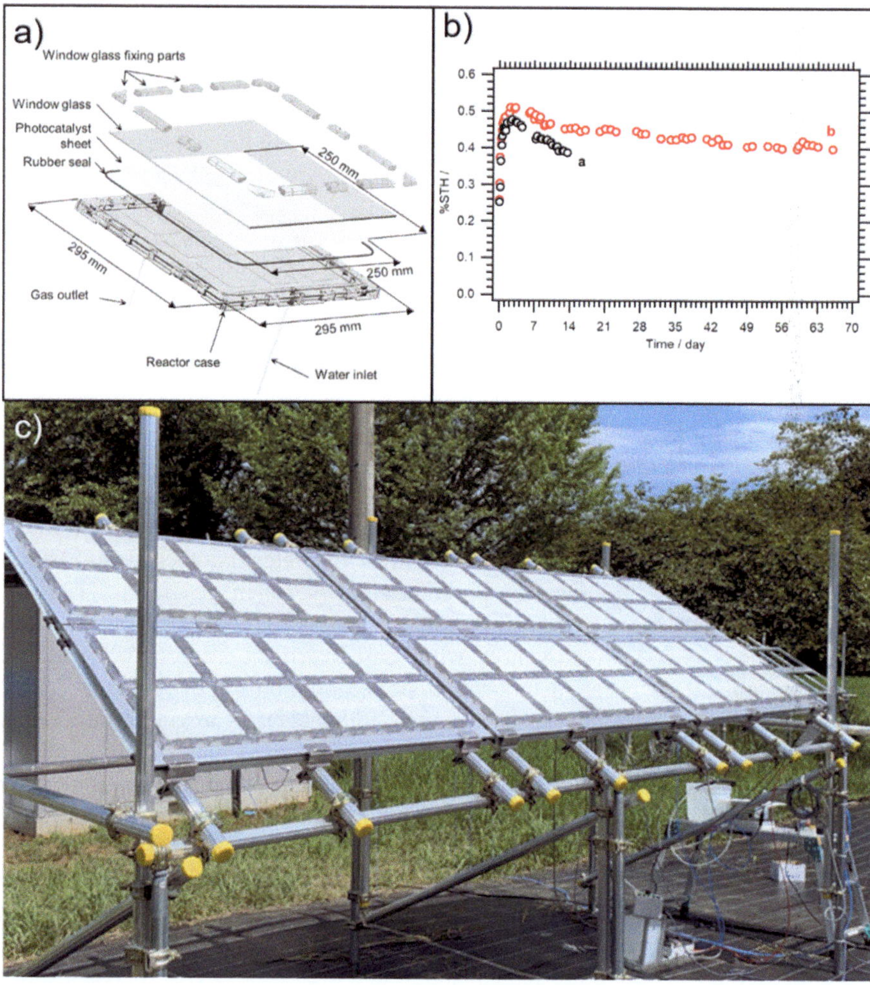

Fig. 8.23 Illustration of an array of photocatalytic reactor panels employing sunlight and a catalytic glass sheet functionalised with aluminium doped strontium titanate photocatalyst; **a** Design of a single panel and its components; **b** STH efficiencies and stability of hydrogen production recorded for the 100 m^2 panel array over a 70-day period; and **c** Image of the array panel reactor under exposure to natural sunlight. Reproduced with permission from Ref. [132]

months [132]. It is necessary to house the photocatalytic sheet catalyst in a reactor so that the oxyhydrogen gas generated by the water splitting reaction can be recovered. The panel reactor design allows efficient collection of both direct and diffused sunlight and presents a simple yet efficient solution for sunlight collection. In addition, panel reactors are easier to handle and cheaper to scale-up. The gap between the window and the photocatalyst sheet is only 0.1 mm and is filled with reactant water. When the photocatalyst sheet is housed in the panel reactor and illuminated with UV light, oxyhydrogen gas bubbles are emitted from the panel reactor despite the narrow headspace of 0.1 mm. The safety of the panel design is most notable since safety is one of the most critical elements in design of hydrogen producing reactors. As demonstrated by the group, oxyhydrogen gas is highly explosive, but the product oxyhydrogen gas confined in the narrow gap cannot explode even if it is intentionally ignited [132]. This achievement marks an important milestone in the production of green hydrogen. The research and developments efforts in green hydrogen, however, still have a long way to go because STH efficiencies near 10% are needed for green hydrogen to be competitive with grey hydrogen. The customization capabilities and design freedom enabled by additive manufacturing can greatly aid the green hydrogen revolution. Microreactors are especially suited in handling hydrogen reactions because of laminar flow and minimal mixing in microchannels which improves safety.

Besides photocatalytic water splitting, additive manufacturing can also relieve the technological restraints and improve cost effectiveness in design of solar photovoltaic electrolyzers and photoelectrochemical cells. For instance, Ambrosi et al. demonstrated a customized design of an electrolyzer for the overall water splitting reaction fabricated using selective laser sintering and fused deposition modelling [133]. SLS was used to fabricate the SS316 mesh electrodes which were subsequently functionalized by Ni-MoS_2 for the hydrogen evolution reaction and NiFe double hydroxide for the oxygen evolution reaction. The electrolyzer body and assembly components were designed in polylactic using FDM. The design flexibility enabled by additive manufacturing allows rapid testing and performance improvement in the lab scale and allows fabrication and testing of sophisticated components at low cost in the pilot scale.

Similarly, Pumera demonstrated the use of custom designed electrodes for bioanalytical applications where surface functionalization of metallic electrodes was used to improve the activity while SLS was used to fabricate the electrode body [134]. Gusmao et al. demonstrated the use of a commercial MoS_2 based lubricant spray in functionalization of a circular electrode additively manufactured using FDM of a graphene-PLA filament and evaluated its performance in the hydrogen evolution reaction [135]. Direct ink writing or robocasting is another popular additive manufacturing method which has been frequently studied for preparation of both electroactive and photoactive materials some examples of which have been elaborated in Chap. 5 of this book. In this field the main focus remains on the optimization of catalyst properties by modification of the composition of the precursors used for robocasting while subsequently maintaining high catalytic activity, uniformity of particle size, improved mass transfer, reduced recombination of holes and photons and optimum exposure to light radiation [137, 138]. Castedo et al. demonstrated an indirect application of additive manufacturing wherein PLA printed microchannels reactor shapes were transferred into PDMS using soft lithography and obtained microreactors were investigated for the photocatalytic production of hydrogen using Au/TiO_2 catalysts [139, 140].

In addition to the conventional approaches of green hydrogen production based on water electrocatalysis and photocatalysis, the scientific community has also investigated the use of concentrated solar power and photo thermal approaches for green hydrogen production. For instance, Camus et al. demonstrated an interesting use of an additively manufactured SS316 microreactor for production of green hydrogen by steam reforming of methane using the solar photothermal approach [136]. The reactor was fabricated using selective laser sintering and consisted of input chambers, an integrated heat exchanging micro channel that allowed to heat the input liquids and cool the outlet gases from the reactor section and a reactor section as shown in Fig. 8.24. Temperatures as high as 840 °C could be achieved using an absorptive black coating on the reactor and a parabolic reflector allowing up to 100% methane conversion under the solar radiation of 240 suns. On the other hand, Chiefari et al. argued in the favor of additively manufactured catalyst supports for green hydrogen

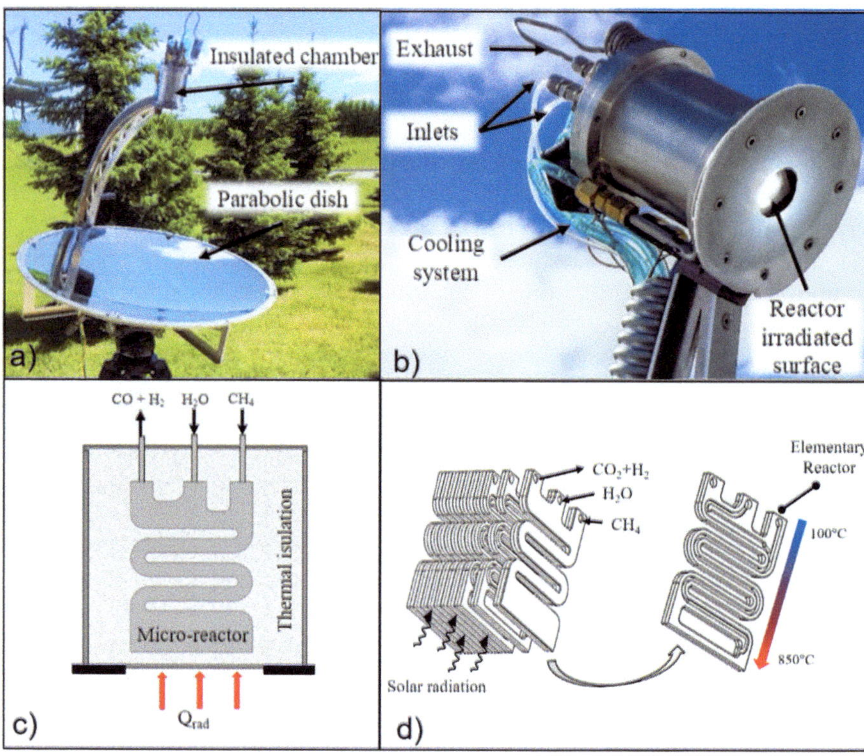

Fig. 8.24 Illustration of a solar photothermal methane reforming reactor for green hydrogen production fabricated using SLM; **a** Overall construction and testing of the photothermal reactor; **b** Reactor section of the device consisting of inlets, outlets, exhaust, cooling system and radiation receiving window; **c** Design of the additively manufactured device depicting inlet, heat exchange and reactor sections; and **d** Functioning of the reforming device with heat flow. Reproduced with permission from Ref. [136]

economy by reforming of liquid hydrogen carriers such as ammonia, methanol, formic acid, methylcyclohexane and perhydrodibenzotoluene [141]. Such production relies on efficient and cyclic catalysis of above chemicals to derive hydrogen and re-hydrogenation for a circular hydrogen economy. The reports clearly highlight the importance of technological improvements in reactor design in terms of sunlight harvesting, microchannel design, hydrogen membranes and hydrogen detection equipment for process control and safety. However, the utilization of additive manufacturing in the green hydrogen sector has only started to emerge and rapid developments are anticipated in the coming years.

8.5 Summary and Outlook

The challenges faced by our modern technological society in the twenty-first century have put extraneous pressure on the existing chemical industry and demanding its systemic reinvigoration. The industry in turn is looking towards wide adoption of modern industrial principles such as automation and digital process control, smart manufacturing using Industry 4.0 principles and process intensification. To meet these demands of the chemical industry, conventional processing equipment and reactors with generic designs are gradually getting replaced by advanced microreactors with sophisticated internal geometries that exhibit high performance, process efficiencies and optimum mass and heat transport. However, conventional manufacturing technologies either fail or struggle in terms of processing cost or manufacturing time for the needed reactor designs. Thus, there is an urgent and critical need to integrate additive manufacturing with the chemical industry which allows rapid prototyping and performance testing, decentralized production, economic viability, and small turnover times in fabrication of microreactors. Adoption of micro reaction technology can potentially alleviate the challenges of modern chemical industry. Reaction rates and process efficiencies can be significantly improved by employing novel geometries that reduce or disrupt boundary layers and increase surface areas. Additive manufacturing can be synergistically employed in conjunction with theoretical and experimental studies for the development of efficient chemical microreactors. Moreover, it acts as a bridge between chemical engineering and applied chemistry by realization of optimal designs derived from in-depth hydrodynamic and transport studies. Additive manufacturing for active reactionware has enormous potential in sustainable, timely, decentralised, and customised manufacturing of healthcare and chemical products. User involvement with the entire design and manufacturing process allows precise control and rapid optimization of reactor dimensions, geometry, and functionality. This level of design freedom is not currently available when using existing microreactor fabrication methods. Additional benefits of AM procedures such as an increased speed of production and reduced manufacturing costs, show that AM processes have significant scope to improve on the existing portfolio of microreactor fabrication processes. Recent research has seen development of a range of complex 3D geometries that can be applied to flow chemistry applications.

AM is currently a rapidly evolving field of research with continuous development of new technologies, systems, and materials. It is anticipated that additive manufacturing will see an increased uptake by the chemical industry through integration of milli-to-micro scale reactor systems in existing production processes.

8.6 Test Your Knowledge

After reading the above chapter you should be able to answer the following questions:

1. *What are the three major drivers that are demanding systemic and technological innovations in existing chemical processes and unit operations?*
2. *What is the fourth Industrial revolution or Industry 4.0 and what are its main indicators?*
3. *What is process intensification and what are some benefits associated with its implementation?*
4. *What are the critical issues which must be paid special attention for full realization of the potential of process intensification?*
5. *What are the five fundamental physical phenomenon that dominate the performance of chemical processes and equipment?*
6. *What is the biggest driver behind miniaturization of chemical processing equipment?*
7. *What are some of the benefits associated with micro reactors?*
8. *What are some limitations of microreactors that are hindering their rapid uptake by the industrial sector?*
9. *Which microscopic forces govern mixing performance in microreactors and micromixers?*
10. *What are some critical design elements which govern the mixing performance in microreactors and micromixers?*
11. *Which reaction or process parameters are directly or indirectly related to mixing efficiency?*
12. *Which dimensionless numbers and quantities become important to define flow regimes in microchannels?*
13. *What are the different types of flow regimes observed in microchannels?*
14. *What is the importance of flow obstruction geometry in microchannels?*
15. *What are the major hurdles in the use of fused deposition modelling and stereolithography for microreactor design and how can this hurdle be circumvented?*
16. *What are the major hurdles in the use of selective laser sintering and selective laser melting for microreactor design and how can this hurdle be circumvented?*
17. *What is the common measure used for comparison of efficiencies in green hydrogen production technologies?*

References

1. Rose J, Lukic V, Milon T, Cappuzzo A (2016) Sprinting to value in industry 4.0: perspective from and implications for US manufacturers. Boston Consulting Group
2. Gerven TV, Stankiewicz A (2009) Structure, energy, synergy, time—the fundamentals of process intensification. Ind Eng Chem Res 48:2465–2474
3. Dautzenberg FM, Mukherjee M (2001) Process intensification using multifunctional reactors. Chem Eng Sci 56:251–267
4. Huang K, Wang S-J, Shan L, Zhu Q, Qian J (2007) Seeking synergistic effect—a key principle in process intensification. Sep Purif Technol 57(1):111–120. https://doi.org/10.1016/j.seppur.2007.03.022
5. SM blank reference
6. Zhakeyev A, Wang P, Zhang L, Shu W, Wang H, Xuan J (2017) Additive manufacturing: unlocking the evolution of energy materials. Adv Sci 4(10):1700187. https://doi.org/10.1002/advs.201700187
7. Parra-Cabrera C, Achille C, Kuhn S, Ameloot R (2018) 3D printing in chemical engineering and catalytic technology: structured catalysts, mixers and reactors. Chem Soc Rev 47(1):209–230. https://doi.org/10.1039/c7cs00631d
8. Jähnisch K, Hessel V, Löwe H, Baerns M (2004) Chemistry in microstructured reactors. Angew Chem Int Ed 43(4):406–446. https://doi.org/10.1002/anie.200300577
9. Kockmann N (2015) Modulare chemische Reaktoren für die Prozessentwicklung und Produktion in kontinuierlichen Mehrzweckanlagen. Chem Ing Tech 87(9):1173–1184. https://doi.org/10.1002/cite.201500028
10. Kulkarni AA (2014) Continuous flow nitration in miniaturized devices. Beilstein J Org Chem 10:405–424. https://doi.org/10.3762/bjoc.10.38
11. Mastronardi F, Gutmann B, Kappe CO (2013) Continuous flow generation and reactions of anhydrous diazomethane using a teflon AF-2400 tube-in-tube reactor. Org Lett 15(21):5590–5593
12. Schwalbe T, Autze V, Wille G (2002) Chemical synthesis in microreactors. Chimia 56:636–646
13. Cao E, Zuburtikudis I, Al-Rifai N, Roydhouse M, Gavriilidis A (2014) Enhanced performance of oxidation of rosalva (9-decen-1-ol) to costenal (9-decenal) on porous silicon-supported silver catalyst in a microstructured reactor. Processes 2(1):141–157. https://doi.org/10.3390/pr2010141
14. Kundu A, Ahn JE, Park S-S, Shul YG, Han HS (2008) Process intensification by microchannel reactor for steam reforming of methanol. Chem Eng J 135(1–2):113–119. https://doi.org/10.1016/j.cej.2007.02.026
15. Hu J, Wang Y, Cao C, Elliott DC, Stevens DJ, White JF (2005) Conversion of biomass syngas to DME using a microchannel reactor. Ind Eng Chem Res 44:1722–1727
16. Liguori F, Barbaro P (2014) Green semi-hydrogenation of alkynes by Pd@borate monolith catalysts under continuous flow. J Catal 311:212–220. https://doi.org/10.1016/j.jcat.2013.11.027
17. Bravo J, Karim A, Conant T, Lopez GP, Datye A (2004) Wall coating of a $CuO/ZnO/Al_2O_3$ methanol steam reforming catalyst for micro-channel reformers. Chem Eng J 101(1–3):113–121. https://doi.org/10.1016/j.cej.2004.01.011
18. Watts P, Haswell SJ (2005) The application of micro reactors for organic synthesis. Chem Soc Rev 34(3):235–246. https://doi.org/10.1039/B313866F
19. Ober TJ, Foresti D, Lewis JA (2015) Active mixing of complex fluids at the microscale. Proc Natl Acad Sci USA 112(40):12293–12298. https://doi.org/10.1073/pnas.1509224112
20. Bakhtiary-Davijany H, Hayer F, Phan XK, Myrstad R, Venvik HJ, Pfeifer P, Holmen A (2011) Characteristics of an Integrated micro packed bed reactor-heat exchanger for methanol synthesis from syngas. Chem Eng J 167(2–3):496–503. https://doi.org/10.1016/j.cej.2010.08.074

21. Tanimu A, Jaenicke S, Alhooshani K (2017) Heterogeneous catalysis in continuous flow microreactors: a review of methods and applications. Chem Eng J 327:792–821. https://doi.org/10.1016/j.cej.2017.06.161
22. Charpentier JC (2005) Process intensification by miniaturization. Chem Eng Technol 28(3):255–258. https://doi.org/10.1002/ceat.200407026
23. Watts P, Wiles C (2007) Recent advances in synthetic micro reaction technology. Chem Commun 5:443–467. https://doi.org/10.1039/B609428G
24. Rahman M, Rebrov E (2014) Microreactors for gold nanoparticles synthesis: from faraday to flow. Processes 2(2):466–493. https://doi.org/10.3390/pr2020466
25. Zhang B (2016) The influence of wall roughness on detonation limits in hydrogen–oxygen mixture. Combust Flame 169:333–339. https://doi.org/10.1016/j.combustflame.2016.05.003
26. Zhang B, Liu H, Yan B (2019) Investigation on the detonation propagation limit criterion for methane-oxygen mixtures in tubes with different scales. Fuel 239:617–622. https://doi.org/10.1016/j.fuel.2018.11.062
27. Gao Y, Zhang B, Ng HD, Lee JHS (2016) An experimental investigation of detonation limits in hydrogen–oxygen–argon mixtures. Int J Hydrogen Energy 41(14):6076–6083. https://doi.org/10.1016/j.ijhydene.2016.02.130
28. Teng H, Ng HD, Jiang Z (2017) Initiation characteristics of wedge-induced oblique detonation waves in a stoichiometric hydrogen-air mixture. Proc Combust Inst 36(2):2735–2742. https://doi.org/10.1016/j.proci.2016.09.025
29. Alshammari YM, Hellgardt K (2017) CFD analysis of hydrothermal conversion of heavy oil in continuous flow reactor. Chem Eng Res Des 117:250–264. https://doi.org/10.1016/j.cherd.2016.10.002
30. Della Torre A, Lucci F, Montenegro G, Onorati A, Dimopoulos Eggenschwiler P, Tronconi E, Groppi G (2016) CFD modeling of catalytic reactions in open-cell foam substrates. Comput Chem Eng 92:55–63. https://doi.org/10.1016/j.compchemeng.2016.04.031
31. Li K, Valla J, Garcia-Martinez J (2014) Realizing the commercial potential of hierarchical zeolites: new opportunities in catalytic cracking. ChemCatChem 6(1):46–66. https://doi.org/10.1002/cctc.201300345
32. Zhou L, Zhang K, Hu Z, Tao Z, Mai L, Kang Y-M, Chou S-L, Chen J (2018) Recent developments on and prospects for electrode materials with hierarchical structures for lithium-ion batteries. Adv Energy Mater 8(6):1701415. https://doi.org/10.1002/aenm.201701415
33. Schwieger W, Machoke AG, Weissenberger T, Inayat A, Selvam T, Klumpp M, Inayat A (2016) Hierarchy concepts: classification and preparation strategies for zeolite containing materials with hierarchical porosity. Chem Soc Rev 45(12):3353–3376. https://doi.org/10.1039/C5CS00599J
34. Roth WJ, Gil B, Makowski W, Marszalek B, Eliášová P (2016) Layer like porous materials with hierarchical structure. Chem Soc Rev 45(12):3400–3438. https://doi.org/10.1039/C5CS00508F
35. Grundemann L, Scholl S (2014) Ecological and economic assessment of micro-/milli-continuous campaign manufacturing: the case of writing ink. Processes 2(1):238–264. https://doi.org/10.3390/pr2010238
36. Karakaya M, Avci AK (2009) Simulation of on-board fuel conversion in catalytic microchannel reactor-heat exchanger systems. Top Catal 52(13–20):2112–2116. https://doi.org/10.1007/s11244-009-9383-1
37. Munera Parra A, Antweiler N, Nagpal R, Agar D (2014) Stability analysis of reactive multiphase slug flows in microchannels. Processes 2(2):371–391. https://doi.org/10.3390/pr2020371
38. Mehla S, Das J, Jampaiah D, Periasamy S, Nafady A, Bhargava SK (2019) Recent advances in preparation methods for catalytic thin films and coatings. Catal Sci Technol 9:3582–3602. https://doi.org/10.1039/C9CY00518H
39. Phan XK, Bakhtiary-Davijany H, Myrstad R, Pfeifer P, Venvik HJ, Holmen A (2011) Preparation and performance of Cu-based monoliths for methanol synthesis. Appl Catal A 405(1–2):1–7. https://doi.org/10.1016/j.apcata.2011.07.005

40. Mehla S, Krishnamurthy KR, Viswanathan B, John M, Niwate Y, Kishore Kumar SA, Pai SM, Newalkar BL (2013) n-Hexadecane hydroisomerization over BTMACl/TEABr/MTEABr templated ZSM-12. Microporous Mesoporous Mater 177:120–126. https://doi.org/10.1016/j.micromeso.2013.05.001
41. Mehla S, Krishsna V, Sriganesh G, Ravishankar R (2018) Mesoporous superacid catalysts for valorisation of refinery naphtha stream. RSC Adv 8(59):33702–33709. https://doi.org/10.1039/C8RA07024E
42. Ganley J, Riechmann K, Seebauer E, Masel R (2004) Porous anodic alumina optimized as a catalyst support for microreactors. J Catal 227(1):26–32. https://doi.org/10.1016/j.jcat.2004.06.016
43. Moharir AS, Shah SS, Gudi RD, Devereux BM, Bussche KV, Venimadhavan G (2001) Generalized reactor model: an object oriented approach to reactor modeling. In: Gani R, Jørgensen SB (eds) Computer aided chemical engineering, vol 9. Elsevier, pp 243–248. https://doi.org/10.1016/S1570-7946(01)80036-5
44. Mandaliya DD, Moharir AS, Gudi RD (2012) Generic reactor modeling and simulation framework for industrial reactors. In: Paper presented at the AIChE annual conference. Pittsburgh, PA, USA
45. Mandaliya DD (2016) Generalized model based analysis approaches for chemical reactors. Indian Institute of Technology Bombay, Mumbai
46. Mandaliya DD, Gudi RD (2021) Multi-scale analysis of chemical reactors using green's function method. In: Paper presented at the AIChE annual conference. Boston, MA, USA
47. Wei Q, Li H, Liu G, He Y, Wang Y, Tan YE, Wang D, Peng X, Yang G, Tsubaki N (2020) Metal 3D printing technology for functional integration of catalytic system. Nat Commun 11(1):4098. https://doi.org/10.1038/s41467-020-17941-8
48. Kuipers JAM, van Swaaij WPM (1998) Computational fluid dynamics applied to chemical reaction engineering. In: Advances in chemical engineering, vol 24. Advances in chemical engineering, pp 227–328. https://doi.org/10.1016/s0065-2377(08)60094-0
49. Salciccioli M, Stamatakis M, Caratzoulas S, Vlachos DG (2011) A review of multiscale modeling of metal-catalyzed reactions: mechanism development for complexity and emergent behavior. Chem Eng Sci 66(19):4319–4355. https://doi.org/10.1016/j.ces.2011.05.050
50. Charpentier J-C (2010) Among the trends for a modern chemical engineering, the third paradigm: the time and length multiscale approach as an efficient tool for process intensification and product design and engineering. Chem Eng Res Des 88(3):248–254. https://doi.org/10.1016/j.cherd.2009.03.008
51. Caygill G, Zanfir M, Gavriilidis A (2006) Scalable reactor design for pharmaceuticals and fine chemicals production. 1: potential scale-up obstacles. Organic Process Research & Development
52. Elvira KS, Casadevall i Solvas X, Wootton RC, deMello AJ (2013) The past, present and potential for microfluidic reactor technology in chemical synthesis. Nat Chem 5(11):905–915. https://doi.org/10.1038/nchem.1753
53. Gutmann B, Cantillo D, Kappe CO (2015) Continuous-flow technology-a tool for the safe manufacturing of active pharmaceutical ingredients. Angew Chem Int Ed Engl 54(23):6688–6728. https://doi.org/10.1002/anie.201409318
54. Hwang G-J, Onuki K (2001) Simulation study on the catalytic decomposition of hydrogen iodide in a membrane reactor with a silica membrane for the thermochemical water-splitting IS process. J Membr Sci 194:207–215
55. Salman BH, Mohammed HA, Munisamy KM, Kherbeet AS (2013) Characteristics of heat transfer and fluid flow in microtube and microchannel using conventional fluids and nanofluids: a review. Renew Sustain Energy Rev 28:848–880. https://doi.org/10.1016/j.rser.2013.08.012
56. Liu RH, Stremler MA, Sharp KV, Olsen MG, Santiago JG, Adrian RJ, Aref H, Beebe DJ (2000) Passive mixing in a three-dimensional serpentine microchannel. J Microelectromech Syst 9(2):190–197

57. Hossain S, Ansari MA, Kim K-Y (2009) Evaluation of the mixing performance of three passive micromixers. Chem Eng J 150(2–3):492–501. https://doi.org/10.1016/j.cej.2009.02.033
58. Nema G, Garimella S, Fronk BM (2014) Flow regime transitions during condensation in microchannels. Int J Refrig 40:227–240. https://doi.org/10.1016/j.ijrefrig.2013.11.018
59. An H, Li A, Sasmito AP, Kurnia JC, Jangam SV, Mujumdar AS (2012) Computational fluid dynamics (CFD) analysis of micro-reactor performance: effect of various configurations. Chem Eng Sci 75:85–95. https://doi.org/10.1016/j.ces.2012.03.004
60. Zhao S, Wang W, Zhang M, Shao T, Jin Y, Cheng Y (2012) Three-dimensional simulation of mixing performance inside droplets in micro-channels by Lattice Boltzmann method. Chem Eng J 207–208:267–277. https://doi.org/10.1016/j.cej.2012.06.098
61. Zhang J, Wang K, Teixeira AR, Jensen KF, Luo G (2017) Design and scaling up of microchemical systems: a review. Annu Rev Chem Biomol Eng 8:285–305. https://doi.org/10.1146/annurev-chembioeng-060816-101443
62. Hessel V, Löwe H, Schönfeld F (2005) Micromixers—a review on passive and active mixing principles. Chem Eng Sci 60(8–9):2479–2501. https://doi.org/10.1016/j.ces.2004.11.033
63. Nguyen NT, Huang X (2005) Mixing in microchannels based on hydrodynamic focusing and time-interleaved segmentation: modelling and experiment. Lab Chip 5(11):1320–1326. https://doi.org/10.1039/b507548c
64. Wang K, Xie L, Lu Y, Luo G (2013) Generating microbubbles in a co-flowing microfluidic device. Chem Eng Sci 100:486–495. https://doi.org/10.1016/j.ces.2013.02.021
65. Panić S, Loebbecke S, Tuercke T, Antes J, Bošković D (2004) Experimental approaches to a better understanding of mixing performance of microfluidic devices. Chem Eng J 101(1–3):409–419. https://doi.org/10.1016/j.cej.2003.10.026
66. Kockmann N, Gottsponer M, Roberge DM (2011) Scale-up concept of single-channel microreactors from process development to industrial production. Chem Eng J 167(2–3):718–726. https://doi.org/10.1016/j.cej.2010.08.089
67. Kockmann N, Kiefer T, Engler M, Woias P (2006) Convective mixing and chemical reactions in microchannels with high flow rates. Sens Actuators B Chem 117(2):495–508. https://doi.org/10.1016/j.snb.2006.01.004
68. Faridkhou A, Larachi F (2014) Two-phase flow hydrodynamic study in micro-packed beds—effect of bed geometry and particle size. Chem Eng Process 78:27–36. https://doi.org/10.1016/j.cep.2014.02.005
69. Dong Z, Yao C, Zhang Y, Chen G, Yuan Q, Xu J (2016) Hydrodynamics and mass transfer of oscillating gas-liquid flow in ultrasonic microreactors. AIChE J 62(4):1294–1307. https://doi.org/10.1002/aic.15091
70. Jung JH, Destgeer G, Ha B, Park J, Sung HJ (2016) On-demand droplet splitting using surface acoustic waves. Lab Chip 16(17):3235–3243. https://doi.org/10.1039/c6lc00648e
71. Abolhasani M, Coley CW, Xie L, Chen O, Bawendi MG, Jensen KF (2015) Oscillatory microprocessor for growth and in situ characterization of semiconductor nanocrystals. Chem Mater 27(17):6131–6138. https://doi.org/10.1021/acs.chemmater.5b02821
72. Oddy MH, Santiago JG, Mikkelsen JC (2001) Electrokinetic instability micromixing. Anal Chem 73:5822–5832
73. Chung Y-C, Wu C-M, Lin S-H (2016) Particles sorting in micro channel using designed micro electromagnets of magnetic field gradient. J Magn Magn Mater 407:209–217. https://doi.org/10.1016/j.jmmm.2016.01.075
74. Sochol RD, Sweet E, Glick CC, Venkatesh S, Avetisyan A, Ekman KF, Raulinaitis A, Tsai A, Wienkers A, Korner K, Hanson K, Long A, Hightower BJ, Slatton G, Burnett DC, Massey TL, Iwai K, Lee LP, Pister KS, Lin L (2016) 3D printed microfluidic circuitry via multijet-based additive manufacturing. Lab Chip 16(4):668–678. https://doi.org/10.1039/c5lc01389e
75. Gong H, Woolley AT, Nordin GP (2016) High density 3D printed microfluidic valves, pumps, and multiplexers. Lab Chip 16(13):2450–2458. https://doi.org/10.1039/c6lc00565a
76. Coleman JW, Garimella S (1999) Characterization of two-phase flow patterns in small diameter round and rectangular tubes. Int J Heat Mass Transf 42:2869–2881

77. Garimella S, Fronk BM (2013) Single- and multi-constituent condensation of fluids and mixtures with varying properties in micro-channels. Exp Heat Transf 26(2–3):129–168. https://doi.org/10.1080/08916152.2012.736841
78. Soleymani A, Yousefi H, Turunen I (2008) Dimensionless number for identification of flow patterns inside a T-micromixer. Chem Eng Sci 63(21):5291–5297. https://doi.org/10.1016/j.ces.2008.07.002
79. Reddy Cherlo SK, Pushpavanam S (2010) Effect of depth on onset of engulfment in rectangular micro-channels. Chem Eng Sci 65(24):6486–6490. https://doi.org/10.1016/j.ces.2010.08.025
80. Fani A, Camarri S, Salvetti MV (2014) Unsteady asymmetric engulfment regime in a T-mixer. Phys Fluids 26(7). https://doi.org/10.1063/1.4885451
81. Bello-Ochende T, Liebenberg L, Meyer JP (2007) Constructal cooling channels for micro-channel heat sinks. Int J Heat Mass Transf 50(21–22):4141–4150. https://doi.org/10.1016/j.ijheatmasstransfer.2007.02.019
82. Salimpour MR, Sharifhasan M, Shirani E (2011) Constructal optimization of the geometry of an array of micro-channels. Int Commun Heat Mass Transfer 38(1):93–99. https://doi.org/10.1016/j.icheatmasstransfer.2010.10.008
83. Kim S-M, Mudawar I (2010) Analytical heat diffusion models for heat sinks with circular micro-channels. Int J Heat Mass Transf 53(21–22):4552–4566. https://doi.org/10.1016/j.ijheatmasstransfer.2010.06.038
84. Leroy F, Muller-Plathe F (2010) Solid-liquid surface free energy of Lennard-Jones liquid on smooth and rough surfaces computed by molecular dynamics using the phantom-wall method. J Chem Phys 133(4):044110. https://doi.org/10.1063/1.3458796
85. Leung WW-F, Ren Y (2013) Crossflow and mixing in obstructed and width-constricted rotating radial microchannel. Int J Heat Mass Transf 64:457–467. https://doi.org/10.1016/j.ijheatmasstransfer.2013.04.064
86. Naher S, Orpen D, Brabazon D, Poulsen CR, Morshed MM (2011) Effect of micro-channel geometry on fluid flow and mixing. Simul Model Pract Theory 19(4):1088–1095. https://doi.org/10.1016/j.simpat.2010.12.008
87. Hsiao K-Y, Wu C-Y, Huang Y-T (2014) Fluid mixing in a microchannel with longitudinal vortex generators. Chem Eng J 235:27–36. https://doi.org/10.1016/j.cej.2013.09.010
88. Kim BS, Kwak BS, Shin S, Lee S, Kim KM, Jung H-I, Cho HH (2011) Optimization of microscale vortex generators in a microchannel using advanced response surface method. Int J Heat Mass Transf 54(1–3):118–125. https://doi.org/10.1016/j.ijheatmasstransfer.2010.09.061
89. Aubin J, Fletcher DF, Xuereb C (2005) Design of micromixers using CFD modelling. Chem Eng Sci 60(8–9):2503–2516. https://doi.org/10.1016/j.ces.2004.11.043
90. Tsai RT, Wu CY (2011) An efficient micromixer based on multidirectional vortices due to baffles and channel curvature. Biomicrofluidics 5(1):14103. https://doi.org/10.1063/1.3552992
91. Jeon W, Shin CB (2009) Design and simulation of passive mixing in microfluidic systems with geometric variations. Chem Eng J 152(2–3):575–582. https://doi.org/10.1016/j.cej.2009.05.035
92. Chuan L, Wang X-D, Wang T-H, Yan W-M (2015) Fluid flow and heat transfer in microchannel heat sink based on porous fin design concept. Int Commun Heat Mass Transfer 65:52–57. https://doi.org/10.1016/j.icheatmasstransfer.2015.04.005
93. Gumuslu G, Avci AK (2012) Parametric analysis of Fischer-tropsch synthesis in a catalytic microchannel reactor. AIChE J 58(1):227–235. https://doi.org/10.1002/aic.12558
94. Kitson PJ, Symes MD, Dragone V, Cronin L (2013) Combining 3D printing and liquid handling to produce user-friendly reactionware for chemical synthesis and purification. Chem Sci 4(8):3099–3103. https://doi.org/10.1039/c3sc51253c
95. Kitson PJ, Glatzel S, Chen W, Lin CG, Song YF, Cronin L (2016) 3D printing of versatile reactionware for chemical synthesis. Nat Protoc 11(5):920–936. https://doi.org/10.1038/nprot.2016.041

96. Kitson PJ, Marshall RJ, Long D, Forgan RS, Cronin L (2014) 3D printed high-throughput hydrothermal reactionware for discovery, optimization, and scale-up. Angew Chem Int Ed Engl 53(47):12723–12728. https://doi.org/10.1002/anie.201402654
97. Kitson PJ, Rosnes MH, Sans V, Dragone V, Cronin L (2012) Configurable 3D-printed millifluidic and microfluidic "lab on a chip" reactionware devices. Lab Chip 12(18):3267–3271. https://doi.org/10.1039/c2lc40761b
98. Rossi S, Porta R, Brenna D, Puglisi A, Benaglia M (2017) Stereoselective catalytic synthesis of active pharmaceutical ingredients in homemade 3D-printed mesoreactors. Angew Chem Int Ed Engl 56(15):4290–4294. https://doi.org/10.1002/anie.201612192
99. Capel AJ, Edmondson S, Christie SD, Goodridge RD, Bibb RJ, Thurstans M (2013) Design and additive manufacture for flow chemistry. Lab Chip 13(23):4583–4590. https://doi.org/10.1039/c3lc50844g
100. Symes MD, Kitson PJ, Yan J, Richmond CJ, Cooper GJ, Bowman RW, Vilbrandt T, Cronin L (2012) Integrated 3D-printed reactionware for chemical synthesis and analysis. Nat Chem 4(5):349–354. https://doi.org/10.1038/nchem.1313
101. Shan Z, Van Kooten WEJ, Oudshoorn OL, Jansen JC, Van Bekkum H, Van Bleek CM, Calis HPA (2000) Optimization of the preparation of binderless ZSM-5 coatings on stainless steel monoliths by in situ hydrothermal synthesis. Microporous Mesoporous Mater 34:81–91
102. Liu Y, Yang Z, Yu C, Gu X, Xu N (2011) Effect of seeding methods on growth of NaA zeolite membranes. Microporous Mesoporous Mater 143(2–3):348–356. https://doi.org/10.1016/j.micromeso.2011.03.016
103. Lefevere J, Mullens S, Meynen V (2018) The impact of formulation and 3D-printing on the catalytic properties of ZSM-5 zeolite. Chem Eng J 349:260–268. https://doi.org/10.1016/j.cej.2018.05.058
104. Konarova M, Aslam W, Ge L, Ma Q, Tang F, Rudolph V, Beltramini JN (2017) Enabling process intensification by 3D printing of catalytic structures. ChemCatChem 9(21):4132–4138. https://doi.org/10.1002/cctc.201700829
105. Zhou X, Liu C-J (2020) Three-dimensional printing of porous carbon structures with tailorable pore sizes. Catal Today 347:2–9. https://doi.org/10.1016/j.cattod.2018.05.044
106. Mehla S, Kandjani AE, Babarao R, Lee AF, Periasamy S, Wilson K, Ramakrishna S, Bhargava SK (2021) Porous crystalline frameworks for thermocatalytic CO_2 reduction: an emerging paradigm. Energy Environ Sci 14(1):320–352. https://doi.org/10.1039/d0ee01882a
107. Hurt C, Brandt M, Priya SS, Bhatelia T, Patel J, Selvakannan PR, Bhargava S (2017) Combining additive manufacturing and catalysis: a review. Catal Sci Technol 7(16):3421–3439. https://doi.org/10.1039/c7cy00615b
108. Lawson S, Li X, Thakkar H, Rownaghi AA, Rezaei F (2021) Recent advances in 3D printing of structured materials for adsorption and catalysis applications. Chem Rev 121(10):6246–6291. https://doi.org/10.1021/acs.chemrev.1c00060
109. Tumbleston JR, Shirvanyants D, Ermoshkin N, Janusziewicz R, Johnson AR, Kelly D, Chen K, Pinschmidt R, Rolland JP, Ermoshkin A, Samulski ET, DeSimone JM (2015) Continuous liquid interface production of 3D objects. Science 347(6228):1349–1352
110. Zhang X, Jiang XN, Sun C (1999) Micro-stereolithography of polymeric and ceramic microstructures. Sens Actuators 77:149–156
111. Chaparro-Garnica CY, Jorda-Faus P, Bailon-Garcia E, Ocampo-Perez R, Aguilar-Madera CG, Davo-Quinonero A, Lozano-Castello D, Bueno-Lopez A (2020) Customizable heterogeneous catalysts: nonchanneled advanced monolithic supports manufactured by 3D-printing for improved active phase coating performance. ACS Appl Mater Interfaces 12(49):54573–54584. https://doi.org/10.1021/acsami.0c14703
112. Knitter R, Liauw MA (2004) Ceramic microreactors for heterogeneously catalysed gas-phase reactions. Lab Chip 4(4):378–383. https://doi.org/10.1039/b403361b
113. Wendel B, Rietzel D, Kühnlein F, Feulner R, Hülder G, Schmachtenberg E (2008) Additive processing of polymers. Macromol Mater Eng 293(10):799–809. https://doi.org/10.1002/mame.200800121

114. Hutter C, Zenklusen A, Lang R, Rudolf von Rohr P (2011) Axial dispersion in metal foams and streamwise-periodic porous media. Chem Eng Sci 66(6):1132–1141. https://doi.org/10.1016/j.ces.2010.12.016
115. Potdar A, Protasova LN, Thomassen L, Kuhn S (2017) Designed porous milli-scale reactors with enhanced interfacial mass transfer in two-phase flows. React Chem & Eng 2(2):137–148. https://doi.org/10.1039/c6re00185h
116. Fonte CM, Leblebici ME, Dias MM, Lopes JCB (2013) The NETmix reactor: pressure drop measurements and 3D CFD modeling. Chem Eng Res Des 91(11):2250–2258. https://doi.org/10.1016/j.cherd.2013.07.014
117. Kenzari S, Bonina D, Marie Dubois J, Fournee V (2014) Complex metallic alloys as new materials for additive manufacturing. Sci Technol Adv Mater 15(2):024802. https://doi.org/10.1088/1468-6996/15/2/024802
118. Amorim FL, Lohrengel A, Neubert V, Higa CF, Czelusniak T (2014) Selective laser sintering of Mo-CuNi composite to be used as EDM electrode. Rapid Prototyp J 20(1):59–68. https://doi.org/10.1108/rpj-04-2012-0035
119. Rombouts M, Kruth JP, Froyen L, Mercelis P (2006) Fundamentals of selective laser melting of alloyed steel powders. CIRP Ann 55(1):187–192. https://doi.org/10.1016/s0007-8506(07)60395-3
120. Tolosa I, Garciandía F, Zubiri F, Zapirain F, Esnaola A (2010) Study of mechanical properties of AISI 316 stainless steel processed by "selective laser melting", following different manufacturing strategies. Int J Adv Manuf Technol 51(5–8):639–647. https://doi.org/10.1007/s00170-010-2631-5
121. Kruth JP, Froyen L, Van Vaerenbergh J, Mercelis P, Rombouts M, Lauwers B (2004) Selective laser melting of iron-based powder. J Mater Process Technol 149(1–3):616–622. https://doi.org/10.1016/j.jmatprotec.2003.11.051
122. Wang Y, Shen Y, Wang Z, Yang J, Liu N, Huang W (2010) Development of highly porous titanium scaffolds by selective laser melting. Mater Lett 64(6):674–676. https://doi.org/10.1016/j.matlet.2009.12.035
123. Louvis E, Fox P, Sutcliffe CJ (2011) Selective laser melting of aluminium components. J Mater Process Technol 211(2):275–284. https://doi.org/10.1016/j.jmatprotec.2010.09.019
124. Scotti G, Kanninen P, Matilainen V-P, Salminen A, Kallio T (2016) Stainless steel micro fuel cells with enclosed channels by laser additive manufacturing. Energy 106:475–481. https://doi.org/10.1016/j.energy.2016.03.086
125. Dawson RJ, Patel AJ, Rennie AEW, White S (2015) An investigation into the use of additive manufacture for the production of metallic bipolar plates for polymer electrolyte fuel cell stacks. J Appl Electrochem 45(7):637–645. https://doi.org/10.1007/s10800-015-0832-1
126. Calignano F, Tommasi T, Manfredi D, Chiolerio A (2015) Additive manufacturing of a microbial fuel cell–a detailed study. Sci Rep 5:17373. https://doi.org/10.1038/srep17373
127. González-Garay A, Frei MS, Al-Qahtani A, Mondelli C, Guillén-Gosálbez G, Pérez-Ramírez J (2019) Plant-to-planet analysis of CO_2-based methanol processes. Energy Environ Sci 12(12):3425–3436. https://doi.org/10.1039/c9ee01673b
128. Path to hydrogen competitiveness: a cost perspective (2020). Hydrogen Council
129. Liu G, Sheng Y, Ager JW, Kraft M, Xu R (2019) Research advances towards large-scale solar hydrogen production from water. EnergyChem 1(2). https://doi.org/10.1016/j.enchem.2019.100014
130. Maeda K, Domen K (2010) Photocatalytic water splitting: recent progress and future challenges. J Phys Chem Lett 1(18):2655–2661. https://doi.org/10.1021/jz1007966
131. Chen S, Takata T, Domen K (2017) Particulate photocatalysts for overall water splitting. Nat Rev Mater 2(10). https://doi.org/10.1038/natrevmats.2017.50
132. Nishiyama H, Yamada T, Nakabayashi M, Maehara Y, Yamaguchi M, Kuromiya Y, Nagatsuma Y, Tokudome H, Akiyama S, Watanabe T, Narushima R, Okunaka S, Shibata N, Takata T, Hisatomi T, Domen K (2021) Photocatalytic solar hydrogen production from water on a 100-m(2) scale. Nature 598(7880):304–307. https://doi.org/10.1038/s41586-021-03907-3

133. Ambrosi A, Pumera M (2018) Multimaterial 3D-printed water electrolyzer with earth-abundant electrodeposited catalysts. ACS Sustain Chem & Eng 6(12):16968–16975. https://doi.org/10.1021/acssuschemeng.8b04327
134. Pumera M (2019) Three-dimensionally printed electrochemical systems for biomedical analytical applications. Curr Opin Electrochem 14:133–137. https://doi.org/10.1016/j.coelec.2019.02.001
135. Gusmao R, Sofer Z, Marvan P, Pumera M (2019) MoS2 versatile spray-coating of 3D electrodes for the hydrogen evolution reaction. Nanoscale 11(20):9888–9895. https://doi.org/10.1039/c9nr01876j
136. Camus P, Dufault JF, Mehanovic D, Braidy N, Fréchette LG, Picard M (2019) Testing of a 3D-printed solar micro-reactor for hydrogen production via natural gas reforming. J Phys Conf Ser 1407(1). https://doi.org/10.1088/1742-6596/1407/1/012005
137. Elkoro A, Soler L, Llorca J, Casanova I (2019) 3D printed microstructured Au/TiO_2 catalyst for hydrogen photoproduction. Appl Mater Today 16:265–272. https://doi.org/10.1016/j.apmt.2019.06.007
138. Martínez L, Soler L, Angurell I, Llorca J (2019) Effect of TiO_2 nanoshape on the photoproduction of hydrogen from water-ethanol mixtures over Au_3Cu/TiO_2 prepared with preformed Au-Cu alloy nanoparticles. Appl Catal B 248:504–514. https://doi.org/10.1016/j.apcatb.2019.02.053
139. Castedo A, Mendoza E, Angurell I, Llorca J (2016) Silicone microreactors for the photocatalytic generation of hydrogen. Catal Today 273:106–111. https://doi.org/10.1016/j.cattod.2016.02.053
140. Castedo A, Uriz I, Soler L, Gandía LM, Llorca J (2017) Kinetic analysis and CFD simulations of the photocatalytic production of hydrogen in silicone microreactors from water-ethanol mixtures. Appl Catal B 203:210–217. https://doi.org/10.1016/j.apcatb.2016.10.022
141. Chiefari J, Hornung C (2021) Mobile hydrogen reformers as a novel approach to decarbonise the transport sector. Curr Opin Chem Eng 34. https://doi.org/10.1016/j.coche.2021.100756

Additive Manufacturing as the Future of Green Chemical Engineering

Digital Process Control, Industry 4.0, Smart Manufacturing, Process Intensification, Micro Reaction Technology, Micro Mixers, Micro Heat Exchangers, Fluid Flow Regimes, FDM, SLA, DIW, SLS, SLM, Green Hydrogen

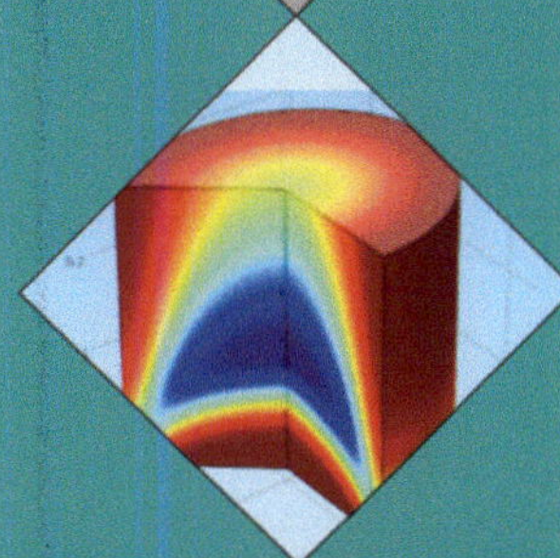

Sunil Mehla, Ravindra D. Gudi, D. D. Mandaliya, Takashi Hisatomi, Kazunari Domen, Suresh K. Bhargava

8

Appendix

Answers to Questions

Chapter 1

1. *What are the three most common manufacturing philosophies used for fabrication of complex shaped objects?*

   Answer: Additive manufacturing, substractive manufacturing (machining) and net-shaping (such as molding, casting and forging)

2. *What is the typical process flow involved in additive manufacturing of an object from a design concept?*

   Answer: Concept > CAD Design > Tessellation (STL) > Slicing > Generation of g-code > printing > post-processing

3. *What are the seven standard additive manufacturing processes according to ASTM F52900-15 committee on additive manufacturing?*

   Answer: Vat photopolymerization, material jetting, material extrusion, binder jetting, powder bed fusion, sheet lamination and directed energy deposition

4. *What is stair-stepping effect in additive manufacturing and what is the consequence of this effect on the surface finish of the printed object?*

   Answer: The surface roughness arising from the layer-by-layer manufacturing in AM is called stair-stepping effect. Larger layer thicknesses and interlayer overlap leads to higher roughness.

S. K. Bhargava et al. (eds.), *Additive Manufacturing for Chemical Sciences and Engineering*, https://doi.org/10.1007/978-981-19-2293-0

Chapter 2

1. *Besides solid object modelling in CAD software what other methods can be used to obtain a CAD file usable for additive manufacturing?*

   Answer: 3D scanning of objects along with suitable software can be used to develop a STL file usable by additive manufacturing.

2. *What is the biggest advantage and limitation of additive manufacturing methodology?*

   Answer: A high degree of freedom in fabrication of solid objects such as in design complexity, material composition and nature of assembly (modular) without tooling requirements is the biggest advantage of AM. Weak inter-layer bonding in AM objects arising from the layered fabrication method is the biggest limitation of AM since such construction impacts the mechanical properties of the object.

3. *Which historical figure can be regarded as the father of additive manufacturing and which historic modelling techniques gave birth to the modern form of additive manufacturing?*

   Answer: François Willème, a French artist, sculptor and photographer who developed the photosculpture technique in 1860s can be considered as the father of additive manufacturing. Photosculpture and topography merged through the works of Morioka in 1930s to gave birth the modern era of additive manufacturing.

4. *Who established the first commercial venture to supply 3D printers and also developed the STL file format?*

   Answer: Charles W. Hull established the first commercial venture '**3D systems**' in 1986 and also developed the STL file format.

5. *What are the different AM standards defined by ASTM and with what objectives?*

   Answer: ASTM has defined 5 key standards for ASTM as given below with their major objectives:

#	Terminology	Standard ID	Objective
1	Standard terminology for additive manufacturing technologies	F2792	Standardize terminology and nomenclature
2	General principles and terminology	ISO/ASTM 52,900	Unify ASTM standards with European Committee for Standardization (CEN)

(continued)

(continued)

#	Terminology	Standard ID	Objective
3	Standard specification for additive manufacturing with titanium-6, aluminum-4 and vanadium with powder bed fusion	F2924	Additive manufacturing of new materials using EBM, SLS, and DMLS
4	General principles and requirements for purchased AM parts	ISO/ASTM 52,901	Standardizing requirements for AM parts
5	Guidelines for design for AM	ISO/ASTM 52,910	Guidelines for design considerations and opportunities and identification of red-flags for designers

6. *Who established the first commercial venture to supply 3D printers and also developed the STL file format?*

   Answer: Charles W. Hull established the first commercial venture '**3D systems**' in 1986 and also developed the STL file format.

7. *The advance of additive manufacturing into modern applications is evident from significant development in which technological sectors?*

   Answer: Biomedical, construction, food and engineering design sectors.

8. *What does the term 3D bioprinting refers to and how does it differ from 3D printing?*

   Answer: By common usage and convention, 3D bioprinting has been used to refer to AM using biological feedstock materials such as cells and tissues whereas 3D printing has been used to refer to AM using metal, polymers, ceramics and composites.

9. *What are the two examples of buildings constructed by 3D printing?*

   Answer: The Apis Cor Printed House in Russia and 3D print Canal House in Amsterdam.

10. *Give some examples AM technologies which have been specifically developed for the engineering design sector, derived benefits and some important clients.*

    Answer: Stereolithography, multi-jet modelling, fused deposition modelling, rapid plasma deposition and selective laser sintering. These technologies have been employed for example by NASA, GE aviation, Bell helicopter and Airbus.

    AM technologies result in benefits of improved sophistication, manufacturing with materials that are difficult to machine, customized production, reduced cost and emissions and high performance at low weights.

11. *What is 4D printing and why is it being regarded as a disruptive technology?*

    Answer: 4D printing refers to 3D printing of 4D objects i.e. objects that change their shape with time. 4D printing allows development of stimuli responsive smart objects and devices and therefore regarded as a disruptive technology with enormous potential in applications such as targeted drug delivery, programming, space, architecture, adaptive structures, personalized medical devices and self-healing implants.

12. *Which industries have benefitted the most from AM in terms of profit.*

    Answer: Consumer products and electronic industries has benefitted the most followed by the transportation and the medical sector.

13. *AM driven customization incentives are rapidly impacting which industrial sectors?*

    Answer: Manufacturing, logistics, services, food and consumers goods and electronics.

Chapter 3

1. *Explain the fundamental principle by which additive manufacturing enables the manufacture of parts with higher geometric complexity then traditional methods*

   Answer: The fundamental principle by which additive manufacturing enables the manufacture of parts with higher geometric complexity then traditional methods is that the layer-by-layer approach to additive manufacture effectively decomposes a complex and difficult to manufacture 3D part geometry, into a series of comparatively simple to manufacture 2D layers.

2. *When additively manufacturing a VP part with an inclined plane, at which plane angle relative to the horizontal build platform would you expect the plane surface roughness to be generally the smallest?*

   Answer: Surface roughness is generally the lowest on vertical or horizontal part surfaces due to the absence of stair-stepping artefacts created by layer-by-layer additive manufacture.

3. *List 3 advantages and 3 challenges associated with FDM and SLA additive manufacturing processes.*

Answer:

<table>
<tr><th>Fused deposition modelling (FDM)</th><th>Stereolithography (SLA)</th></tr>
<tr><td>Advantages
• Comparatively low feedstock material and machine cost
• Feedstock material can be standard engineering plastics like Acrylonitrile butadiene styrene (ABS), polystyrene (PS), polylactic acid (PLA) and polycarbonate (PC) with well understood and high-performance mechanical properties
• Parts require minimum level of post-processing (only support structure removal if present)
• Internal part cavities do not contain unfused feedstock material
Challenges
• Lower resolution and higher surface roughness compared to SLA
• Higher tendency for warping and delamination due to thermomechanical stress
• Higher anisotropy in mechanical properties compared to SLA produced parts due to increased presents of inter-layer and intra-layer porosity
• Generally slow build rate</td><td>Advantages
• Ability to produce high resolution parts with small minimum feature sizes
• Parts exhibit low surface roughness
• Low tendency for warping and delamination due to absence of significant thermomechanical stresses
Challenges
• Feedstock materials can be more expensive than for FDM
• Feedstock materials may exhibit lower mechanical properties compared to FDM materials
• Secondary post-processing required to remove excess resin and conduct final curing
• Material changes requires changing entire vat of feedstock
• Internal part cavities will contain unfused feedstock resin material
• Generally slow build rate</td></tr>
</table>

4. *Why do parts manufactured by Fused Deposition Modelling (FDM) typically exhibit anisotropic behavior?*

   Answer: FDM parts are formed by the fusion of many individual materials track and layers which can result in anisotropy in the part mechanical properties due to:

   1. Reduced inter-layer bond strength compared to within layer strength
   2. Accumulation of voids/porosity between material tracks and layers
   3. Stair-stepping resulting in inter-layer stress concentration which reduce part strength

5. *Suggest 3 components relevant to chemical engineering application that could be good candidates for redesign for additive manufacture using mex or vp processes? what would be the advantages and disadvantages of making those parts using additive manufacturing?*

   Answer: Additive manufacturing processes such as MEX and VP can be used to: Economically manufacture customized components in low batch sizes with reduced waste, increase performance or provide new functionality with complex geometry not manufacturable with other methods, simplify manufacturing by reducing process steps or by consolidating parts together, or reduce inventory with just-in time manufacture. In general, good candidate parts would exploit these benefits while adhering to MEX and VP process constraints associated with cost, part size, and production quantity as detailed in this chapter. Specifically, good candidates would be parts high in complexity or with the potential to benefit from customization, small in size, and manufactured in a small production quantity. Poor candidates would be large parts in large production volumes with relatively simple shapes that are well suited to conventional manufacturing process.

Chapter 4

1. *How do the 'part cost versus part complexity' and 'part cost versus production volume' relationships differ between additive manufacturing and conventional manufacturing processes? What are the reasons for these differences?*

   Answer: Traditional Manufacturing (TM) processes typically incur high initial costs due to the outlay for tooling, fixtures, and process setup. As these costs need to be amortized per component, small batch sizes result in higher part costs compared to larger batch sizes. Furthermore, the cost of traditionally manufactured components typically increases with higher part complexity, as the number and intricacy of required manufacturing processes, equipment, and tooling rise. The part complexity can also exceed the capability of the traditional manufacturing processes due to constraints associated with tool access in subtractive machining, or material flow in formative processes.

   Conversely, as AM processes do not require special tooling or machinery to produce varying geometric features, the associated set-up costs are substantially reduced. Similarly, the number and intricacy of manufacturing processes and equipment required for additive manufacture is mostly independent of part complexity, and the constraints on manufacturable part geometry are low. As such the key advantages of AM over conventional methods are in the economic manufacture of components: in low batch sizes; with complex geometry where conventional manufacture is expensive; or where the geometric complexity is beyond the capabilities of traditional methods. However, traditional manufacture has significant economic advantages at higher production volumes and lower part complexities, primarily due to the substantially higher speed of manufacture compared to AM methods. The three unique areas of opportunity for AM, as well as the areas of advantage of tradition methods, are highlighted in Fig. 4.1 where typical cost relationships of both manufacturing processes are shown in relation to batch size (Fig. 4.1a) and product complexity (Fig. 4.1b).

2. *List 3 advantages and 3 challenges associated with the L-PBF additive manufacturing processes.*

Answer:

Advantages presented by L-PBF	*Challenges associated with L-PBF*
• Ability to manufacture fully functional high density (>99%) metal parts with very high design freedom and part complexity • High resolution and small minimum achievable feature size • No custom tooling required resulting is short lead times and very high level of part customization • Broad range of compatible materials including steel, titanium, aluminium, nickel and copper alloys etc • Reusable feedstock material	• Cost of equipment and feedstock material can be • Generally slow process compared to machining/casting/forging requiring hours to days in cycle time to complete a part • Significant number of process, laser, material, powder and geometry parameters influence the quality of parts • Part can require post-processing including, powder and support structure removal, heat treatment and machining • Internal part cavities will contain unfused powder feedstock material which requires drain holes

3. *State three possible effects of part build orientation on the L-PBF process and on part quality*

   Answer: The orientation of the part during L-PBF manufacture can influence the following characteristics:

   1. The quantity of support structures required and the ease of their removal.
   2. The build time, as taller parts with more layers typically require longer build times.
   3. Surface roughness, due to effects such as stair-stepping.
   4. Internal stress built up.
   5. Strength of the part, due to the influence of layer orientation on material microstructure material strength (Sect. 4.3.1)
   6. The number of parts that can be built concurrently (influenced by the base area of a boundary box enclosing the part).

4. *How do the mechanical properties of L-PBF parts typically compare to components manufactured by conventional manufacturing processes?*

   Answer: In general, the mechanical properties of L-PBF materials under static loading are superior to conventionally cast materials and similar to wrought materials produced by rolling/forging/extrusion. However, under dynamic loading the fatigue life of as-built L-PBF parts can be inferior to conventionally manufactured parts. As fatigue life is dependent on the presence of surface or internal stress concentrations, as-built L-PBF components exhibit low fatigue lives (similar to cast parts) due to comparatively high surface roughness and the presence of residual internal gas pores. Machining and polishing the surface of the parts improves the fatigue life significantly due to reduced surface roughness. Hot isostatic pressing (HIP) treatment which closes internal pores further improves the fatigue lives of the samples and the results are comparable to wrought materials.

5. *Suggest 3 components relevant to chemical engineering application that could be good candidates for redesign for additive manufacture using L-PBF? What would be the advantages and disadvantages of making those parts using additive manufacturing?*

   Answer: Additive manufacturing processes such as L-PBF can be used to: Economically manufacture customized components in low batch sizes with reduced waste, increase performance or provide new functionality with complex geometry not manufacturable with other methods, simplify manufacturing by reducing process steps or by consolidating parts together, or reduce inventory with just-in time manufacture. In general, good candidate parts would exploit these benefits while adhering to L-PBF process constraints associated with cost, part size, and production quantity as detailed in this chapter. Specifically, good candidates would be parts high in complexity or with the potential to benefit from customization, small in size, and manufactured in a small production quantity. Poor candidates would be large parts in large production volumes with relatively simple shapes that are well suited to conventional manufacturing process.

Chapter 5

1. *What would be the major difference between Robocasting printed structure to the other 3D printed structures?*

   Answer: Robocasting is a slurry-based extrusion process of pseudoplastic paste or ink con-taining ceramic particles to print a green body containing the ceramic particles and other additives present in the ink. The ceramic green bodies fabricated by the robocasting technique need to be dried and cured in order to remove the binder, and further sintered at high temperature to obtain a high-density ceramic object.

2. *What is the role of solvent in controlling the viscosity of the pseudoplastic ink? Is water a good solvent for making the pseudoplastic ink?*

   Answer: Viscosity of pseudoplastic ink is a key parameter in robocasting. After the extrusion process, viscosity will start increases when the liquid content decreases as a consequence of drying. The majority of the robocasting slurry formulations use water as the solvent (continuous phase) due to its ability to form stable suspen-sions with ceramic particles. Water has a low viscosity which can be increased with additives, but the green pre-sintered ceramics produced from aqueous slurry formulations are Viscosity of pseudoplastic ink is a key parameter in robocasting. After the extru-sion process, viscosity will start increases when the liquid content decreases as a consequence of drying. The majority of the robocasting slurry formulations use water as the solvent (continuous phase) due to its ability to form stable suspen-sions with ceramic particles. Water has a low viscosity which can be increased with additives, but the green pre-sintered ceramics produced from aqueous slurry formulations are not as mechanically strong as those with resin or polymer based green bodied ceramic. The drying of the water from the green-body can also lead to cracks and voids if drying is not sufficiently rapid.

3. *What will be the general composition of pseudoplastic ink? Comment about the volume fraction of solids and its role in printing the ceramic green body*

   Answer: The pseudoplastic paste consist of ceramic materials, solvents, binders and plasti-cizers in a specific ratio to control its rheological properties in order to be printed effectively and smoothly. Solid content must be maintained and regulated around 70% or above to obtain better ceramic paste for the appropriate fluidity and drying kinetics.

4. *How post-processing of ceramic green body is significant in getting a dense ce-ramic structure?*

   Answer: The ceramic green bodies fabricated by the robocasting technique need to be dried and cured in order to remove the binder, and further sintered at high tem-perature to obtain a high-density ceramic object. Adversely, sintering at high tem-peratures may lead to cracks and less dense structures, hence the amount of total shrinkage needs to be minimized during the post-processing of the green-body. This is why it is ideal for the slurry to have a high weight percentage of fine ceramic particles and conversely a low weight percentage of organic binders and other additives.

5. *How multi-ceramic printing process can be operated?*

   Answer: Robocasting has the potential to produce multi-ceramic materials by including multi-ceramic particles in the ink preparation stage. This can be achieved in two ways; First approach uses a single mixing nozzle that is equipped with multiple feeding systems, while the second approach involves the use of discrete multi-nozzle systems. In the first approach, the single mixing nozzle blends the materi-als received from the multiple feeding systems at specific feed-stock proportions to form a final mixture, which can be extrudates as single filament. In the case of discrete multi-nozzle systems, they have the advantage of allowing automatic in-terchangeable printing heads during printing [41], that allow multi-material printing.

Chapter 6

1. *What would be the methodology, to functionalize the lattice structure printed us-ing Selective Laser Melting?*

   Answer: Thermal activation of the lattice will be the first post-processing step prior to the surface functionalization. Formation of metal oxides as a result of thermal acti-vation enhance the adsorption of functional layer.

2. *Stereo lithographically printed structures swell in organic solvents. What would be the best way to use these printed objects in organic medium?*

   Answer: Stereo lithographically printed polymer materials are known to react with any or-ganic solvent, therefore a modified approach of using a resin containing isocya-nate groups as a monomer in vat photopolymerization can be followed. The as-printed object with this modified resin can be resistant against organic solvents.

3. *FDM printed objects are made up of thermoplastics. How to deposit metals to make electronic circuits on the surface of FDM printed objects?*

   Answer: Through electroless deposition approach, metals can be deposited on the surface of FDM printed thermoplastics. Electronic circuits made up of copper, silver and other metals can be deposited on these FDM printed materials.

4. *What are the post-processing physical methods available for reducing the surface roughness of the SLM printed metal objects?*

   Answer: Stress relief, heat treatment, powder removal, hot isostatic pressing, surface treatment, machining, ultrasonic treatment, and laser based post-processing tech-niques are the post-processing physical methods available for reducing the sur-face roughness of SLM printed objects.

5. *Why chemical or thermal activation of 3D printed objects are required for func-tionalization?*

   Answer: 3D printed objects are chemically inert on their surface and structurally aniso-tropic. To functionalize their surface, chemical and thermal treatments need to be followed to enhance their surface reactivity and structural reorganization to ena-ble surface functionalization.

Chapter 7

1. *Which additive manufacturing technology has the highest resolution in XY and Z direction as well as allows unsupported fabrication of overhanging parts?*

   Answer: Two-photon lithography (~100 nm can be achieved in both XY and Z directions). Since TPL fabricates at extremely small scales, overhanging parts are usually not an issue.

2. *Which additive manufacturing technology allows the highest print speeds for fabrication?*

   Answer: Dynamic Light Projection Stereolithography (DLP SLA allows up to 720 mm/hour).

3. *How can the fabrication times be drastically reduced for photopolymerization based AM technologies?*

   Answer: A common method employs printing of only side walls of designed objects as solids leaving unpolymerized liquid in the bulk of the solid. A UV flood exposure post-printing is used to cure the liquid in the bulk and improve mechanical stability.

4. *Identify the only correct statement from the following:*

   (a) Direct ink writing (robocasting) provides highest resolution and lowest fabrication times.
   (b) Selective laser sintering can be used to fuse the powdered form of any material
   (c) Resolution in the Z direction and fabrication time are inversely proportional for any additive manufacturing technology.
   (d) Additive manufacturing cannot be used for printing of elastic materials

   Answer: c

5. *Identify the only correct statement from the following:*

   (a) Microreactors are used because they increase the rate of reactions.
   (b) High pressure reactions cannot be performed in microreactor because of safety issues.
   (c) Microreactors are only suitable for lab testing and not commercial production as they involve small volumes of reactants and products.
   (d) Physical processes like mass transfer, heat transfer, temperature gradients, maldistribution and channelling are reduced at small scales involved in microreactors.

   Answer: d

6. *Identify the only incorrect statement from the following:*

   (a) Nano sized powders are most suited for chemical reactors owing to their high specific surface areas and high chemical reactivities.
   (b) Catalyst pellets and extrudates also serve the dual purpose of improvement in mixing for continuous flow reactors.
   (c) Robocasted catalyst bodies can potentially replace the use of catalyst pellets and extrudates.
   (d) Foam packed reactors are use in diffusion limited reactions to enhance mass and heat transfer.

   Answer: a

7. *Identify the only incorrect statement from the following:*

   (a) Additive manufacturing enables better integration of optoelectronic device components in three dimensions compared to conventional assembly techniques.
   (b) Additive manufacturing can be used to fabricate flexible and elastic 3D objects suited for biotic/abiotic interfaces with body organs
   (c) DLP SLA is the most promising AM technology for applications of optoelectronic devices in biosensing, tissue engineering, cell culturing and organ printing because its high resolution, high print speeds, smooth finish to objects and capability to print optically transparent objects.
   (d) Optoelectronic devices can greatly benefit by integration with microfluidics because of reduced consumption of analytes arising from the small scale.

   Answer: c

8. *Identify the only correct statement from the following:*

   (a) Batteries and capacitors are identical energy storage devices composed of two electrodes and an electrolyte and exhibiting same charging/discharging characteristics.
   (b) Lithium-ion batteries are the most promising energy storage solutions because of the use of low-weight and low-cost lithium which is abundantly found in earth crust.
   (c) Supercapacitors are going to completely replace batteries soon because of their capabilities for fast charging and discharging.
   (d) Two metal plates separated from each other can be used as a capacitor.

   Answer: d

9. *Identify the only incorrect statement from the following:*

   (a) Extraction of organics from an aqueous mixture can be performed using 3D printed mixers and phase separators.
   (b) Membrane filtration is based on sieving as well as retention by chemical interactions.
   (c) Separation using zeolites and crystalline frameworks is based on their chemical interaction with molecules to be retained.
   (d) Throughput in membrane filtration can be improved by using 3D printed formed bodies without significant deterioration of separation performance.

   Answer: c

10. *Match the Following:*

S. no	AM technology	Mechanism of fabrication	Application	Classification of Process
1	Inkjet Printing	Extrusion of ink from a nozzle in three dimensions as per design coordinates	(a) Contact-less fabrication of conducting interconnects on 3D objects	Form and shape sensitive process
2	Selective Laser Sintering	(b) Deposition of atomised liquid particles carried with carrier gas through a nozzle	(b) Rapid prototyping using material sheets	Bulk sensitive process
3	Laminate Object Manufacturing	(c) Melting and extrusion of a thermoplastic filament through a nozzle	(c) Cheap 3D objects using ABS	Fusion-based additive manufacturing
4	(d) Aerosol Jet Printing	(d) On-demand ink droplet generation and deposition	(d) Batteries and capacitors	High throughput experimentation
5	(e) Direct Ink Writing	(e) Selective and localised sintering of powders using an IR laser	(e) Rapid preparation of composite materials	Surface sensitive process
6	(f) Stereolithography	(f) Photopolymerization of photocurable resin in a 3D voxel using an IR laser	(f) 3D objects with sub-micron features	(f) Energy storage materials
7	(g) Two-photon lithography	(g) Photopolymerization of photocurable resin using a UV light source	(g) Rapid prototyping, single use molds and microfluidics	(g) Photopolymerization-based additive manufacturing
8	(h) Fused Deposition Modelling	(h) Laser machining of a material sheet in 2D as per CAD design and binding of cut laminates	(h) Chemical Reactors	(h) Extrusion-based additive manufacturing

Answer: 1-a-d-e-d, 2-b-e–h-a, 3-c-h-b-c, 4-d-b-a-e, 5-e-a-d-f, 6-f-g-g-g, 7-g-f-f-b, 8-h-c–c-d.

Chapter 8

1. *What are the three major drivers that are demanding systemic and technological innovations in existing chemical processes and unit operations?*

   Answer:

   (1) The rise of new feedstocks that must be processed with existing infrastructure.
   (2) The rise of expectations around energy efficiency, process efficiency, sustainability, and environmental footprint
   (3) Dynamic shifts in customer demands

2. *What is the fourth industrial revolution or industry 4.0 and what are its main indicators?*

   Answer: Industry 4.0 or the fourth industrial revolution (4IR) refers to the reorganization, modernization, and digitization of the entire industrial value chain and characterized by elements such as internet of things, artificial intelligence, data analytics, advanced robotics, additive manufacturing, process automation, augmented reality, cyber security and cloud systems.

3. *What is process intensification and what are some benefits associated with its implementation?*

   Answer: Process intensification refers to increase in intensity or severity of process parameters with the goal to improve process efficiency. It can lead to several benefits such as reduction in equipment size, energy consumption, or waste formulation, improved product quality and improvement in sustainability, efficiency, economics, safety of the process.

4. *What are the critical issues which must be paid special attention for full realization of the potential of process intensification?*

   Answer: Control over intentional or unintentional reactive chemistry issues, waste generation and disposal, uncontrolled release to the environment, safe storage, shipping and transport.

5. *What are the five fundamental physical phenomenon that dominate the performance of chemical processes and equipment?*

   Answer: Mass transfer, heat transfer, maldistribution, temperature gradients and process fluid composition homogeneity.

6. *What is the biggest driver behind miniaturization of chemical processing equipment?*

   Answer: Physical and chemical phenomenon that govern the performance of chemical processed and reactor are scale dependent and become much easier to control when the processes happen at a small scale.

7. *What are some of the benefits associated with micro reactors?*

   Answer: Improved heat and mass transfer, safe operation under severe process conditions, safer handling of hazardous chemicals, efficient mixing, enhanced catalyst-reactant interactions, improved catalyst lifetimes and improved activity and selectivity leading to a reduced E-factor.

8. *What are some limitations of microreactors that are hindering their rapid uptake by the industrial sector?*

   Answer: Lack of agility and adaptability in existing chemical process and practical considerations such as periodic maintenance, space availability, shape, and size constraints, geometrical (performance) uniformity in parallel microreactors and catalyst related issues such as stability and lifetime of catalyst coating.

9. *Which microscopic forces govern mixing performance in microreactors and micromixers?*

   Answer: Different physico-chemical factors and governing forces such as size effect, rarefaction effect, viscous effect, hydrodynamic and electrostatic forces govern mixing performance in microreactors and micromixers.

10. *What are some critical design elements which govern the mixing performance in microreactors and micromixers?*

    Answer: Microchannel geometry, path variations and flow obstruction geometry are the three most important design elements which govern mixing performance.

11. *Which reaction or process parameters are directly or indirectly related to mixing efficiency?*

    Answer: Back pressure, number, and distribution of dead zones, reaction kinetics and thermodynamics, mass transfer and heat transfer are all directly or indirectly related to mixing efficiency.

12. *Which dimensionless numbers and quantities become important to define flow regimes in microchannels?*

    Answer: Reynolds number, Weber number, identification number, Bonds number, Froude number, Martinelli parameter and hydraulic diameter.

13. *What are the different types of flow regimes observed in microchannels?*

    Answer: Stratified flow regime consisting of stratified smooth and stratified wavy flow, intermittent flow regime consisting of the elongated bubble or plug flow and slug flow, annular flow regime consisting of the annular and wavy annular flow and dispersed flow regime consisting of the bubble and dispersed flows.

14. *What is the importance of flow obstruction geometry in microchannels?*

    Answer: Flow obstruction geometry in microchannel improve mixing efficiency and heat transfer by generation of different types of vortices, flow churning and chaotic advective while also providing addition surface area for catalyst integration and fluid contacting.

15. *What are the major hurdles in the use of fused deposition modelling and stereolithography for microreactor design and how can this hurdle be circumvented?*

    Answer: The chemical, thermal and mechanical stability of polymeric microreactors fabricated using fused deposition modelling and stereolithography under reaction conditions is the major hurdle in their wide use for microreactors. New resins with good stability under reaction conditions or integration of stable materials into resins and removal of resin post fabrication can resolve this issue for stereolithography. Similarly, for fused deposition modelling use of new thermoplastic materials which are stable under reaction conditions or indirect use of FDM for fabrication of ceramic or metallic microreactors such as in lost polylactic acid (PLA) casting techniques can resolve this stability issues for FDM.

16. *What are the major hurdles in the use of selective laser sintering and selective laser melting for microreactor design and how can this hurdle be circumvented?*

    Answer: In both SLM and SLS, resolution along the object walls is limited by laser spot in the range of 0.3 – 0.6 mm. In addition, high surface roughness and removal of powder from microchannels are also some of the important shortcomings of these technologies. Further improvement of these techniques is anticipated to resolve these issues.

17. *What is the common measure used for comparison of efficiencies in green hydrogen production technologies?*

    Answer: Solar-to-hydrogen (STH) and Qauntum Efficiencies (QE)

Suresh K. Bhargava, Member of the Order of Australia, Centre for Advanced Materials and Industrial Chemistry, RMIT University, Melbourne, VIC, Australia

Milan Brandt, Centre for Additive Manufacturing, RMIT University, Melbourne, VIC, Australia

Seeram Ramakrishna, Centre for Nanofibers and Nanotechnology, National University of Singapore, Singapore

Kazunari Domen, Department of Chemical System Engineering, University of Tokyo, Japan

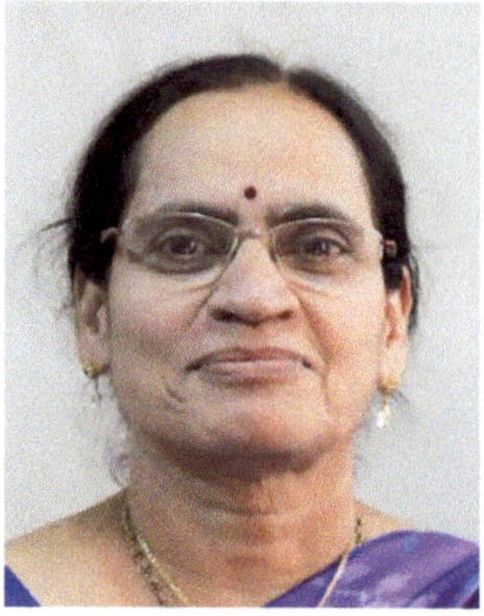

Lakshmi Kantam Mannepalli, Department of Chemical Engineering, Institute of Chemical Technology Mumbai, India

Ravindra D. Gudi, Department of Chemical Engineering, Indian Institute of Technology Bombay, India

Yonxiang Li, School of Engineering, Electronic and Communication Engineering, RMIT University, Melbourne, VIC, Australia

Takashi Hisatomi, School of Science and Technology, Institute of Engineering, Shinshu University, Japan

Dharmendra D. Mandaliya, Department of Chemical Engineering, L.D. College of Engineering Ahmedabad, India

Maciej Mazur, Centre for Additive Manufacturing, RMIT University, Melbourne, VIC, Australia

PR. Selvakannan, Centre for Advanced Materials and Industrial Chemistry, RMIT University, Melbourne, VIC, Australia

Sunpreet Singh, Centre for Nanofibers and Nanotechnology, National University of Singapore, Singapore

Uzma Malik, Centre for Advanced Materials and Industrial Chemistry, RMIT University, Melbourne, VIC, Australia

Sunil Mehla, Centre for Advanced Materials and Industrial Chemistry, RMIT University, Melbourne, VIC, Australia

Roxanne Hubesch, Centre for Advanced Materials and Industrial Chemistry, RMIT University, Melbourne, VIC, Australia

Xiaochen Sun, Centre for Additive Manufacturing, RMIT University, Melbourne, VIC, Australia

What peers say

"*Additive manufacturing (AM) has already established itself as a new discipline and with manifold and now routine applications. What is exciting about this book is its comprehensive approach to developing and game changing applications of AM ranging from green process engineering to the intriguing next dimension of surface modified AM*"

Professor Robin Batterham, AO, FREng, FAA, FTSE, Chief Scientist of Australia (1999-2006), President of Institute of Chemical Engineers (2004)

"*This book is a powerful guide that shows how the promise of 3D printing is finally showing performance in terms of 3Ds - the power of 'Digitalisation' that is leading 'Decentralisation', which in turn is contributing to 'Decarbonisation'*"

Dr. R A Mashelkar, FRS, FAA, FTSE, FREng, FNAE(US), President of Institute of Chemical Engineers (2007), President-Indian National Science Academy (INSA) (2005)

"*The Additive Manufacturing revolution is gathering pace in chemical science and engineering to make manufacturing eco-sustainable. This book is a unique initiative by Distinguished Professor Bhargava and his team to guide both students and scientists interested in the interdisciplinary research*"

Yasuhiro Iwasawa, Professor at the University of Electro-Communications, Emeritus Professor at the University of Tokyo Japan, President - Japan Union of Chemical Science and Technology (JUCST) (2018)

"*Wonderful guide to how additive manufacturing is accelerating human Endeavor, from the arts to zero emissions energy systems*"

Dr. Alan Finkel, AC, FAA, FTSE, Chair of Australia's Low Emissions Technology Investment Advisory Council, Chief Scientist of Australia (2016-2020), President - Australian Academy of Technological Science and Engineering (ATSE) (2013)

MIX
Papier aus verantwortungsvollen Quellen
Paper from responsible sources
FSC® C105338

If you have any concerns about our products,
you can contact us on
ProductSafety@springernature.com

In case Publisher is established outside the EU,
the EU authorized representative is:
Springer Nature Customer Service Center GmbH
Europaplatz 3, 69115 Heidelberg, Germany

Printed by Libri Plureos GmbH
in Hamburg, Germany